Catalysis from Waste

Scrivener Publishing
100 Cummings Center, Suite 541J
Beverly, MA 01915-6106

Publishers at Scrivener
Martin Scrivener (martin@scrivenerpublishing.com)
Phillip Carmical (pcarmical@scrivenerpublishing.com)

Catalysis from Waste

Transforming Solid Residues into Value

Edited by

Deepti Goyal

Sakshi Kabra Malpani

and

Renuka Jain

Scrivener Publishing

WILEY

This edition first published 2026 by John Wiley & Sons, Inc., 111 River Street, Hoboken, NJ 07030, USA and Scrivener Publishing LLC, 100 Cummings Center, Suite 541J, Beverly, MA 01915, USA

For more information about Scrivener publications please visit www.scrivenerpublishing.com.

Wiley Global Headquarters
111 River Street, Hoboken, NJ 07030, USA

For details of our global editorial offices, customer services, and more information about Wiley products visit us at www.wiley.com.

The manufacturer's authorized representative according to the EU General Product Safety Regulation is Wiley-VCH GmbH, Boschstr. 12, 69469 Weinheim, Germany, e-mail: Product_Safety@wiley.com.

Library of Congress Cataloging-in-Publication Data

ISBN 9781394289301

Cover image: Generated with AI using Adobe Firefly
Cover design by Russell Richardson

Set in size of 11pt and Minion Pro by Manila Typesetting Company, Makati, Philippines

Printed in the USA

10 9 8 7 6 5 4 3 2 1

Dedicated to our beloved mothers — Nirmala Kabra and to the memory of Smt. Sarla Jain whose unwavering encouragement and endless blessings have inspired every step of our paths.

From waste to wealth: unlocking the planet's overlooked assets.

Contents

Foreword

The generation of solid waste is increasing rapidly due to the growth of the global population. The volume and composition of solid waste have surged because of urbanization and industrialization. This increase creates not only economic and environmental challenges but also significant global risks to sustainable development. Poor management of waste poses high risks to ecosystems and human health because of its toxic nature. To address these problems, it is essential to find solutions for solid waste management or to utilize this waste sustainably without harming human health or the environment. This book mainly focuses on the critical role of waste-derived catalysts in sustainable solutions, highlighting the transformative potential of ingredients from solid waste. Notably, carbon-based catalysts derived from farming, municipal, and industrial waste can be converted into porous activated carbon and biochar/hydrocarbons, which have many applications as flocculants or adsorbents for water purification. Additionally, transition metal-based catalysts from used batteries, electronic waste, and manufacturing byproducts also demonstrate effectiveness in environmental remediation. Calcium and other metal-based catalysts obtained from food waste, including seashells, bones, and construction debris, play an important role in biofuel production, proving their promising place in sustainable and eco-friendly practices. In summary, by transforming waste into valuable catalysts, this book advocates a model for an economically viable, highly sustainable, and resource-efficient future.

The current book (*Catalysis from Waste: Transforming Solid Residues into Value)* includes 15 chapters written by various scientists in this field and covers key features of solid waste-derived catalysts. The content of this book bridges the gap between laboratory research and industrial applications, not only in the field of catalysis but also for sustainable development. We appreciate the efforts of the authors in presenting this book, which can benefit researchers learning these methods for their research and academic purposes.

Dr. Priyanka Sharma
Department of Chemical and Paper Engineering, Western Michigan University, Kalamazoo, MI, USA

Preface

The global demand for sustainable development has prompted scientists and researchers to explore innovative approaches for resource recovery and waste valorization. With the increasing accumulation of solid residues from agricultural waste and industrial byproducts to municipal refuse there arises both a challenge and an opportunity. Rather than viewing these residues merely as waste, recent advancements in catalysis have opened pathways to transform them into valuable materials, chemicals, and fuels. This edited volume, *Catalysis from Waste: Transforming Solid Residues into Value*, emerges from this vital intersection of environmental concern, technological innovation, and sustainable chemistry.

The primary objective of this edited volume is to illuminate the transformative potential lying dormant within waste materials and their conversion through catalytic processes into valuable resources. It seeks to bridge the gap between environmental stewardship and economic sustainability by exploring innovative catalytic methodologies that can harness the latent value embedded in solid residues. Through a meticulous examination of case studies, cutting-edge research, and practical applications, the book serves as a vital resource for researchers, practitioners, and policymakers alike, fostering a deeper appreciation for the pivotal role catalysis can play in reshaping our approach towards waste management and resource utilization.

The chapters in this volume, contributed by distinguished experts from academia, research institutions, and industry, collectively provide a thorough exploration of the field. They cover fundamental principles of catalyst development, detailed methodologies for the synthesis and characterization of catalysts from diverse waste sources, and their practical applications across a spectrum of processes—including organic transformations, energy conversion, and environmental remediation. By combining theoretical perspectives with experimental evidence and case studies, the book offers a balanced and comprehensive view of current progress and future possibilities.

As our global population continues to surge, so does the generation of waste, presenting a pressing environmental crisis. This book endeavors to illuminate the innovative potential of catalysis as a transformative force, addressing the dual challenge of waste accumulation and the quest for sustainable resource utilization. Unprecedented levels of waste necessitate novel solutions, and through the lens of catalysis, we explore pathways to convert seemingly burdensome residues into valuable resources. This work seeks to inspire a paradigm shift, fostering a deeper understanding of the intrinsic value embedded in waste streams, and encouraging the adoption of catalytic processes to unlock this latent potential. A particular strength of this volume lies in its multidisciplinary approach, bridging chemistry, materials science, environmental engineering, and process technology. It aims to serve as a reference resource for researchers, academicians, industrial practitioners, policymakers, and students who are keen to understand the transformative potential of waste valorization through catalysis. The examples and methodologies presented herein are not only technically significant but also aligned with the broader vision of a circular economy, where waste is redefined as a valuable feedstock for sustainable production systems.

The book is divided into two parts: Introduction to Catalysis and Various Applications of Solid Waste-derived Catalysts. Part I comprises eight chapters while part II contains seven chapters. It covers a wide spectrum of agricultural, municipal, industrial solid waste like rice husk ash, fly ash, eggshells, corn cob ash, plastic waste, red mud, e-waste and many more. Part I gives a glimpse of different types of solid waste, their status and conversion into heterogeneous catalysts. Further it covers various techniques in catalyst synthesis as well as in catalyst characterization. Part II illustrates multiple applications of waste-derived catalysts in green chemical organic reactions, energy generation and environmental remediation. It also sketches strategies used in solid waste management, challenges and opportunities in this sector. Each chapter is self-consistent and illustrates a reasonable proportion of imperative basics and relevant recent results.

Our aim is to provide researchers, academicians, industrial practitioners, or students with insights into cutting-edge methodologies and real-world applications that showcase how waste materials can be repurposed into high-performance catalysts. We believe that this book will serve as a valuable reference for advancing the vision of a circular economy, where waste is not discarded but reimagined as a resource. We hope it will inspire further research, policy development, and industrial innovation aimed at building a cleaner, greener, and more resource-efficient future. We extend our sincere gratitude to all the contributing authors for their thoughtful chapters and to the reviewers for their constructive feedback. The enormous workmanship of the Wiley publishing team is appreciated. Special thanks also go to the editorial and publishing team for their unwavering support throughout the compilation of this book. We would also like to thank our colleagues and friends for helping us understand and develop various aspects of solid waste valorization. Last but not the least, we are thankful to our family members for their continued support to complete this work. We wish the readers an enjoyable and advantageous time while reading this book and we anticipate that it serves as a regularly unwrapped reference textbook.

Deepti Goyal
Sakshi Kabra Malpani
Renuka Jain
Editors

Part I
INTRODUCTION TO CATALYSIS

1

Foundations of Catalysis: Understanding the Key Concepts and Mechanisms

Deepti Goyal[1], Rajesh Kumar Meena[2] and Sakshi Kabra Malpani[3]*

[1]Department of Applied Chemistry, School of Vocational Studies & Applied Sciences, Gautam Buddha University, Greater Noida, UP, India
[2]Department of Chemistry, Kalindi College, University of Delhi, New Delhi, Delhi, India
[3]A-12, 102, Florida Estate Co-op Housing Society, Pune, Maharashtra, India

Abstract
Chemical transformations are pivotal to manufacture products like fuels, chemicals, polymers, plastics, and pharmaceuticals. The materials needed for our daily lives would not be produced without using catalysts and catalytic processes. The foundation of catalysis is based on understanding the fundamental concepts and mechanisms of catalysts and their interactions with reactants. This book chapter provides a comprehensive overview of the key principles and theories of catalysis, giving insights into the various applications in chemistry. The chapter begins with the historical development of catalysis, followed by contributions of pioneering scientists. Subsequently, the chapter focuses on the fundamental principles of catalysis, including the concept of activation energy, transition state theory, and the role of catalysts in lowering reaction barriers. It discusses the various types of catalysts, ranging from homogeneous and heterogeneous catalysts, and their mechanisms of action in facilitating chemical transformations. Furthermore, the chapter explores the factors influencing catalytic activity and selectivity. It also examines the importance of reaction kinetics and thermodynamics for careful optimization and characterization of catalysts to achieve desired reaction outcomes. It discusses the usage of modern tools like artificial intelligence, computational models in catalyst design and synthesis. Some examples of new age catalysts have also been discussed in brief. The chapter concludes with the need for continued research and collaboration to unlock new catalytic processes and advance scientific knowledge.

**Corresponding author*: sakshikmalpani@gmail.com

Deepti Goyal, Sakshi Kabra Malpani and Renuka Jain (eds.) Catalysis from Waste: Transforming Solid Residues into Value, (3–30) © 2026 Scrivener Publishing LLC

Keywords: Chemical transformations, catalysis, activation energy, heterogeneous catalysts, deactivation; computational models, new age catalysts

1.1 Introduction

Catalysts are the substances which when added to a chemical reaction, increases the rate of the reaction without being consumed in the reaction. Indeed, a catalyst accelerates the reaction by reducing the activation energy or by providing the alternate reaction path. This makes it convenient for reactant molecules to reach the transition state and get converted into product in a lesser time. Catalysis is essential for life and has been used since ancient times, particularly in fermentation. In ancient times, food and beverages were used to synthesize using catalysts. After this, catalysts were used in the making of dairy and baking products for 1000 years which is confirmed by archaeological evidence. In the second half of the 18th century, scientist Berzelius introduced the term "catalysis" and proposed the term "catalytic force" beyond chemical affinity. After the speedy research in inorganic chemistry, the modern form of catalysis has been entered in the scientific literature. Several concentrated mineral acids were found to be efficient to catalyse organic reactions like dehydration of ethanol and hydrolysis of starch. The first commercial catalytic process, performed in the 18th century was the synthesis of sulfuric acid *via* lead chamber process using vanadium catalyst. Thereafter, several metal catalyzed organic reactions were performed such as decomposition of ethanol and combustion of methane. After a few years, Davy, Dobereiner and others have performed platinum catalyzed combustion of hydrocarbons and explained the concept of reaction mechanism. In the 19th century, Sabatier got the Nobel Prize for his discovery on catalytic hydrogenation. He introduced the concept of stability of intermediates formed at the catalyst surface. If intermediates are too stable, they would not decompose but if they are too unstable, they would not be formed. Sabatier's study formulated the molecular basis of catalytic reactions [1]. A historical timeline of catalytic technologies is represented in Figure 1.1 [2].

Catalyst is a substance which facilitates a chemical reaction by lowering the activation energy and thus providing an alternate path for formation of new molecules. It reduces the energy barrier which is an essential requirement for a chemical reaction to occur. Thus, a catalyst helps in breaking old bonds and forming new bonds to form new compounds or molecules, making the process fast and energy efficient. Catalysts also help in increasing selectivity of desired product by decreasing the amount of side products [3].

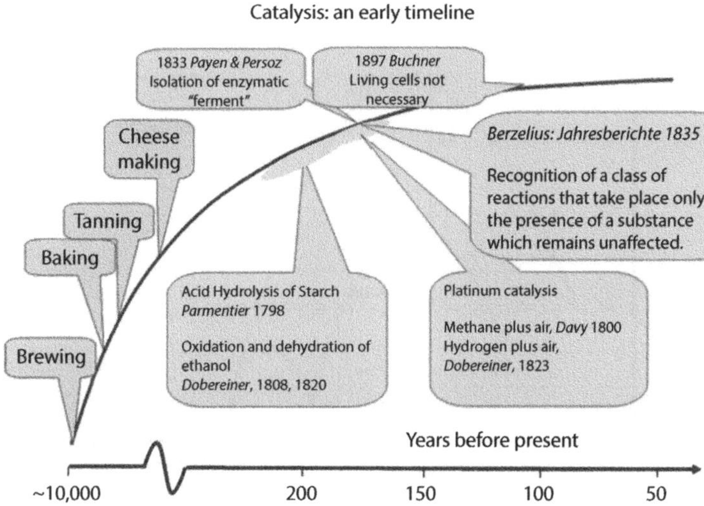

Figure 1.1 Historical timeline for catalyst technologies [2].

The most important property of a catalyst is specificity of action; that is, each chemical reaction, or a group of chemical reactions can be accelerated only by very specific catalysts. Over the past several decades, scientists have developed increasingly specialized catalysts for essential real-world applications. In particular, powerful catalysts have transformed the chemical industry. These advances have led to biodegradable plastics, new pharmaceuticals, and environmentally safer fuels and fertilizers.

1.2 Fundamental Concepts

1.2.1 Activation Energy and Reaction Pathways

Catalysis is a foundation of modern chemistry supporting a wide range of industrial operations, environmental applications and biological activities. For an effective and long-lasting catalytic system, it is important to understand the fundamental concepts of catalysis. These fundamental concepts provide a comprehensive framework for comprehending and advancing the catalytic reactions.

The activation energy of a chemical reaction strongly affects the rate of reaction. By lowering activation energy and providing alternative reaction pathways, catalysts enhance reaction rates and selectivity (Figure 1.2).

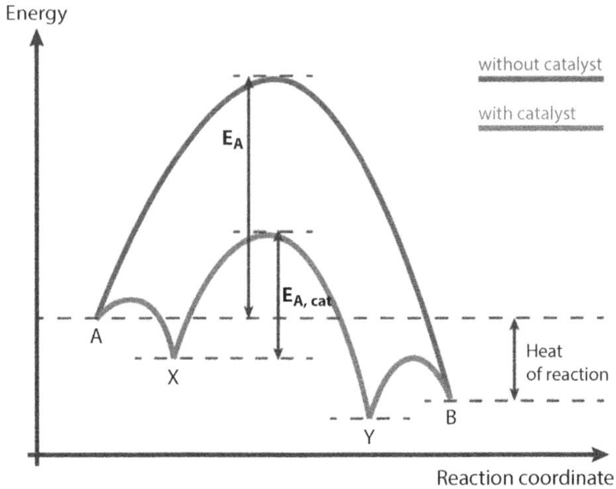

Figure 1.2 Potential energy diagram of a chemical reaction with or without catalyst.

Catalysts do this by providing an alternative path that requires comparatively less energy.

The benefit of the catalyst presented in Figure 1.2, is thus intimated to the reduction of activation energy (Ea) of the reaction in the presence of the catalyst. The increase in the rate of reaction using catalyst can be well understood using Arrhenius equation (Equation 1.1):

$$k = Ae^{\frac{-Ea}{RT}} \tag{1.1}$$

Where: k = rate constant, A = Pre factor, E_a = activation energy, R = gas constant and T = absolute temperature.

Understanding reaction rates begins with the origin of activation energies, making the potential energy diagram (PED) for surface chemical transformations a crucial starting point. The value of activation energy largely varies with changing catalyst or the reactants, or with change in the reaction mechanism while the value of thermal energy (RT) varies with a small value. Consequently, from the exponential dependence of k on E_a, we can conclude that the reduction in activation energy in the presence of a catalyst is responsible for increase in the rate of reaction. From PED, it is also clear that the nature of intermediate species also plays an important role behind the action of catalyst. At this point, the understanding of surface behavior of a heterogeneous catalyst plays a crucial role since the intermediate species get adsorbed the catalyst's surface.

1.2.2 Catalyst Mechanism

The mechanism of heterogeneous catalysts can be explained on the basis of the hypothesis that the reactants and the catalysts are in different phases. Most of the heterogeneous catalysts are solids, while the reactants are gases or liquids. This theory assimilates the traditional adsorption model with the intermediate compound formation. Adsorption is a basic phenomenon of heterogeneous catalysis where the heat is liberated during adsorption of reactants on catalysts surface. This heat is used to break the bond between the reactants and thus favor the formation of products.

In adsorption process, reactants accumulate on the surface of the catalyst instead of getting absorbed in a solid or liquid. This is because of the unbalanced attractive forces. The molecules that adhere to the surface are known as adsorbates, while the substance on which adsorption occurs is called the adsorbent. Desorption is just the reverse of adsorption in which adsorbate get separated from the surface of adsorbents [4].

Some examples of adsorption process are as follows:

- Adsorption of O_2, Cl_2, H_2 and NH_3 gas on charcoal surface
- Adsorption of moisture on silica gel

The mechanism of adsorption on the surface of heterogeneous catalyst is performed in the following five steps (Figure 1.3):

Step 1: Reactant molecules get diffused on the surface of the catalyst.

Step 2: Diffused reactant molecules physically adsorbed in initial phase and then chemically adsorbed at catalyst surface *via* bond formation between reactant and the catalyst.

Step 3: Thus, adsorbed reactant molecules react with each other to form the product. When reactants are attached to the surface, they have a greater chance of interacting and forming an intermediate molecule.

Step 4: This step is the reverse of step 1, where the product formed is desorbed from the surface by breaking of bonds between catalyst and product. The surface is now available for further adsorption of other reactant molecules.

Step 5: In this step product molecules diffuse from the catalyst surface.

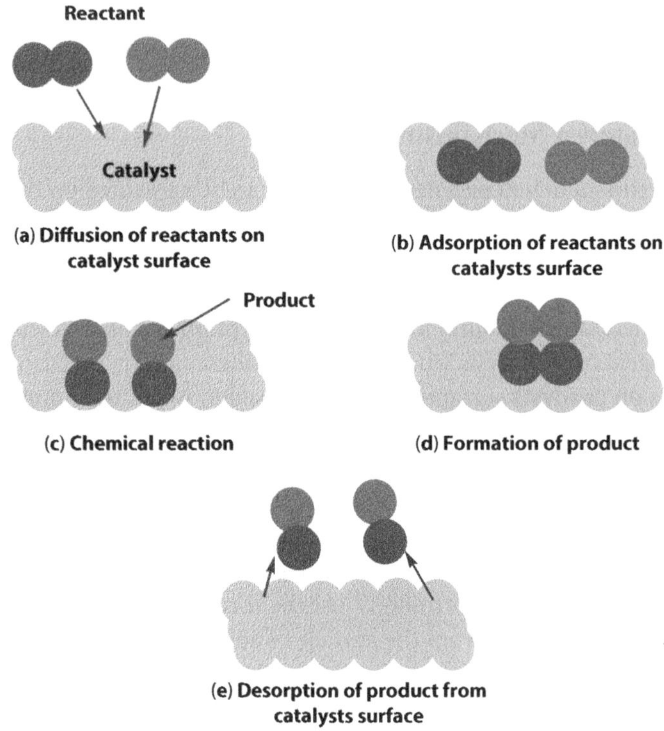

Figure 1.3 Mechanism of heterogeneous catalysis.

1.2.3 Catalyst Design

Various factors affect the catalytic efficiency including temperature, pressure, and surface area. Their effects should be taken into account while designing effective catalysts. Catalysts do not possess uniform active sites on the entire surface, but only specific active sites are involved in the catalytic reaction. The number of active sites available on catalysts surface depends upon the surface area of the catalyst. For this reason, catalyst designing is performed in a way so that the resulting catalyst will be of high surface area with superior porosity. Most of the synthesized solid catalysts for industrial applications possess surface area in the range between 50-450 m^2/g which make the process economical. Along with this, such catalysts also enhance the catalytic efficiency. For further improvement in catalytic efficiency, several supported solid catalysts with increased surface area and stability have been synthesized. Such support materials are less reactive with high melting point and porous. Some commonly used support materials are silica, alumina, zirconia and zeolite etc. The essential requirement for a support material is the free movement of reactants and products on the catalyst surface. Furthermore, to increase the catalyst performance, some promoters are added to the catalyst like alumina which is used as a promotor for iron catalyst in synthesis of ammonia. The selection of a suitable promotor is essential as it enhances the activity, selectivity and stability of the catalyst as well as the support material. The addition of promoters can contribute to the following aspects [5]:

- ✓ Enhancing thermal stability.
- ✓ Preventing unwanted secondary reactions that lead to catalyst deactivation
- ✓ Improving the dispersion of the active phase metal.

To design an efficient catalyst, a balance between mechanical, physiochemical and catalytic properties are most required. Figure 1.4 illustrates how these interconnected properties are affected by factors such as process type, reactor design, operating conditions, and economic considerations. For example, selecting an appropriate catalyst particle size ensures optimal flow distribution and minimizes pressure drop in the reactor while maintaining sufficient mechanical strength and resistance to attrition.

High activity and selectivity are essential for catalysts, which are primarily determined by different factors, namely, chemical composition, preparation method and synthesis conditions. Both the catalytic materials and the synthesis process are crucial for achieving the desired specific surface area and ensuring good dispersion of the metal active phase on the support. Additionally, the trade-off between porosity and mechanical strength must be considered; a highly porous catalyst may provide higher activity but can compromise mechanical strength [7]. Beyond activity, stability is also critical. Catalyst lifespan depends on its resistance to deactivation

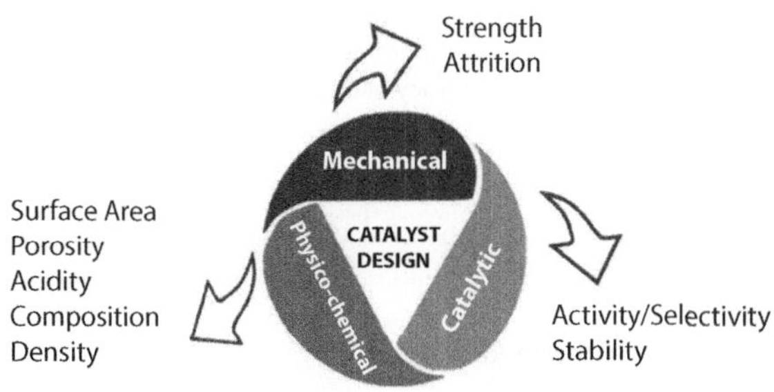

Figure 1.4 Mechanical, physio-chemical and catalytic properties of catalyst [6].

mechanisms, particularly metal sintering and coke deposition in reforming reactions. While thermal stability and coke resistance are influenced by process conditions, these properties are often enhanced by the addition of a promoter [8].

1.2.4 Thermodynamics and Kinetics of Catalysis

The thermodynamic study of a catalytic system provides basic comprehensions about the spontaneity and feasibility of that particular chemical reaction. It also plays a crucial role in designing a stable catalyst. Some key concepts of thermodynamics (Scheme 1.1) are shown below:

Thermodynamics is the study of the relationship between free energy (G), enthalpy (H) and entropy (S). The first and second law of thermodynamics and certain formulas explain the relationship between G, H and S (Figure 1.5).

Figure 1.5 shows the thermodynamic parameters of the system.

$$G = H - TS$$

First Law of Thermodynamics	Second Law of Thermodynamics	Gibbs Free Energy
It states that the energy neither be created nor destroyed but can converted into one form to another form.	It state that the state of entropy of entire universe as an isolated system increases over the time.	Gibbs free energy is related to the enthalpy and entropy as in the given equation: $\Delta G = \Delta H - T\Delta S$
Concerning a catalytic system, this law emphasized on conversion of energy during a chemical reaction.	Concerning a catalytic system, this law helps in predicting the direction of spontaneous reactions and the efficiency of energy conversion.	The value of free energy helps in determining spontaneity of the chemical reaction.
		If AG is negative, reaction will be spontaneous.
		If AG is positive, reaction will be non-spontaneous

Scheme 1.1 Key concepts of thermodynamics.

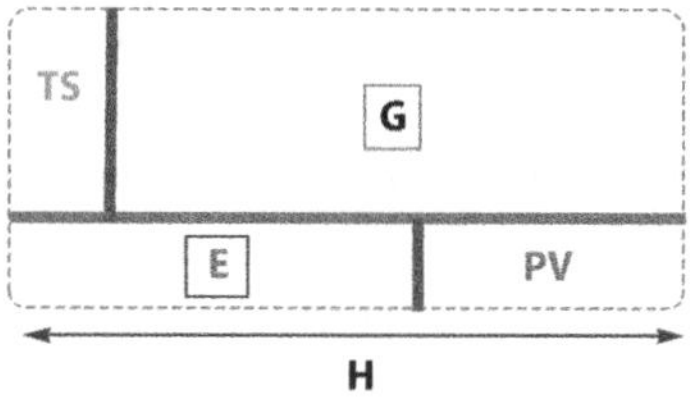

Figure 1.5 Thermodynamic parameters of a system.

The change in free energy of system can be given as:

$$\Delta G = \Delta H - T\Delta S$$

Where

- A negative value of ΔH revealed that the heat is released during the reaction (Exothermic reaction), while the positive value of ΔH revealed that the heat is absorbed (endothermic reaction). The role of a catalyst is to reduce the activation enthalpy, thus increasing the rate of reaction.
- The change in entropy (ΔS) measures the disorder or randomness of a system. For an efficient catalytic process catalysts are designed in a way to give optimized values of enthalpy and entropy, to achieve favorable reaction conditions.

The optimized values of S and H, helps in determining the free energy change (ΔG), which further decides the spontaneity and equilibrium position of a reaction. Catalysts do not alter ΔG but provide a pathway with a lower activation free energy (ΔG), thus increasing the reaction rate. Indeed, the activation energy is reduced to a change in the Gibbs free energy when the initial substances are converted into reaction products. Catalysts lower the activation energy (E_a) required for a reaction, thereby increasing the reaction rate. The design of effective catalysts often involves understanding and manipulating the thermodynamic parameters to minimize E_a. Along with this, thermodynamics also helps identify the most favorable reaction pathways. Catalysts are designed to provide alternative pathways with lower activation energies and favorable thermodynamic properties. Thermodynamic stability of catalysts and intermediates is crucial for sustained catalytic activity. Catalysts must also be selective, promoting the desired reaction over side reactions. Thermodynamic analysis aids in designing catalysts that achieve these objectives.

In terms of thermodynamics, catalysis helps in speeding-up the reaction, which alters the reaction path by lowering the activation energy. This phenomenon is important in terms of saving energy and increasing the efficiency of the process. The presence of a catalyst does not alter the Gibbs free energy, confirming that the catalyst only speeds up the reaction but does not alter the thermodynamic equilibrium of the reaction.

Kinetics, the study of the rates of chemical reactions and the factors affecting them, is a critical aspect of catalytic processes. Kinetics helps in evaluating catalyst activity, which is the ability of the catalyst to accelerate the reaction rate. Understanding the kinetic behavior, allows the selection of catalysts that offer the highest activity for a given reaction. Selectivity refers to the catalyst's ability to direct the reaction towards a specific product. Kinetic studies help in understanding how reaction conditions and catalyst properties influence selectivity, enabling the design of catalysts that minimize unwanted by-products. Catalyst stability is crucial for long-term performance. Kinetic analysis can identify factors leading to catalyst deactivation, such as sintering, coking, or poisoning, and help in developing strategies to enhance catalyst longevity. Understanding the kinetics of catalytic processes provides insights into the reaction pathways and mechanisms. This knowledge is essential for optimizing reaction conditions and improving catalyst performance.

1.2.5 Catalyst Deactivation

In the last ten years, extensive analysis has been conducted on the mechanism and causes of catalyst deactivation in order to understand the basis of catalysis, thus establishing the foundation for modeling deactivation processes and improving catalyst design [7, 9, 10].

The sustainability of a catalytic process mainly depends on the activity and longevity of catalyst. While the decline in catalytic activity over the time is inevitable, various strategies can be implemented to enhance catalytic performance. The basic cause of catalyst deactivation is mainly divided into three factors: i) mechanical ii) thermal and iii) chemical. Only a single factor does not affect the deactivation, but all three factors affect altogether [11].

Deactivation of a catalyst means a loss in activity of the catalyst after a certain period. Deactivation of the catalyst is due to deposition of some substances called poisons on its surface [12]. These poisons get chemically adsorbed on the catalyst surface and reduce the number of active sites present on the surface of the catalyst, which is responsible for decrease in reaction rate. Some of the common elements like sulfur, oxygen, phosphorus, and chlorine (non-metal contaminants) reduce the catalyst activity of metal catalysts generally by adsorption on the catalyst surface. Metal catalysts are used in hydrogenation reactions. In addition, alkali and alkali earth metals (metal contaminants) reduce the activity of acidic catalysts like zeolites by neutralizing strong acid sites. Along with this, in some cases, the interaction between non-metal anions and metal cations is also responsible for catalyst poisoning due to deposition of salt, for example calcium phosphate [13]. One example of catalyst poison is sulfur which reduces the production of methanol by poisoning the Cu/ZnO catalyst [14].

The proposed mechanism of catalytic poisoning by sulfur on metal surface in hydrogenation of ethylene shows that sulfur atoms are strongly adsorbed on the metal surface and block at-least one-three or four-fold active sites. Additionally, due to strong chemical bonding, it accumulates a large surface of catalyst which affects the adsorption of reactant molecules on metal surface as in case of hydrogen and ethylene molecules (Figure 1.6) [9].

The mechanism of catalyst deactivation can be broadly divided into six categories namely (i) sintering (ii) fouling (iii) coking (iv) vapor compound formation *via* transportation (v) vapor-solid and/or solid-solid reactions (vi) attrition/crushing. In sintering, catalytic metal particles transfer on the surface of support material on heating and get deposited which results in reduction of surface area. In fouling, the catalytic material deposits from liquid phase to solid surface, and thus becomes responsible for blockage of surface-active sites of the catalyst. Coking is the deposition of carbon particles on the catalyst surface. In vapor-solid and/or solid-solid reactions, an inactive layer of volatile compound forms on the catalyst surface which results in reduced surface area. In vapor formation, gas reacts with catalysts to produce volatile compounds. In attrition or crushing, loss of catalytic material/internal surface area occurs due to abrasion or crushing

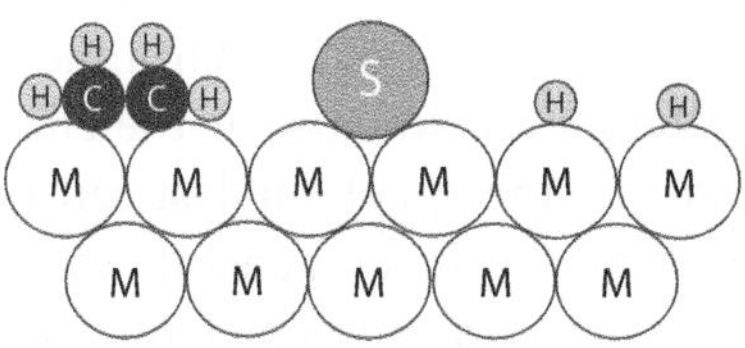

Figure 1.6 Proposed model of poisoning by sulfur atoms of a metal surface during ethylene hydrogenation.

of catalyst particles. In industry, catalyst deactivation costs billions every year due to process shutdown and catalyst replacement [9].

Despite poisons, there are some substances which increase the activity of the catalyst called promoters. For example, alkali metals that act as promoter in ammonia synthesis increases the rate of N_2 dissociation [15]. The use of catalyst poisons or promoters alter the activation energy of the reaction and thus affect the percent selectivity of the final product. The amount of catalyst poison and promoter also influences the rate of a chemical process. For example, less amount of chlorine acts as a promoter for Ag catalyst in the production of ethylene while high amounts of chlorine act as poison.

Poisoning of catalysts can be mitigated by the following methods:

- By catalyst regeneration
- By designing contaminant-resistant catalysts
- By removing impurities before introducing inside the catalyst bed.

1.3 Types of Catalysts

Catalysts are the substances which when added to a chemical reaction, increases the rate of the reaction without being consumed in the reaction. Indeed, a catalyst accelerates the reaction by reducing the activation energy or by providing the alternate reaction path. This makes it convenient for reactant molecules to reach the transition state and get converted into product in a lesser time. Conventionally, catalysts are divided into three categories: homogeneous, heterogeneous and biocatalysts (Figure 1.7).

Unlike heterogeneous catalysts, homogeneous catalysts are soluble in reaction medium and offer better catalytic performance due to uniform particle size. However, the major problems with the use of homogeneous catalysts are their separation from the reaction mixture, which makes the process costly as it needs extra steps to separate the catalysts from the mixture. Also, the use of homogeneous catalysts creates some toxic chemicals as by-products which can cause severe environmental issues. Therefore, the scientists started doing experiments for the effective recycling of the remaining solution, so that the spent catalyst can be reused in the next reaction cycle. In this way the heterogeneous catalysts which are present in a different phase than from the reactants, are the best alternatives of homogeneous catalysts as they can be easily separated from the reaction mixture by simple filtration method. This property of heterogeneous

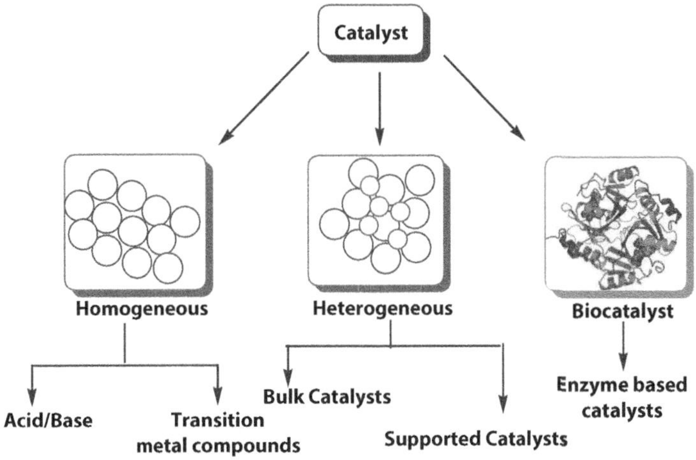

Figure 1.7 Types of catalysts.

catalysts makes them preferable in industries for synthesis of value-added chemicals. In heterogeneous catalysis, catalysts adsorb at the surface of the reactant, for that reason heterogeneous catalysts should possess high surface area.

The basic components of heterogeneous catalysts are summarized in Figure 1.8 [6]:

- ➢ Active phase consists of metal representing catalytic active sites responsible for a chemical reaction.
- ➢ A solid support with high surface area, required to disperse and stabilize the active phase.
- ➢ The promoter, an additive that enhances the catalyst's performance by improving its activity, selectivity, and lifespan.

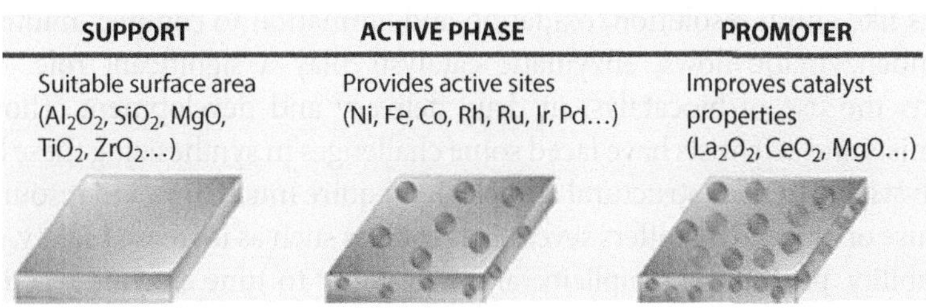

Figure 1.8 Basic components of a heterogeneous catalyst [6].

The summary of homogeneous and heterogeneous catalysts is given below:

Homogeneous catalysis	Heterogeneous catalysis
Liquid phase Low temperature High selectivity High diffusivity Easy heat transfer Difficult catalyst separation Expensive recycling Well defined active sites Easy catalyst modification Well understood reaction mechanism	Solid/liquid/gas phase High temperature Low selectivity Low diffusivity Heat transfer can be an issue Easy catalyst separation Simple recycling Poorly defined active sites Difficult catalyst modification Poorly understood reaction mechanism

Apart from homogeneous and heterogeneous catalysts, biocatalysts are also a widely known category of catalysts. Biocatalysts are the natural substances like enzymes (trypsin and pepsin) received from biological sources. These enhance the rate of a biochemical reaction. Biocatalysts play an important role in modern organic synthesis, both in academia and industries due to quick expansion of a variety of biochemical reactions. This is possible due to advanced tools and optimization techniques available for biocatalyst optimization. In recent years a variety of highly efficient and selective tuneable enzymes have been prepared in large quantities with the aim to ease accessibility of these biocatalysts to the scientific communities [16].

In the past few decades, biocatalysts have been significantly utilized in fine chemicals and pharmaceutical industries. Latest developments in DNA sequencing, bioinformatics and protein engineering ensure the successful implementation of specific and selective enzymes in the synthesis processes [17]. Moreover, nowadays in every aspect from conventional organic synthesis like chiral resolution, oxidation and amination to complex multistep chemoenzymatic flows, enzymatic catalysts play a significant role. This uplifts the use of biocatalyst in drug delivery and development. Though chemists and scientists have faced some challenges in synthesizing these biocatalysts due to their structural complexity require much time and resources. The use of biocatalysts offers several advantages such as increased safety, sustainability, procedural simplicity, and the ability to tune enzyme reactivity or selectivity through protein engineering. The sustainability of enzymatic transformations is particularly motivating their adoption in industrial processes. Unlike organometallic catalysts developed from precious metals, which are depleting rapidly, enzymatic catalysts can be produced sustainably. These advantages position bio-catalysis for mainstream adoption, making it accessible to enzyme novices and global biotech companies alike.

In the early 1800s, researchers and scientists appreciated the use of enzymes such as yeast in the fermentation process [18, 19]. Earlier it was believed that the entire microorganism is taking part in the reaction, but after the discovery of diastase enzyme, this belief was changed because only the specific part of the organism was involved in the reaction. This specificity of enzymes can be categorized as follows:

- **Bond specific:** Relatively specific about the bond within the substrate. For example, lipase enzyme hydrolyses only ester bonds in fats.
- **Group specific:** Very specific to the type of bond and surrounding groups, like endopetidases and exopeptidases.
- **Substrate specific:** Highly specific towards the substrate, like lactose, sucrose, maltose has a specific function and does not perform any other reaction.
- **Stereo specific:** Enzymes are very specific about the arrangement of atoms in the space, like L-amino acid only acts as L-amino acid not D-amino acid.
- **Geometric specific:** Less specific about the size and shape of the substrate, like alcohol dehydrogenase can react with any alcohol functional group like methanol, ethanol and propanol etc.

The latest research in enzymology, helped in understanding the reaction mechanism of enzymes. These milestones include the development of the lock and key model, the achievement of cell-free fermentation, the realization that enzymes are proteins, and the elucidation of DNA structure all of which have paved the way for modern bio-catalysis.

1.4 Modern Tools in Catalysis Research

Research in heterogeneous catalysis has become a transdisciplinary field and is facing challenges from multiple directions, like disparities in composition and design of catalyst, experimental conditions, deactivation, formation of undesired side products, optimizing product yield and selectivity, and many more [20, 21]. Conventional approaches like gaining conceptual knowledge about reactants, reaction atmosphere, reaction mechanism and doing experiments cannot solve the above discussed problems completely [22]. Heterogeneous catalyst discovery using 21st century tools: a tutorial]. Blending modern tools like mathematical, computational modelling, advanced characterization techniques, artificial intelligence (AI), machine learning (ML) can provide efficient solutions in this area. Some of these modern tools are discussed below-

1.4.1 Mathematical Modeling

Mathematical modeling techniques like response surface methodology (RSM), artificial neural network (ANN), central composite design (CCD) use different algorithms to predict reaction conditions, yield, selectivity of end products, kinetics of the reaction before actual experimentation. These statistical models and analysis not only provide convenience in handling multiple variables simultaneously but also reduce consumption of chemicals, time, energy and cut overall cost of the chemical process. Khudsange and Wasewar used RSM and ANN for modeling butyl butyrate synthesis *via* esterification reaction [23]. They also compared both models and found out that ANN model predicted yield of butyl butyrate and optimized reaction conditions better than RSM. Error values and accuracy of both models were determined by applying different validation tests. Similarly, Soltani and co-workers optimized biodiesel production by ultrasound-assisted transesterification of sesame oil. They also compared two methods, RSM and a combined model of ANN-particle swarm algorithm (ANN-PSO) [24]. They studied the effects of many variables like molar ratio of methanol: oil, reaction temperature, time, NaOH catalyst % on biodiesel yield. The modeling analysis showed good harmony with experimental results, but optimal points were more accurately predicted by combined ANN-PSO model. Irfan and team extracted bio-oil from rice husk ash (RHA) using pyrolysis [25]. They also utilized the RSM model to optimize reaction variables for getting maximum bio-oil yield %. The results concluded that agricultural waste can be a potential source of bio-oil extraction, and the process can be conveniently carried out by integrating RSM before experimentation. A better overview of RSM predicted results vs experimental results is shown in Figure 1.9.

The reaction parameters of one more esterification reaction was also optimized using RSM, ANN-Genetic Algorithm (ANN-GA) mathematical models [26]. In ferric sulfate catalyzed esterification of neem oil, apart from optimizing reaction parameters, the amount of free fatty acids was reduced to a recommended level of $\leq$ 1% using mathematical models. Results demonstrated that the combined ANN-GA model outperformed the RSM model in this study.

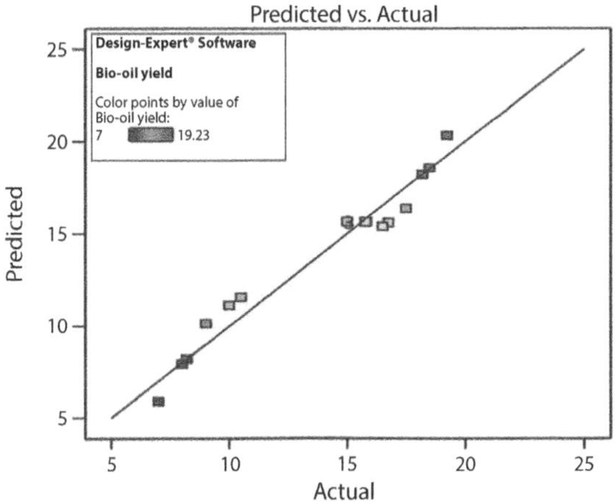

Figure 1.9 Comparison between RSM predicted vs actual bio-oil yield [25].

1.4.2 Computational Chemistry

Computational chemistry is an advanced technique to study reaction mechanisms at atomic scale. It helps in designing complex catalysts and determining multi-step reaction pathways. It can be done by using different computational models, simulations, for example, density functional theory (DFT), molecular dynamics (MD), kinetic modeling, quantum mechanics, computational software, and many more as shown in Figure 1.10 [27]. These models also give us insights about choice of solvent, reaction medium (microwave, ultrasound, vapor phase, photo irradiations etc.), choice of transition metals/ions, quadratic energy potential (QEP), etc. Free energy barrier can be calculated with the help of computational chemistry methods which gives us the probability of applying transition state theory related to use of catalysts for a specific reaction. Hosseini and co-workers reviewed the recent work done on designing different catalysts with the help of computational chemistry and its experimental validation [28]. Barona *et al.* applied DFT calculations to check the structure and performance of meta-organic frameworks (MOFs) as catalysts in alkane C-H bond activation reactions. For experimental validation, they synthesized many catalysts using combinations of different metals like Mn, Fe, Co and Ni. The lab results showed that MOFs catalyst with Fe, Mn metals were more active than other catalysts which was an absolute agreement with DFT calculations and predictions. Gonzalez and team used molecular dynamics to study different aspects of biodiesel production, along with complex features of fatty acids [29]. Experimental determination of these features is quite challenging. Murphy and Xu also analyzed the synthesis of multifunctional molecules from

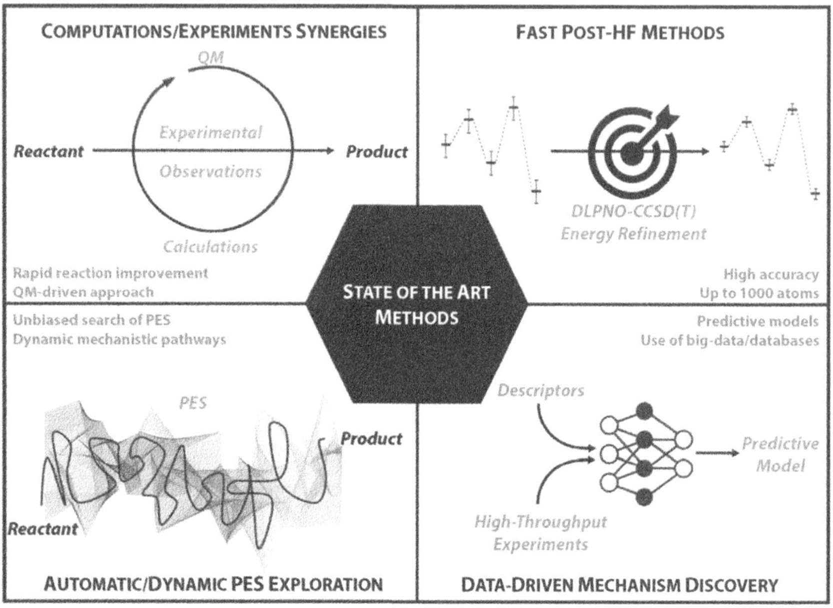

Figure 1.10 Different methods used in computational chemistry. Adapted with permission from [27].

biomass waste rich in lignin and cellulose, using computational methods. They used kinetic information and results of molecular spectroscopies to validate the data provided by computational methods [30].

Felice and group worked designing a computational quantum chemistry software by integrating principles of chemistry, mathematical calculations and computer science [31]. They proposed the utility of this software in numerous applications like battery production, catalyst design, advancing clean energy, fuel production, gaseous chemical reactions, heavy metal chemistry, etc. Literature also shows the use of computational simulation tools in solid waste management by thermal incineration [32]. Molecular dynamics has also been used to determine viscosity and density of bio-aviation fuel blends [33]. The computational results of this study were then verified by using theoretical models and lab experiments. All the results coincided with each other proving the utility of modern tools in chemistry.

1.4.3 Artificial Intelligence and Machine Learning

The onset of 2020 marked the increasing applications of data processing tools- AI and ML in catalysis research in combination with computational methods. ML is a set of advanced data analysis tools and processes which when fed with datasets, determine patterns among various variables. Based on these patterns, they can predict experimental results, optimized reaction parameters even when the input variables get changed. It is a blend of concepts of statistics, mathematics and probability [34]. In a typical catalyzed organic reaction, input variables constitute reaction time, temperature, catalyst: substrate ratio, catalyst concentration while yield %, selectivity of products are main output parameters. In AI methods, machines are simulated to think like the human brain. They can solve problems, learn new things and make decisions on their own. With the help of ML, natural language processing and deep learning, AI can analyze complex, vast datasets, identify patterns, predict output and make self-decisions to solve the problems. It has been greatly used in drug delivery applications [35]. AI and ML can help in most of the steps of catalyst synthesis including theoretical catalyst design, optimization of conditions of synthesis, automated catalyst preparation and analysis of characterization results [36]. Gu *et al.*, used ML to design the electrocatalyst. Later they also utilized this tool in application of as-synthesized electrocatalyst [37]. In this work, ML helped in catalyst design, reaction processes, optimization of reaction energy barriers. A review deeply analyzed the role of ML in studying reaction mechanism, kinetics of catalyzed reactions [38]. In this review, Araujo *et al.*, studied in detail and presented an elaborated overview of theory-driven and data-driven approaches used in catalysis. Figure 1.11 depicts the possible applications of AI and ML in the field of catalysis.

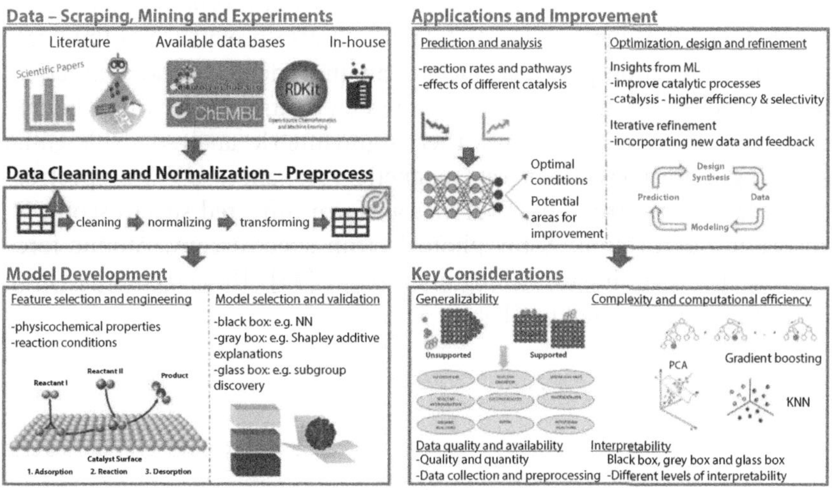

Figure 1.11 Possible applications of AI and ML in catalysis [38].

Xing and team used combined DFT and ML in CO_2 electroreduction reaction to determine modifications in metals and ligands of MOF catalyst [39]. DFT analyzes catalytic performance while ML analyzes atomic features like electron affinity, first ionization energy, electronegativity, radius of metal atom etc. They examined about 37 MOFs and found 6 most compatible ones. ML has also proven its application in wastewater treatment. Models based on ML were used to predict catalytic performance of silver-carboxymethyl cellulose in adsorption of nitrophenols and azo dyes from wastewater [40]. Both experimental and ML-model predictions coincide in this work. Mohammadpour and team synthesized photocatalytic nanocomposites using transition metals like Fe, Cu, Zn, Ga with spent coffee ground extract and gum arabica and utilized in nitrate removal from water samples [41]. In this work, RSM is integrated with ML models. Kohzadi *et al.*, assessed photocatalytic performance of neodymium doped on TiO2 along with their toxicity. It was found out that the photocatalyst can efficiently remove about 95.4% Rhodamine B dye in presence of sunlight while 98.12% dye is removed under UV light [42]. In waste management, AI and ML methods are used to sort waste, improve pyrolysis process, conserve energy, estimate carbon emission, cost savings. They can also handle logistics, resource recovery, illegal dumping, monitoring and tracking, process efficiency and help in building smart cities [43, 44].

1.4.4 Advanced Characterization Techniques

Our chemical processes have been equipped with cutting-edge characterization techniques helping in assessing structural, morphological, chemical properties of heterogeneous catalysts. It also provides real time data to monitor reaction progress. Such techniques include X-ray diffraction, Scanning electron microscopy, Transmission electron microscopy, Energy dispersive X-ray spectroscopy, Thermogravimetric analysis, Differential scanning calorimetry, Nuclear magnetic resonance, UV-Vis spectroscopy, Raman spectroscopy, Fourier transform infra-red spectroscopy, N_2 adsorption-desorption, X-ray photoelectron spectroscopy, Atomic force microscopy, Scanning tunnelling microscopy [45]. These techniques have been discussed in detail in Chapter 5.

1.5 New Age Catalysts

Modernisation of chemical processes, changing customer requirements, updated environmental regulations, rise of computer-aided techniques compelled researchers to explore new avenues in catalyst synthesis. Now-a-days not only the process of synthesizing catalysts has been changing but also new substrates have been discovered for making heterogeneous catalysts. They are termed 'new age catalysts' such as catalysts derived from wastes, renewable resources, biocatalysts, environmental catalysts, nano catalysts, MOFs and many more. Velusamy reviewed a lot of nanocatalysts and catalysts derived from biochar in production of biofuels. They presented different ways to synthesize nanocatalysts as well as modification of biochar to form heterogeneous catalysts [46]. Damian *et al.* also investigated a number of nanocatalysts used in biodiesel production. [47]. They concluded that nanocatalysts can be conveniently recovered from reaction mixture and easily regenerated to be reused in next reaction cycles, thus saving overall process cost. Also, nanocatalysts gave better catalytic performance as compared to traditional heterogeneous catalysts. In another study, Damian and Devrajan assessed catalytic performance of different nanocatalysts based on metal oxides, carbon, magnetic nanoparticles being used in biodiesel production from waste cooking oil, animal fats etc. [48]. Nanocatalysts have applications in diverse fields like drug delivery, synthesis of organic molecules [49], wastewater treatment and many more.

The use of enzymes/biocatalysts in carrying out chemical reactions have been emerging as another interesting avenue. They have major applications in pharmaceutical research, drug delivery, synthesis of metabolites, intermediates, discovery chemistry [50]. Experimental protein engineering, computational designing are widely known methods used to produce such catalysts. ML and computational modelling are highly beneficial in designing biocatalysts at industrial scale [51]. Biocatalysts were initially used to study enzyme-mediated kinetic resolutions. Higher yield, targeted product formation and higher enantioselectivity are some of the prominent features of using biocatalysts [17].

Environmental catalysts represent a broad class of materials which catalyse the reactions and processes which can improve environment, sustainability and mitigate pollution. Isahak and Al-Amiery explored different catalysts such as zeolites, transition metals, metal oxides etc. to drive thermal reactions [3]. These thermal reactions have huge potential in the petrochemical sector. Zhu *et al.* studied environmental catalysts at molecular level with application in water purification. They demonstrated effects of microenvironment engineering on catalytic performance [52].

Higher surface area, greater porosity, tunable structural properties and diversity are the major features of MOFs which make them ideal candidates for catalysing organic reactions, adsorption, photocatalysis etc. They can behave as hosts for metal nanoparticles, forming complex catalytic solids which have been used in photocatalysis and electrocatalysis [53]. MOFs can be used in green hydrogen production by water splitting [54]. In this application, MOFs act as photocatalysts. As photocatalysts they are also used in removal of organic pollutants like antibiotics, dyes, phthalates, pesticides and so on [55, 56]. They also have applications in polymerisation reactions, drug-delivery, wastewater treatment [57, 58].

Waste-derived catalysts have been considered as a major class of new age catalysts. They have been derived from industrial, agricultural, municipal solid waste materials. Their catalytic performance is at par with conventional heterogeneous catalysts prominently used in reactions at industrial scale. This book is based on such catalysts, and the forthcoming chapters will provide a detailed study of these catalysts.

1.6 Conclusion

The global demand for sustainable and green processes has prompted researchers to put their efforts in synthesis of catalysts. Catalysts are responsible for the production of over 60 % of all chemicals and are used in about 90 % of all chemical processes worldwide. Catalyst manufacturing alone accounts for over 100-billion-dollar sales globally in major sectors namely: refining, fine chemicals, polymerization, exhaust emission catalysts, pharmaceuticals etc. Modern catalysis is centred on the development of more efficient catalysts, which must be used under both sustainable and eco-friendly conditions, hence envisaging all possible principles of green chemistry, with special attention to strategies to approach a renewable chemical industry. This chapter outlines the fundamental concepts of catalysis like activation energy, reaction pathways, reaction mechanisms, kinetics of reactions, kinetics of catalysts. Further it also tells us about the clogging of catalysts and their deactivation, ageing. The chapter also shows the basis of the classification of catalysts into different classes like homogeneous, heterogeneous, acid, base, oxidation, reduction etc. Modern tools like mathematical models, computational methods, AL, ML, advanced characterization techniques used in catalysis research are also described. The last section gives an overview of trending catalysts used in organic reactions, photocatalysis, wastewater treatment, drug discovery and lot more. This work is a stepping-stone for prospective researchers diving into the world of organic reactions.

AI disclosure statement: The author acknowledges the use of AI-based tools for minor language editing or grammatical correction. However, all intellectual contributions, analyses, and conclusions presented in this chapter are entirely the author's own.

References

1. Wisniak, J., The History of Catalysis. From the Beginning to Nobel Prizes. *Educ. Quim.*, 21, 1, 60–69, 2010.
2. Adams, C., Applied Catalysis: A Predictive Socioeconomic History. *Top. Catal.*, 52, 8, 924–934, 2009, https://doi.org/10.1007/s11244-009-9251-z.
3. Wan Isahak, W.N.R. and Al-Amiery, A.A.H., Catalysts driving efficiency and innovation in thermal reactions: A comprehensive review. *Green Technol. Sustainability*, 2, 2, 100078, 2024.
4. Qu, X., Hu, Q., Song, Z., Sun, Z., Zhang, B., Zhong, J., Cao, X., Liu, Y., Zhao, B., Liu, Z., Shen, Y., Bao, Y., Wang, Z., Zhang, Y., Niu, L., Adsorption and desorption mechanisms on graphene oxide nanosheets: Kinetics and tuning. *Innovation*, 2, 3, 100137, 2021.
5. Nogueira, F.G.E., Assaf, P.G.M., Carvalho, H.W.P., Assaf, E.M., Catalytic steam reforming of acetic acid as a model compound of bio-oil. *Appl. Catal.*, 160–161, 188–199, 2014.
6. Santamaria, L., Lopez, G., Fernandez, E., Cortazar, M., Arregi, A., Olazar, M., Bilbao, J., Progress on Catalyst Development for the Steam Reforming of Biomass and Waste Plastics Pyrolysis Volatiles: A Review. *Energy Fuels*, 35, 21, 17051–17084, 2021.
7. Argyle, M.D. and Bartholomew, C.H., Heterogeneous catalyst deactivation and regeneration: A review. *Catalysts*, 5, 145–269, 2015.
8. Kimura, T., Miyazawa, T., Nishikawa, J., Kado, S., Okumura, K., Miyao, T., Naito, S., Kunimori, K., Tomishige, K., Development of Ni catalysts for tar removal by steam gasification of biomass. *Appl. Catal., B*, 68, 160–170, 2006.
9. Bartholomew, C.H. and Farrauto, R.J., *Fundamentals of Industrial Catalytic Processes*, 2nd ed., John Wiley & Sons, Inc., Hoboken, NJ, 2010.
10. Denny, P.J. and Twigg, M.V., Factors determining the life of industrial heterogeneous catalysts. *Stud. Surf. Sci. Catal.*, 6, 577–599, 1980.
11. Bartholomew, C.H., Mechanisms of catalyst deactivation. *Appl. Catal., A*, 212, 17–60, 2001.
12. Cimino, S. and Lisi, L., Catalyst Deactivation, Poisoning and Regeneration. *Catalysts*, 9, 8, 668, 2019.

13. Rezania, S., Din, M.F.M., Taib, S.M., Sohaili, J., Chelliapan, S., Kamyab, H., Saha, B.B., Review on fermentative biohydrogen production from water hyacinth, wheat straw and rice straw with focus on recent perspectives. *Int. J. Hydrogen Energy*, 42, 20955–20969, 2017.
14. Lazzarotto, I.P., Ferreira, S.D., Junges, J., Bassanesi, G.R., Manera, C., Perondi, D., Godinho, M., The role of CaO in the steam gasification of plastic wastes recovered from the municipal solid waste in a fluidized bed reactor. *Process Saf. Environ. Prot.*, 140, 60–67, 2020.
15. Wang, Q., Guo, J., Chen, P., The impact of alkali and alkaline earth metals on green ammonia synthesis. *Chem*, 7, 12, 3203–3220, 2021.
16. Kuo, C.H., Huang, C.Y., Shieh, C.J., Dong, C.D., Enzymes and Biocatalysis. *Catalysts*, 12, 9, 993, 2022.
17. Pyser, J.B., Chakrabarty, S., Romero, E.O., Narayan, A.R.H., State-of-the-Art Biocatalysis. *ACS Cent. Sci.*, 7, 7, 1105–1116, 2021.
18. Reetz, M.T., Biocatalysis in Organic Chemistry and Biotechnology: Past, Present, and Future. *J. Am. Chem. Soc.*, 135, 34, 12480–12496, 2013.
19. Heckmann, C.M. and Paradisi, F., Looking Back: A Short History of the Discovery of Enzymes and How They Became Powerful Chemical Tools. *ChemCatChem*, 12, 24, 6082–6102, 2020.
20. Mukhtar, A., Saqib, S., Lin, H., Ul Hassan Shah, M., Ullah, S., Younas, M., Rezakazemi, M., Ibrahim, M., Mahmood, A., Asif, S., Bokhari, A., Current status and challenges in the heterogeneous catalysis for biodiesel production. *Renewable Sustainable Energy Rev.*, 157, 112012, 2022, doi: doi.org/10.1016/j.rser.2021.112012.
21. de Maria, P.D., Grand challenges in industrial catalysis: let´s put academia and industry on the same page! *Front. Catal.*, 3, 2023, 2024, https://doi.org/10.3389/fctls.2023.1359527.
22. Ras, E.J. and Rothenberg, G., Heterogeneous catalyst discovery using 21st century tools: a tutorial. *RSC Adv.*, 4, 5963–74, 2014, https://doi.org/10.1039/C3RA45852K.
23. Khudsange, C.R. and Wasewar, K.L., Experimental and Modeling Study of Esterification Reaction for Synthesis of Butyl Butyrate: Desirability Function Approach for Optimization and Prediction Comparative Study of RSM and ANN. *Int. J. Chem. Reactor Eng.*, 15, 3, 20160176, 2017, https://doi.org/10.1515/ijcre-2016-0176.
24. Soltani, H., Karimi, A., Falahatpisheh, S., The optimization of biodiesel production from transesterification of sesame oil *via* applying ultrasound-assisted techniques: comparison of RSM and ANN–PSO hybrid model. *Chem. Prod. Process Model.*, 17, 1, 55–67, 2022, https://doi.org/10.1515/cppm-2020-0076.
25. Irfan, M., Ghalib, S.A., Waqas, S., Khan, J.A., Rahman, S., Mursal, S.N.F., Ghanim, A.A.J., Response Surface Methodology for the Synthesis and Characterization of Bio-Oil Extracted from Biomass Waste and Upgradation

Using the Rice Husk Ash Catalyst. *ACS Omega*, 8, 20, 17869–79, 2023, https://doi.org/10.1021/acsomega.3c00868.

26. Okpalaeke, K.E., Ibrahim, T.H., Latinwo, L.M., Betiku, E., Mathematical Modeling and Optimization Studies by Artificial Neural Network, Genetic Algorithm and Response Surface Methodology: A Case of Ferric Sulfate–Catalyzed Esterification of Neem (Azadirachta indica) Seed Oil. *Front. Energy Res.*, 8, 2020, 2020, doi: 10.3389/fenrg.2020.614621.
27. Ardoiz, I.F. and Schoenebeck, F., Established and Emerging Computational Tools to Study Homogeneous Catalysis—From Quantum Mechanics to Machine Learning. *Chem*, 6, 8, 1904–13, 2020, https://doi.org/10.1016/j.chempr.2020.07.008.
28. Hosseini, H., Herring, C.J., Nwaokorie, C.F., Sulley, G.A., Montemore, M.M., Computational Design of Catalysts with Experimental Validation: Recent Successes, Effective Strategies, and Pitfalls. *J. Phys. Chem. C*, 128, 43, 18144–57, 2024, https://doi.org/10.1021/acs.jpcc.4c04949.
29. González, K.R.A., Perez, J.C.A., Núñez, E.V., Cárdenas, J.C., Mendoza, A., Aguilera, J.A.R., Gerstenmaier, S.F., Determination of liquid–vapor equilibrium and critical properties of fatty acids for biodiesel production through molecular dynamics. *J. Phys.: Condens. Matter*, 34, 21, 2022. 10.1088/1361-648X/ac5996.
30. Murphy, B.M. and Xu, B., Foundational techniques for catalyst design in the upgrading of biomass-derived multifunctional molecules. *Prog. Energy Combust. Sci.*, 67, 1–30, 2018, https://doi.org/10.1016/j.pecs.2018.01.003.
31. Felice, R.D., Mayes, M.L., Richard, R.M., Williams-Young, D.B., Chan, G.K.L., de Jong, W.A., Govind, N., Head-Gordon, M., Hermes, M.R., Kowalski, K., Li, X., Lischka, H., Mueller, K.T., Mutlu, E., Niklasson, A.M.N., Pederson, M.R., Peng, B., Shepard, R., Valeev, E.F., Schilfgaarde, M.V., Vlaisavljevich, B., Windus, T.L., Xantheas, S.S., Zhang, X., Zimmerman, P.M., A Perspective on Sustainable Computational Chemistry Software Development and Integration. *J. Chem. Theory Comput.*, 19, 20, 7056–76, 2023, https://doi.org/10.1021/acs.jctc.3c00419.
32. Lemieux, P., Boe, T., Tschursin, A., Denison, M.K., Davis, K., Swensen, D., Computational simulation of incineration of chemically and biologically contaminated wastes. *J. Air Waste Manage. Assoc.*, 71, 4, 462–476, 2021, https://doi.org/10.1080/10962247.2020.1853627.
33. Yang, X., Liu, Q., Ma, Y., Xie, J., Cao, B., Molecular dynamics study on the kinematic viscosity, density and structure of fuel blends containing n-decane and biofuel compound of ethyl decanoate or ethyl dodecanoate. *J. Mol. Liq.*, 379, 121680, 2023, https://doi.org/10.1016/j.molliq.2023.121680.
34. Osman, A.I., Nasr, M., Farghali, M., Rashwan, A.K., Abdelkader, A., Muhtaseb, A.H.A., Ihara, I., Rooney, D.W., Optimizing biodiesel production from waste with computational chemistry, machine learning and policy insights: a review. *Environ. Chem. Lett.*, 22, 1005–71, 2024, https://doi.org/10.1007/s10311-024-01700-y.

35. Choudhary, N., Bharti, R., Sharma, R., Role of artificial intelligence in chemistry. *Mater. Today: Proc.*, 48, 5, 1527–33, 2022, https://doi.org/10.1016/j.matpr.2021.09.428.
36. Zhang, L., Bing, Q., Qin, H., Yu, L., Li, H., Deng, D., Artificial intelligence for catalyst design and synthesis. *Matter*, 8, 5, 102138, 2025, https://doi.org/10.1016/j.matt.2025.102138.
37. Gu, Y., Zhang, H., Xu, Z., Ren, R., Kong, X., Wang, Y., Zhu, H., Xue, D., Zhang, Y., Ma, Y., Zhao, D., Zhang, J., Design and Application of Electrocatalyst Based on Machine Learning. *Interdiscip. Mater.*, 4, 3, 456–479, 2025, https://doi.org/10.1002/idm2.12249.
38. de Araujo, L.G., Vilcocq, L., Fongarland, P., Schuurman, Y., Recent developments in the use of machine learning in catalysis: A broad perspective with applications in kinetics. *Chem. Eng. J.*, 508, 160872, 2025, https://doi.org/10.1016/j.cej.2025.160872.
39. Xing, G., Liu, S., Sun, G.Y., Liu, J.Y., Modification of metals and ligands in two-dimensional conjugated metal-organic frameworks for CO_2 electroreduction: A combined density functional theory and machine learning study. *J. Colloid Interface Sci.*, 677B, 111–119, 2025, https://doi.org/10.1016/j.jcis.2024.08.069.
40. Khan, M.S.J., Sidek, L.M., Kumar, P., Alkhadher, S.A.A., Basri, H., Zawawi, M.H., Shafie, A.E., Ahmed, A.N., Machine learning based-model to predict catalytic performance on removal of hazardous nitrophenols and azo dyes pollutants from wastewater. *Int. J. Biol. Macromol.*, 278, 3, 134701, 2024, https://doi.org/10.1016/j.ijbiomac.2024.134701.
41. Mohammadpour, A., Kazemi, A., Baghapour, M.A., Samaei, M.R., Kannan, K., Khaneghah, A.M., Bioengineered FeZn/GA@Cu nanocomposite utilizing spent coffee ground extract and gum arabic: Enhanced nitrate removal *via* (RSM) and machine learning optimization. *Int. J. Biol. Macromol.*, 277, 2, 134060, 2024, https://doi.org/10.1016/j.ijbiomac.2024.134060.
42. Kohzadi, S., Bundschuh, M., Rezaee, R., Marzban, N., Vahabzadeh, Z., Johari, S.A., Shahmoradi, B., Amini, N., Maleki, A., Integrating machine learning with experimental investigation for optimizing photocatalytic degradation of Rhodamine B using neodymium-doped titanium dioxide: a comprehensive approach with toxicity assessment. *Environ. Sci. Pollut. Res.*, 31, 55301–16, 2024, https://doi.org/10.1007/s11356-024-34843-0.
43. Fang, B., Yu, J., Chen, Z., Osman, A.I., Farghali, M., Ihara, I., Hamza, E.H., Rooney, D.W., Yap, P.S., Artificial intelligence for waste management in smart cities: a review. *Environ. Chem. Lett.*, 21, 1959–89, 2023, https://doi.org/10.1007/s10311-023-01604-3.
44. Alsabt, R., Alkhaldi, W., Adenle, Y.A., Alshuwaikhat, H.M., Optimizing waste management strategies through artificial intelligence and machine learning - An economic and environmental impact study. *Cleaner Waste Syst.*, 8, 100158, 2024, https://doi.org/10.1016/j.clwas.2024.100158.

45. Anekwe, I.M.S., Akpasi, S.O., Enemuo, E.M., Ashiegbu, D., Mustapha, S.I., Isa, Y.M., Innovations in catalytic understanding: A journey through advanced characterization. *Mater. Today Catal.*, 7, 100061, 2024, https://doi.org/10.1016/j.mtcata.2024.100061.
46. Velusamy, K., Devanand, J., Kumar, P.S., Soundarajan, K., Sivasubramanian, V., Sindhu, J., Vo, D.V.N., A review on nano-catalysts and biochar-based catalysts for biofuel production. *Fuel*, 306, 121632, 2021, https://doi.org/10.1016/j.fuel.2021.121632.
47. Damian, C.S., Devarajan, Y., R.J., R.T., Nanocatalysts in Biodiesel Production: Advancements, Challenges, and Sustainable Solutions. *ChemBioEng Rev.*, 12, 2, 2025, https://doi.org/10.1002/cben.202400055.
48. Damian, C.S. and Devarajan, Y., A Comprehensive Review of the Impact of Nano-Catalysts on Biodiesel Production. *J. Biosyst. Eng.*, 49, 277–290, 2024, https://doi.org/10.1007/s42853-024-00234-z.
49. Jangir, N., Bagaria, S.K., Jangid, D.K., Nanocatalysts: applications for the synthesis of N-containing five-membered heterocycles. *RSC Adv.*, 12, 30, 19640–66, 2022, https://doi.org/10.1039/d2ra03122a.
50. France, S.P., Lewis, R.D., Martinez, C.A., The Evolving Nature of Biocatalysis in Pharmaceutical Research and Development. *JACS Au*, 3, 3, 715–35, 2023, https://doi.org/10.1021/jacsau.2c00712.
51. Bell, E.L., Hutton, A.E., Burke, A.J., O'Connell, A., Barry, A., O'Reilly, E., Green, A.P., Strategies for designing biocatalysts with new functions. *Chem. Soc. Rev.*, 53, 2851–62, 2024, https://doi.org/10.1039/D3CS00972F.
52. Zhu, Z.S., Zhong, S., Cheng, C., Zhou, H., Sun, H., Duan, X., Wang, S., Microenvironment Engineering of Heterogeneous Catalysts for Liquid-Phase Environmental Catalysis. *Chem. Rev.*, 124, 20, 11348–434, 2024, https://doi.org/10.1021/acs.chemrev.4c00276.
53. Bavykina, A., Kolobov, N., Khan, I.I.S., Bau, J.A., Rmairez, A., Gascon, J., Metal–Organic Frameworks in Heterogeneous Catalysis: Recent Progress, New Trends, and Future Perspectives. *Chem. Rev.*, 120, 16, 8468–535, 2020, https://doi.org/10.1021/acs.chemrev.9b00685.
54. Shi, Y., Yang, A.F., Cao, C.S., Zhao, B., Applications of MOFs: Recent advances in photocatalytic hydrogen production from water. *Coord. Chem. Rev.*, 390, 50–75, 2019, https://doi.org/10.1016/j.ccr.2019.03.012.
55. Khan, M.S., Li, Y., Li, D.S., Qu, J., Xu, X., Yang, H.Y., A review of metal–organic framework (MOF) materials as an effective photocatalyst for degradation of organic pollutants. *Nanoscale Adv.*, 5, 6318–48, 2023, https://doi.org/10.1039/D3NA00627A.
56. Patial, S., Raizada, P., Hasija, V., Singh, P., Thakur, V.K., Nguyen, V.H., Recent advances in photocatalytic multivariate metal organic frameworks-based nanostructures toward renewable energy and the removal of environmental pollutants. *Mater. Today Energy*, 19, 100589, 2021, https://doi.org/10.1016/j.mtener.2020.100589.

57. Goetjen, T.A., Liu, J., Wu, Y., Sui, J., Zhang, X., Hupp, J.T., Farha, O.K., Metal–organic framework (MOF) materials as polymerization catalysts: a review and recent advances. *Chem. Commun.*, 56, 10409–18, 2020, https://doi.org/10.1039/D0CC03790G.
58. Yusuf, V.F., Malek, N.I., Kailasa, S.K., Review on Metal–Organic Framework Classification, Synthetic Approaches, and Influencing Factors: Applications in Energy, Drug Delivery, and Wastewater Treatment. *ACS Omega*, 7, 49, 44507–31, 2022, https://doi.org/10.1021/acsomega.2c05310.

2

Introduction to Solid Waste-Derived Catalysts

Avinash K. Srivastava[1*†], Munsaf Ali[2†] and Outi M. Keinänen[1]

[1]Department of Chemistry, University of Alabama at Birmingham, Birmingham, AL, USA
[2]Department of Nanoscale Science and Engineering, University at Albany, Albany, New York, USA

Abstract

In the realm of catalysis, using waste materials to create catalysts has shown promise and offers financial and environmental advantages. With an emphasis on the synthesis, characterization, and catalytic performance of waste-derived catalysts, this chapter examines current advancements and uses of these materials. Numerous waste materials are investigated as possible precursors for catalyst production, including post-consumer waste, industrial byproducts, and agricultural leftovers. Using leading-edge techniques like pyrolysis, hydrothermal treatment, and chemical modification, these waste materials may be transformed into catalysts that can be used in catalytic processes. Additionally, the efficacy of waste-derived catalysts is assessed in various applications, including energy production, biomass conversion, and wastewater treatment. Waste-derived catalysts' sustainable nature lessens reliance.

***Keywords*:** Waste-derived catalysts, recycling, sustainable development, heterogeneous catalysts, value-added applications, industrial waste recycling, agriculture waste recycling

**Corresponding author*: aircmd_avinash@yahoo.com
†All the authors contributed equally

Deepti Goyal, Sakshi Kabra Malpani and Renuka Jain (eds.) Catalysis from Waste: Transforming Solid Residues into Value, (31–58) © 2026 Scrivener Publishing LLC

2.1 Introduction

All production processes e.g., industrial and agricultural produce waste of different forms: solids, liquids and gaseous. Several regulations and policies control the accumulation and production of liquid and gaseous waste to protect the air we breathe and the water we consume. In addition, a lot of ongoing research focuses on repurposing the produced waste. The focus of these research efforts has been on liquid and gaseous waste, and comparatively limited attention has been given to the reduction and repurposing of solid waste. It is important to note that solid waste is not only produced from industrial or agricultural processes, it also originates from the worldwide urbanization of the human population.

Most commonly solid waste ends up into the landfills, although some of it is incinerated or recycled. These treatments can create further challenges, such as contamination of our soil and water sources. Other methods to process solid waste are usually energy intensive and consume a lot of resources. However, some recent technological advancements like optical sorting and recycling, bioreactor landfills and energy recovery from incineration, have shown efficiency in repurposing of solid waste. Currently, there are five main ways to sustainable waste management: *Recycling, Composting, Waste to Energy, Construction and Infrastructure*, and *Upcycling and Repurposing*. The most innovative and practical approach to sustainable waste management, *upcycling and repurposing*, involves transformation of discarded waste into higher value products. This "waste-to-value" strategy promotes resource efficiency as well as economic development [1, 2]. In this chapter, we will discuss catalysts that have been repurposed from solid waste, i.e., *solid waste-derived catalysts*.

Catalysts are the cornerstone of modern chemistry and engineering. They drive advancements, but also contribute to cleaner, efficient, environmentally benign and sustainable chemical processes. With the onset of industrial revolution, the use of catalysts has increased enormously, both in terms of types and quantity. Various research groups worldwide are working on different types of catalysts and catalyst production methodologies. Waste-derived catalysts offer innovative solutions to both environmental and economic challenges. The utilization of waste materials for catalytic purposes presents a compelling strategy to reduce waste accumulation while simultaneously driving advancements in chemical processes [3–5]. Even the precious or rare metals that traditional catalysts utilize can be found in industrial waste.

The properties of waste-derived catalysts can be tailored to fit specific applications. Moreover, the abundance and variety of waste feedstock provides immense possibilities of customization and optimization. As far as methodologies are concerned, the development of waste-derived catalysts encompasses a diverse array of approaches, such as *Pyrolysis*, *Hydrothermal Treatment*, and *Chemical Modification* [6]. In this chapter we will briefly describe the sources of solid waste and methodologies through which the waste can be upcycled to yield solid waste-derived catalysts.

2.2 Agricultural and Agro-Industrial Waste

Agricultural waste can be categorized into two types: field residue and process residue. Field residue consists of dried organic waste that usually comprises stems, stalks, leaves, and pods. Process residue comprises husks, seeds, roots, and bagasse. Agro-industrial waste is produced when agricultural products are industrially processed, and it can be generally categorized as industrial waste. The agro-industrial waste consists of peels, cassava, and pulp.

The available literature describes numerous methods for alteration of agro-related waste to catalysts. The synthesis methodology for transforming agro-related waste to catalyst precursors comprises four steps: i) collection and sorting, ii) cleaning and drying, iii) grinding and sieving, and iv) modification (Figure 2.1). The obtained materials have been modified to catalysts through various methodologies that include sol-gel method, wet impregnation, physical mixing, calcination, co-precipitation, bi-functional modification, and hydration – dehydration (Figure 2.2). Alarmed by waste management policies, various research groups have started efforts to transform agricultural or agro-industrial waste into valuable catalysts. There are several factors such as type of waste, the method of catalyst preparation, particle size, and morphology of prepared catalyst, density of active sites, acidic or basic nature of prepared catalysts, and the reusability that must be considered to achieve the benefits of the derived catalyst. Agricultural or agro-industrial waste have been effectively employed as precursors and

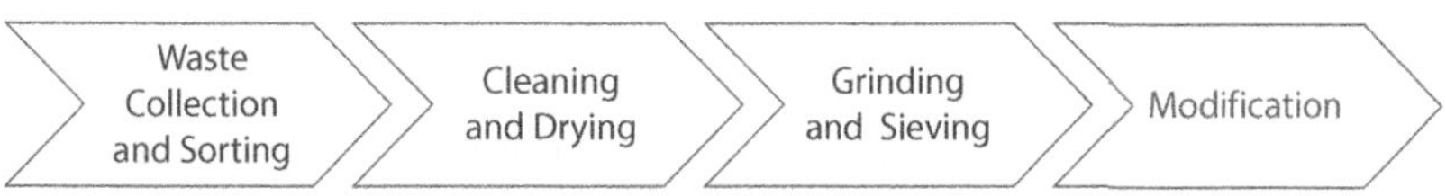

Figure 2.1 Generic procedure of agro-related waste transformation into catalyst precursors.

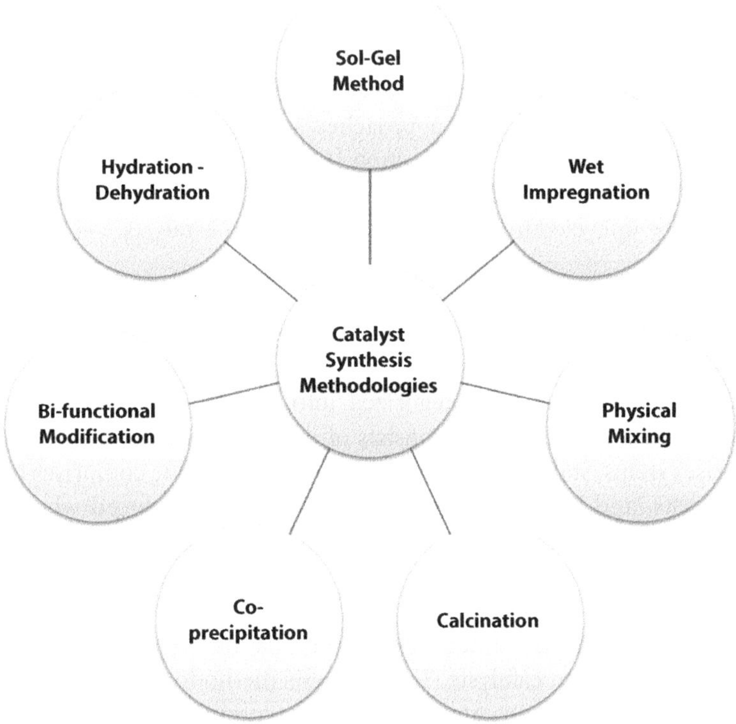

Figure 2.2 Methodologies for catalyst preparation from solid waste.

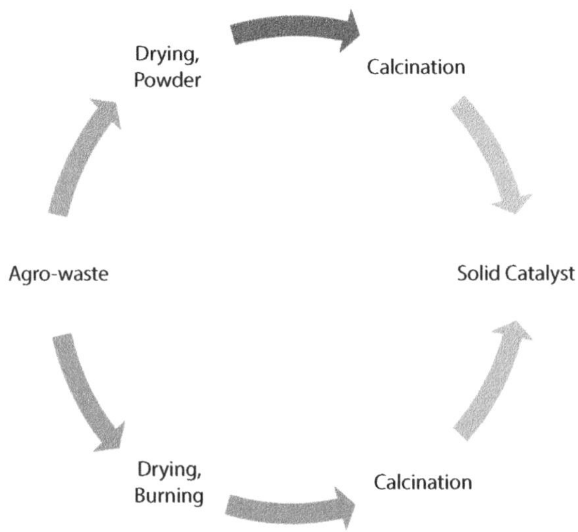

Figure 2.3 A general scheme to produce solid catalysts from agricultural wastes.

materials for catalyst development [7]. Based on some reported examples, a general flow chart can be formed (Figure 2.3) [8, 9].

2.2.1 Calcination

Freitas and co-workers, reported the synthesis of catalysts by using tucuma peels. The synthetic protocol involves the **calcination** of waste at 800 °C that is then directly used for the biodiesel production [10]. XRD revealed that the formed catalyst was a mixed phase of potassium chloride, potassium carbonate, calcium phosphate, and calcium magnesium silicate. All these components are active ingredients for various chemical transformations. The catalytic ability of the prepared material was evaluated through a transesterification reaction of soybean oil and methanol at different catalyst loading of 1%, 5%, and 10%. High yields were recorded (up to 97%), and the catalyst showed a substantial catalytic activity for several catalytic cycles. Similar methodologies of catalyst preparation have been used for different types of banana peels [11, 12]. In both of these methods, the catalyst preparation initiates with burning followed by calcination at 700°C for 4 h. Several other reports in recent years have used very similar methods to produce catalysts from agricultural waste. For example, *sesamum indicum* waste plants, pineapple leaf ash, tectona grandis leaves, acai seeds were transformed into catalysts either by preheating or burning followed by calcination at different temperatures [13–15].

A slightly different method has been reported by Wang and co-workers [16]. In their approach, the preparation of solid catalyst from the waste is followed by the activation of the solid catalyst using Li_3CO_3 and further calcination in air for 4 h. The addition of bases into solid catalysts increases catalytic activity, specifically for biodiesel synthesis by avoiding the saponification of free fatty acids, therefore, increasing the reusability. In their previous study, the same research group used Li_2SiO_3 [17]. Both approaches are excellent ideas to overcome the formation of by-products, like fatty acid methyl esters (FAME). The separation of FAME from the catalyst is the only way to regenerate the catalyst for further use. However, the addition of cost-effective bases like CaO, K_2CO_3, SiO_2, MgO [18–21], Li_3CO_3, and Li_2SiO_3 can overcome this issue.

2.2.2 Co-Precipitation

Co-precipitation is a process frequently used in synthesis methods. In this method, an impurity or another substance is separated by precipitation. Co-precipitation can be used when two or more substances not only have similar solubility but also have different precipitation characteristics. Researchers and chemists often need to use co-precipitation for the purification of bulk materials because most of the other purification techniques are only suitable for small quantities. Recently, a magnetically retrievable solid acid catalyst was developed using Fe_3O_4 and SiO_2 nanoparticles [22]. Rice husk was used to produce SiO_2 nanoparticles by calcination (650°C, 3 h). The SiO_2 nanoparticles were co-impregnated with Fe_3O_4 in a mixture of ethanol and water to produce SiO_2 coated Fe_3O_4 nanoparticles. This was followed by surface functionalization using sulfuric acid. The prepared catalyst was used for the hydrolysis of the corncob [22]. Such acidified catalysts have been reported for various reactions like condensation, esterification, and hydrolysis. The use of agricultural related waste as a source of SiO_2 makes it a cost-effective alternative to tetraethoxysilane (TEOS) and mercaptopropyltrimethoxysilane (MPTPMS).

In another study, Rokhum and co-workers reported a method where banana peel is used to produce extract that is co-precipitated with a mixture of $FeSO_4$ and $FeCl_3$ to yield Fe_3O_4 composite as a catalyst [23]. The methodology incorporates the collection of banana peel (*Musa acuminata*) washed thoroughly with distilled water to remove impurities. After that, the peels were dried and burned in an oven to collect ash, which was dissolved in excess of water at 80°C with continuous stirring. In a separate setup, Fe_3O_4 nanoparticles were prepared through a co-precipitation of mixture of $FeSO_4$ and $FeCl_3$ (1:2 molar ratio) at 90°C in deionized water. The dissolved banana peel ash was then drop wise added to the iron mixture followed by addition of 1 M ammonia solution while vigorously stirring the mixture. The mixture was allowed to settle down and solvent was removed in vacuo. The separated nanoparticles were dried and could be used as catalyst for Henry reaction.

Another co-precipitation report with transition metals (Ni and Co) and eggshell derived CaO and anthill lumps produced heterogeneous catalysts [24]. The preparation begins with a manual pulverization of anthill lumps and sieving through a 0.3 mm mesh. The obtained powder was dried in a hot air oven at 105°C for 4 h to remove moisture and organic content. Separately the collected eggshells were thoroughly washed in hot water to remove the white membrane and dried in hot air oven at 110°C overnight. The dried eggshells were mechanically ground and sieved to similar particle size to the anthill lumps. Later the anthill and eggshell powder were mixed in an aqueous media containing 0.1 M Na_2CO_3 and stirred for 5 h at 80°C. The resulting slurry was aged for 2 h allowing components to settle down followed by drying in oven at 110°C for 12 h. In a recent report, Lee and co-workers showed a synthesis of agricultural waste and manganese ferrite co-precipitation [25]. The procedure involved feedstock of pine-cone, white popinac, and sugarcane bagasse that were dried and pyrolyzed under inert atmosphere at 550°C for 4 h, sieved to separate particles less than 200 mesh, and finally the collected biochar was washed with water, filtered and dried. The manganese-ferrite-biochar catalyst was prepared by ultra-sonicating pristine biochar mixed with iron and manganese sulphates in aqueous media for 30 min. After ultra-sonication, 3 M NaOH was added, and the solution was stirred for 3 h. The solid catalyst was separated through centrifugation and washed with water and ethanol.

2.2.3 Sol-Gel

The **sol-gel method** is another versatile technique that can be used to produce valuable products from agricultural waste. The sol-gel represents suspensions (sol) that undergo gelation to form a solid network (gel). When applied to agricultural waste, such as crop residues or by-products, the sol-gel method offers several potential advantages like resource utilization, controlled processing, economic viability, and enhanced properties. Sol-gel methods involve several steps that include sol preparation (mixing precursors in solvent to form a colloidal solution) and gelation (forming a solid network through a chemical or physical process). Sometimes it involves aging or drying that are needed to get the desired properties for the target material. Overall, the sol-gel method offers a promising route to process agricultural waste to value-added products, including catalysts. Sol-gel method is an excellent method to produce green silica. Green silica refers to silica prepared or extracted by an environmentally friendly or sustainable method.

Despite being discarded as waste, eggshells have few industrial applications. For instance, the calcined eggshell waste – mainly composed of $CaCO_3$ – is used in the production of biodiesel and as a raw material for ceramic products. To further improve the applications of eggshell waste, several authors have investigated the adsorption of water contaminants [26]. In the study, the investigators were able to use eggshell waste to adsorb transition and heavy metals from aqueous solutions. In addition, the amount of adsorption of metal ions to the eggshell residues can be controlled by changing the temperature and pH. For example, Yeddou and co-workers reported that the capacity of Cr(III) and Fe(III) adsorption was reduced when the temperature was reduced in a range between 20–60°C, while the Cr(III) adsorptions increased when pH of the solution was increased from 3 to 5 [27]. Possibility of modification and fine tuning of metal concentrations onto the carrier makes eggshells an excellent support material to develop catalysts. In a recent report by Cree and Rutter, a comparison of the calcite extraction processes from eggshell bio-waste was performed [28]. The aim of the study was to identify the cheapest process to produce highly crystalline calcite preferably without using bleach. Because of the excellent adsorption properties of the calcite produced from eggshell waste, several reports have discussed its applications for water purification [29]. Similarly, Ahn and co-workers reported synthesis of CaO nanoparticles from sol-gel methods [30]. Here, the researchers followed the process reported by Mok and co-workers to produce calcite and later treated the calcite with 1 M HCl to produce $CaCl_2$ followed by treatment with 1 M NaOH to produce $Ca(OH)_2$ [30, 31]. Due to the drop-wise addition of NaOH to the solution of $CaCl_2$, a $Ca(OH)_2$ gel is formed which is then left for aging overnight followed by drying, washing, and calcination at 900°C, 1 h.

2.2.4 Hydration-Dehydration

Hydration-dehydration is a complementary process that is used to manage the moisture content of agricultural residues. Hydration methods involve water addition to the residues, composting or biological processes like anaerobic digestion for biogas production. All these processes are usually followed up by a dehydration process. Dehydration usually involves sun drying, mechanical drying, or sometimes chemical treatments. The common applications of hydrated agricultural residue include composting, biofuel production, animal feed, soil erosion control, bioremediation, and biopolymer and chemical production [32]. The modified residues can be used for diverse applications; however, for agricultural residues the hydration-dehydration method is never directly used for the catalyst development because it is an energy intensive process if used at bulk scale. It also requires calcination before using any residue to catalytic applications, and as we know from our previous discussion, calcination cannot be done directly to the residues. Nevertheless, some reports in the last few years have discussed this method to develop solid catalysts from eggshell, oyster shell, and clam shell bio-waste [33–35]. In comparison to sol-gel method, hydration-dehydration method provides better alternative for catalyst development, as it increases the surface area and active sites when subjected to calcination and offers enhanced possibility to process several other types of waste along with eggshells. Lee and co-workers reported a process for synthesis of clamshells derived catalyst, where the clamshells were washed with hot water several times and dried overnight in oven at 100°C [36]. Later clamshells were crushed and sieved to 200 μm followed by calcination at 900°C, 4 h under atmosphere conditions to produce active oxides. Further, the as-formed precursor underwent hydration-dehydration treatment and was refluxed in water for different time durations of 1, 3, 6, 9, or 12 h at 60°C. The formed slurry was dried at 100°C, 24 h in hot air oven, followed by grinding and sieving. The obtained powder was calcined at 600°C for 3 h before using the mixed phase as a catalyst. In another report from Vital and co-workers, a slightly modified method to produce solid catalysts from eggshells, oyster shells, and clam shells was developed [37]. All the shells were subjected to base hydrolysis with 10% NaOH to remove all the organic impurities, after which the shells were washed with water before drying in hot air oven at 100°C for 24 h. The shells were crushed using a hammer mill followed by ball milling for 24 h, the formed fine powder was calcined in muffle furnace at 900°C for 4 h. In the following steps, powdered eggshells were treated differently in comparison to oyster and clam shells. Eggshells were ultra-sonicated in excess of DMF, for 1, 3, 5 and 7 h, while oyster and clam shells underwent hydration-dehydration – oyster and clam shells samples were suspended in water for 4 h and recovered through centrifugation. The eggshell particles suspended in DMF were also separated through centrifugation followed by washing with water and ethanol and dried in hot air oven at 100°C for 24 h. Similar report by Rathod and co-workers also discussed the solid catalyst preparation using eggshell waste [38]. In brief, the eggshells were washed and dried in a hot air oven, calcined in muffle furnace (850°C for 3 h), and crushed to fine powder. The powdered calcined eggshells were suspended in water and subjected to sonication for 3 h at 60°C. Later, the pulverized samples were again calcined at 600°C to produce the desired catalyst. Among the most recent reports of eggshell derived catalysts, Mekonnen and co-workers have reported a modified methodology to synthesize nano-catalysts [39]. The collected eggshells were first washed by boiling in water for 10 min, then washed with running water, and finally dried in hot air oven for 24 h at 105°C. After drying, the eggshells were grounded and sieved with 63 μm sieve and calcined at 900°C for 3 h. The calcined eggshell powder was suspended in water at 60°C for 6 h and dried at 105°C overnight in a hot air oven. The formed powder was again calcined at 800°C for 3 h to get calcium oxide nanoparticles.

2.2.5 Wet Impregnation

Wet impregnation refers to a process generally used in chemistry and material sciences to dope active substances (metal ions or active catalytic species) onto the surface or in the pores of any porous material. Typically, wet impregnation produces active material that could have exceptional applications as catalysts [40]. The advantages of wet impregnation are simple execution procedures and low amount of waste produced during the process. The common precursors for wet impregnation require metal salts, such as the easily available chlorides, sulfates, carbonates, acetates, and nitrates. In addition to metal salts, generally a support material is required for wet impregnation. Both materials – metal salts and support material are mixed in solvent (usually water), to obtain desired catalysts [41]. Recently, Ji and co-workers reported synthesis of Mo catalyst supported on carbon-derived from agro-waste [41]. The carbon support – called biochar was prepared by carbonization of corn husk with 98% H_2SO_4, after which water was slowly added, and the mixture was stirred for 12 h. After washing the solid residue with deionized water, the biochar was formed as a paste. Molybdenum catalyst was prepared by mixing sodium molybdate with the biochar and stirring the mixture for 5 days at 90°C. The obtained mixture was hydrothermally treated at 550°C for 5 h followed by separation using suction filtration. The obtained material was calcined at 550°C for 10 h, washed with NH_4Cl to remove unreacted metal ions, and finally dried to yield the solid catalyst.

2.2.6 Physical Mixing

Physical mixing is a method that involves combining different components through mechanical means, without any chemical reactions or reagents required to produce final products. This method is typically used when the components are non-reactive to each other and can be mixed homogeneously to get a uniform catalytic activity. This method has several advantages over the other known methodologies because of its simplicity of production, and lower energy consumption, making it a cost-effective methodology compared to the other methods [42]. Also, it is the best way to control the composition and distribution of the desired material. However, some limitations exist. Most prominently, physical mixing is not the best way to maximize the surface area compared to the methods involving chemical treatments or other high-energy processes. Another challenge is to find materials that are compatible with each other to yield optimal catalytic performance. In the past decade, there have been several reports of catalyst preparation through this method. Lee and co-workers reported a method utilizing cocoa pod husk for catalyst preparation [43]. The method involves washing the cocoa pod husk with running water and cutting it into small cubes followed by drying under the sun for 2 weeks. Later, the dried husk was milled to fine powder using a grinder and calcined in a muffle furnace at 650°C for 4 h. The obtained calcined powder was mixed with potash (K_2CO_3) in aqueous medium and later filtered and washed with water to remove all the excess potash. The filtrate was collected and dried in a hot air oven. The collected oven dried powder – the unsupported catalyst was mixed with MgO in ratio of 1:0.6 and calcined at 550°C for 3 h to produce the supported catalyst. In another report, Boro and co-workers developed a Ba doped CaO catalyst, where the CaO was derived from the *Turatia striatula* waste shells [44]. The waste shells were washed with flowing water and dried in a hot air oven. Later, the waste shells were grinded and calcined at 900°C in a muffle furnace. In the second step, $BaCl_2$ was dissolved in water and dropwise added to the mixture of calcined powder with continuous stirring for 10 h. The obtained mixture was filtered under vacuum, dried in a hot air oven, and calcined again at 900°C. Joshi and co-workers used a similar strategy, where the eggshell waste derived CaO was mixed with different metal oxides to yield catalysts [45]. The synthesis protocol involves thorough washing of collected eggshell waste and drying in hot air oven at 100°C for 24 h. The dried eggshells were then grinded and calcined at different temperatures of 500, 700 and 900°C for 4 h, with slow ramp rate of 2°C/min. The formed powder was suspended into aqueous media, and metal (Fe, Al, Zn and Mn) nitrates were added. Afterwards, the solution was stirred at room temperature for 4 h. The formed mixture was dried in a hot air oven at 120°C, and after grinding, the mixture was again calcined at their respective temperatures (500, 700 and 900°C) for 4 h to obtain mixed phase metal oxides. In a recent report, Tomano and co-workers developed CaO supported Sr catalyst using cuttlebone (*Sepia officinalis*) bone waste [46]. The bone bio-waste was first washed with 10% NaOH and DI water, then grinded and sieved through 120 mesh. In the next step, the obtained powder was dissolved into 0.5 M HNO_3 followed by addition of $Sr(NO_3)_2$ and the reaction mixture was stirred for 2 h at room temperature. Na_2CO_3 was added to the reaction mixture, and the mixture was dried and calcined at 1000°C for 5 h.

2.2.7 Bi-Functional Modification

Bi-functional modification in catalyst preparation refers to improvements in catalytic activity and selectivity. This is achieved by incorporating two or more different catalytic active sites within the same material. This approach synergistically affects different functionalities and improves overall performance to facilitate chemical reactions. These modifications offer possibilities to form acid-base pairs, redox pairs, and metal and promoter pairs. Using the aforementioned pairs, researchers have improved the selectivity and stability of the catalysts. The bi-functional modification is recently introduced methodology for heterogeneous catalyst preparation from agricultural waste. Some reports from previous decade have discussed synthesis of catalysts using bi-functional modification. For example, Taufiq-Yap and co-workers [47] have used angel wing shell (AWS) for the catalyst preparation. The method begins with washing of AWS with DI water to remove organic impurities before drying in an oven at 150°C for 24 h. The dried shells were ground and sieved using 250 mesh to achieve homogeneity of the particle size. Later the particles were calcined at 900°C for 2 h. To form the sulfated CaO catalyst, the particles were treated with diluted sulfuric acid for 2 h at room temperature. The formed slurry was filtered, washed, and dried in a hot air oven at 150°C overnight. Another report from the same research group showed the synthesis of bi-functional catalyst using eggshell and phosphotungstic acid/zirconia [48]. First, the eggshells were washed with hot water and dried in hot air oven at 120°C overnight. Then the eggshells were crushed using a mortar pestle, sieved to 250 mesh and calcined in furnace at 900°C for 3 h. The obtained pure pristine CaO was suspended in water and stirred with appropriate quantity of phosphotungstic acid/zirconia at room temperature for 3 h. The formed material was oven dried and calcined at 650°C for 5 h. Other than eggshells, obtuse horn shell has also been reported as a CaO source by Tan and co-workers [49]. The collected waste from the local restaurants was used for the catalyst preparation. The preparation protocol is very similar to what was previously discussed.

2.3 Industrial Waste

Industrial waste refers to any waste generated by industrial activities that include mining, commercial productions, and urbanization. Such waste could be broadly categorized into solid, liquid, and gaseous waste. As this chapter only encompasses solid waste repurposing, we will specifically discuss the types of solid industrial waste and methods of catalyst development from such waste. The industrial solid waste can be further categorized into – glass waste, slag and sludge, ceramic waste, ash waste, construction and demolition waste, stone waste, chemical waste, electronic waste, and polymer waste. The most viable option to repurpose industrial waste is mortar production [50]. However, there are techniques – such as pyrolysis and cracking to repurpose industrial waste into catalytically active species as well.

2.3.1 Pyrolysis

Pyrolysis refers to a process that involves thermal decomposition of organic material in the absence of air or oxygen. The purpose of pyrolysis is to break down complex organic materials into simpler molecules that can be easily further processed. Plastics, rubber, polymers, most of the refinery waste, and biomass can be pyrolyzed to produce simpler products that could be used in an array of applications. Pyrolysis is an energy intensive and technologically sophisticated process that requires a heat source (ranges 400-800°C) and an anaerobic atmosphere. The obtained products of pyrolysis usually have direct applications as char and bio-oil. In addition, pyrolysis also produces gaseous products which can be used in various applications. The only solid residue obtained from pyrolysis is char. As an outcome of thermally decomposed products, char has high surface area and mineral content that makes it an interesting catalyst candidate. Most commonly known form of char is biomass char or biochar that is prepared from the agricultural or agro-industrial waste, when the waste is pyrolyzed at elevated temperatures. The produced charcoal or amorphous carbon has immense applications in various fields of research [51]. In this part of the chapter, we will only discuss the char produced from other than agricultural waste. Because of the wide scope of applications and dependency on polymers, we produce huge amounts of plastic waste. In addition, currently only a few of the polymer types can be recycled effectively and economically. To reduce the plastic waste in the environment, the investigators are hoping to find alternatives for plastics. The very high carbon content in all the polymeric materials make them excellent candidates for pyrolysis. In the last four years, four different reviews have comprehended the various applications of char produced from plastic waste [52–58]. In comparison to the biomass-derived char, synthetic polymers have well-defined structures which make the polymer char unique. Polymer-derived char with well-defined micro-structures has seized the attention of material scientists and substantial research has been published on various applications. However, catalytic applications are rarely reported. Among the few reports in the last decade on catalytic applications, Granados and coworkers [59], reported the development of two different waste-derived catalysts from polystyrene. A water-soluble variant poly-(styrene sulphonic acid) (PSSA) and a heterogeneous solid SiO_2-poly-(styrene sulphonic acid) (SiO_2 - PSSA) were formed. The catalytic properties of these two catalysts were explored in three different reactions of biomass valorization: biodiesel synthesis, xylose dehydration to furfural and furfural oxidation to maleic and succinic acids. Method of water-soluble variant PSSA preparation follows two earlier reports [60, 61]. The protocol includes collection of polystyrene (PS) waste from yoghurt packaging; compact disc covers and expanded polystyrene sources like thermoplastic foam or solid beads of polystyrene. The collected PS waste was dissolved in organic solvent followed by sulphonation of polymer by H_2SO_4 and acetyl sulfate and separation through dissolution-precipitation. The preparation of SiO_2 - PSSA was as follows – 18% of polystyrene- sulphonating acetyl sulfate agent (a mixture of H_2SO_4 and acetyl sulfate), mixed with tetraethylorthosilicate (TEOS, 99%) and 3-aminopropyltriethoxysilane (APTES, 96%) in a round bottom flask to achieve Si/N = 11 and S/N = 4 (Figure 2.4). The mixture was refluxed at 348 K while stirring for 17 h. After completion, solids were separated through centrifugation, to remove the unreacted polymer. The collected solids were washed with water under constant stirring, centrifuged again and washed until neutral pH. Finally, the collected solids were dried at 383 K overnight to get the heterogeneous catalyst.

Another recent report by Lim and coworkers [62], developed homogeneous and cross-linked heterogeneous catalysts derived from waste polystyrene. During the synthesis of polystyrene acidic catalyst, initially, polystyrene styrofoam was thoroughly washed with detergent to remove grease and dried overnight. After cutting the polymer samples, small random sizes were dissolved in DCM, that required a stirring for several hours at 50°C. To a separate round bottom flask immersed in ice bath, acetyl sulfate was prepared by mixing acetic anhydride and conc. H_2SO_4 [63]. This mixture was stirred for 5 min. Later, the freshly prepared acetyl sulfate was gradually added to the polymer DCM solution and stirred vigorously for 17 h at 50°C. After completion, reaction was quenched with excess of ethanol. The formed light brown polymer catalyst was obtained by dialysis and

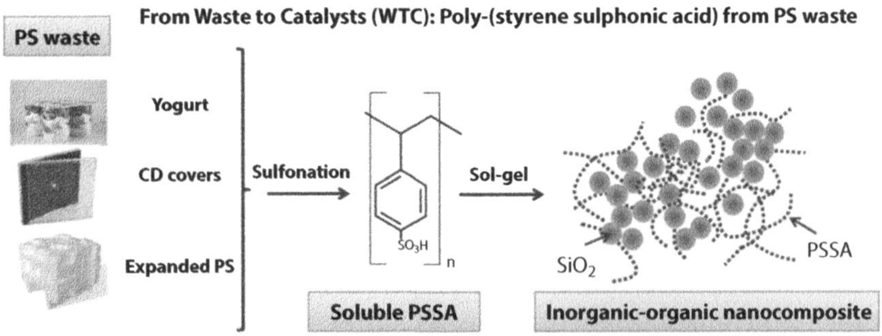

Figure 2.4 Preparation of solid SiO_2 - poly-(styrene sulphonic acid) (SiO_2 - PSSA) (Adapted with permission from the version that appears in Granados *et al.* [59]).

lyophilization. Authors observed that the synthesis of homogenous catalyst requires copious amount of conc. H_2SO_4; to further improve the recovery and reuse of the homogenous catalyst, they were crosslinked with poly (propylene glycol) bis (2-aminopropyl ether) (PPGBA, Mn ¼ 4000 Da). The use of long chain cross linked polymer enables the flexibility to swell and form organogels allowing adequate porosity to various reactants to penetrate; although the excessive cross linking eases recovery but reduces the number of accessible active sites for catalysis. Another interesting report by Souza and coworkers [64], reported the use of the cigarette filters for catalyst preparation. Cigarette filters were collected from the city streets thoroughly washed with water and dried in oven at 80°C for 4 h. Initially 1 g of filter was mixed with 5 mL of conc. H_2SO_4, later transferred to Teflon lined stainless steel autoclave. Using the similar mixture, various sets were produced by heating the autoclave in oven at 100, 150 and 190°C for various durations i.e., 2, 4, 6, and 8 h. The collected samples were washed with water to remove excess nitrate ions. The formed acid catalysts were used for esterification of oleic acid and methanol.

2.3.2 Cracking

Pyrolysis is a broad term that refers to the thermal decomposition of large molecules into small fragments in absence of oxygen. While **cracking** is similar in process typically more specific to long chain polymeric alkanes. The key difference between the two methods is the substrate types and use of catalyst. Pyrolysis can be done with or without catalyst depending upon the nature of the substrate, whereas cracking generally requires specific catalyst due to inherent stability of alkane polymers. Cracking is more controlled degradation, often targeting the breakdown of alkane plastics and polymers yielding smaller and useful hydrocarbons. These products are gaseous or liquid primarily which are commercially useful as fuel or precursors for chemical synthesis. However, the direct application of cracking products as catalysts has not been reported in literature to date. Instead, one promising route is to convert these cracking derived hydrocarbons into carbon-based materials like – carbon nanotubes (CNTs), multi-walled carbon nanotubes (MWCNTs), graphitic carbon, fullerenes and carbon quantum dots (CQDs). Several carbonization methods are available for transformation like - hydrothermal carbonization, template-assisted carbonization, additive-assisted carbonization, microwave-assisted carbonization, ionothermal carbonization, and CO_2 laser carbonization. However, these carbonization techniques typically require further activation steps to enhance their surface area and porosity for catalytic applications. Two main types of activations are physical and chemical activations [65]. While a detailed discussion of these techniques is beyond the scope of this chapter, it is important to note that the resulting activated carbon materials exhibit excellent physiochemical properties. These characteristics make them highly promising as catalyst or catalyst support in wide range of industrial applications.

During pyrolysis of polymer waste and cracking of plastic waste, a solid carbonaceous residue called coke can also form. It is formed usually under high temperature and catalyst driven conditions. Based on the source, these two types of coke are named as – pyrolytic coke and catalyst coke. Pyrolytic coke is formed during pyrolysis under thermal conditions; it is the carbonaceous material which is pyrolyzed in the absence of oxygen. It results from secondary reactions such as polymerization, aromatization and condensation of heavier volatile species generated during thermal decomposition. The quantity and properties of pyrolytic coke depends upon the factors like feedstock composition, temperature, time, and presence or absence of catalyst. The pyrolytic coke holds significant potential as precursor for high value carbon material which can be engineered for use in energy storage devices, water purification and catalyst support, whereas catalyst coke is a carbonaceous deposit that forms on the surface of catalyst during catalytic cracking. It originates from undesired side reactions like condensation, dehydrogenation, and polymerization of heavy hydrocarbons or aromatics. The coke gradually accumulates on the catalyst active sites and within the pores; unlike pyrolytic coke, catalyst coke is typically considered a contaminant rather than a useful byproduct. Catalyst coke is routinely removed through regeneration, which involves controlled combustion in presence of air to restore catalytic activity. Some recent research has begun exploring the potential of spent catalyst containing coke as precursor for functional carbon materials after appropriate treatment [66].

2.4 Fly Ash

Fly ash is a solid waste residue produced from the burning of coal, oil, or biomass. Fly ash waste is a major environmental concern. Fly ash not only creates problems with the soil, but also with the air due to its airborne particulates. Figure 2.5 lists the reported applications of fly ash [67]. Alterary and co-workers [68], categorized fly ash into three types based on the formation time; i.e., primary, secondary, and tertiary phases. The primary phase is generated during the combustion of coal in power plants. The particles of the primary phase fly ash are fine, spherical, light weighted and can be carried away with flue gas composed of silica, alumina, calcium

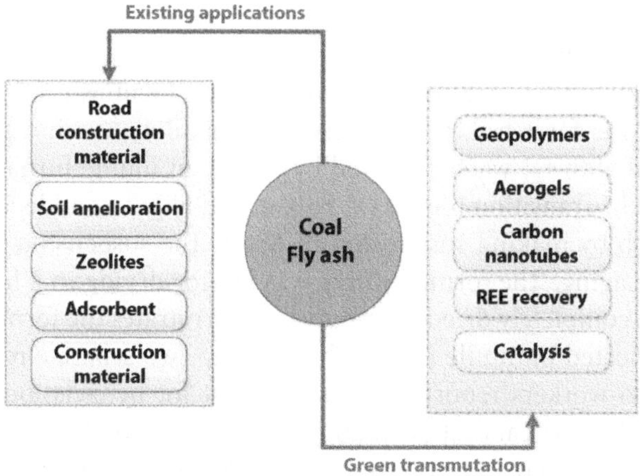

Figure 2.5 Various applications of fly ash (Adapted with permission from the version that appears in Shu *et al.* [67]).

oxide and iron oxide. The secondary phase fly ash forms when primary phase is exposed to moisture and calcium hydroxide. The reactions of primary phase with moisture and calcium hydroxide produce calcium silicate hydrate and calcium aluminate hydrate. The tertiary phase fly ash forms when primary phase fly ash undergoes long term pozzolanic reactions, contributing to the development of strength to the construction material.

Fly ash can be classified into five types based on the source and chemical composition - class F, class C, high calcium, low calcium, and pozzolanic fly ash [68]. Class F fly ash is primarily produced by burning anthracite or bituminous coal, whereas class C fly ash originates from sub-bituminous or lignite coal. The high and low calcium classification are determined by the calcium oxide (CaO) content - the low calcium fly ash contains less than 7% CaO, while high calcium fly ash contains more than 20% CaO. In contrast, class F typically contains 7-10% CaO and class C contains 10-20% CaO. Pozzolanic fly ash is characterized by its significant reactivity with calcium hydroxide, making it valuable as a cement additive to enhance strength and durability. Typically, a single batch of fly ash consists of a mixture of crystalline and amorphous phases. The compositional and structural features contribute to their effectiveness and versatility as a catalyst in various chemical processes.

Xuan and co-workers [69], reported the reduction of nitric acid (NO) to ammonia, where the catalyst preparation includes the collection of fly ash and compressing it to molded bricks. The prepared bricks were kept in the steam chest at 900°C under high pressure (90 atm) for activation.

After cooling, the fly ash blocks were again crushed to different particle sizes, and the different particle sizes were separated through sieving. The powdered fly ash was treated with conc. HNO_3 (acid/ash = 5 mL/g) and stirred at 70°C for 1 h. Later the mixture was diluted with water, filtered and washed with DI water followed by drying in muffle furnace at 200 – 400°C for 3 h. Transition metals (V, Fe, Cu, and Ni) were added to the activated fly ash by making suspension in water. The suspension was stirred and left for aging. The formed slurry was left for drying in a hot air oven until it gets completely dried; to remove excess nitrates the formed powder was again heated in muffle furnace before use. Two separate reports from Rani and co-workers reported fly ash catalyst for benzylation by fly ash supported zirconia catalyst [70], and Knoevenagel condensation reaction by fly ash supported CaO catalyst [71]. Two different fly ash supported catalysts were prepared through a simplified protocol. The collected fly ash was directly treated with conc. H_2SO_4 in 1:2 ratio, followed by drying in hot air oven at 110°C and calcination in a muffle furnace at 450°C for 4 h. To prepare the fly ash supported zirconia catalyst, zirconium propoxide was then added to the fly ash suspended in n-propanol. Dropwise addition of water while continuously stirring, produced a gel which was aged for 24 h and later dried in oven at 110°C for 24 h. The dried gel was mechanically crushed to powder and treated with 1 N H_2SO_4 for 1 h followed by drying (110°C, 24 h in hot air oven) and calcination (550°C, 2 h in muffle furnace) prior to use. In the other method, the calcined fly ash (900°C, 3 h in muffle furnace) was stirred and refluxed with $CaCO_3$ in water for 2 days. The collected paste was dried (120°C, 24 h in hot air oven) and calcined (700°C, 2 h in muffle furnace) before use. A further shortened protocol was reported by Zhou and co-workers [72], where the collected fly ash was first sieved to less than 150 μm, then mixed with 1 M nitric acid at room temperature for 4 h, filtered, washed, and dried (105°C in hot air oven) to be directly used as catalyst. The most recent report by Zhang and co-workers [73], suggests a similar protocol for catalyst preparation. First the collected fly ash was treated with 5 M H_2SO_4 and stirred at 100°C for 2 h. Later the reaction was allowed to cool, filtered, and washed until a neutral pH was observed. The slurry was then dried and crushed mechanically to get powdered catalyst. Another report from Malpani and co-workers [74], discussed preparation and catalytic application of coal fly ash based phosphomolybdic acid catalysts. In this report, class F fly ash is collected from local thermal power plant washed with deionized water and dried in hot air oven at 110°C for 12 h. Later, it was mechanically activated for 10, 20 and 30 h, the specific surface area was observed highest in sample mechanically active for 30 h which is 30.02 m^2/g. The mechanically activated fly ash was later thermally activated at 800°C for 3 h to remove excess of water and other organic impurities. Further the activated fly ash was treated with phosphomolybdic acid in methanol with various loading percentages. This method is known as wetness impregnation method. After overnight stirring, the impregnated solids were separated through filtration and dried at 110°C for 12 h followed by calcination at 400°C for 3 h. The prepared catalyst successfully catalyzed a range of microwave assisted acylation reactions. A similar method for catalyst preparation was reported by Katara and co-workers [75]. In this work, the collected fly ash was mechanically activated using a planetary ball mill for 15 h at 250 RPM and later thermally activated at 800°C for 3 h. The activated fly ash was treated with NaOH (6 N, aq.) for 3 days under continuous stirring at 90°C, followed by precipitation by dropwise addition of H_2SO_4 (5 N, aq.) up to pH 7 at room temperature. Later, the slurry was filtered and washed with deionized water to remove sodium sulfate and dried in hot air oven at 110°C for 24 h. To finally get the solid base catalyst, Ca $(NO_3)_2$ (15% solution) was stirred with activated catalyst for 1 h at 90°C. Again, precipitation was done by dropwise addition of mixture of NaOH (3 M) and Na_2CO_3 (2 M) in 1:1 ratio up to pH 8 and resultant slurry was aged for 24 h. The collected slurry was filtered and washed with deionized water, dried (110°C, 24 h) and calcined (500°C, 3 h) to produce fly ash based base catalyst.

2.5 Zeolites

Zeolites are a group of minerals or synthetic materials with distinctive crystalline structure. Primarily zeolites are composed of aluminum, silicon, and oxygen in a tetrahedral framework. This type of structure provides channels and cavities that can accommodate ions; and because of that, the applications of zeolites have increased to a great extent. Zeolites have very high surface area and porosity; therefore, they have well-known applications in ion exchange, adsorption, catalysis, agriculture, and industries. Based on the ratio of two main elements, Si and Al, zeolites are categorized in three classes: low silica ($Si/Al < 2$), intermediate silica ($2 < Si/Al < 5$), and high silica ($Si/Al > 5$) [76]. The conventional synthesis of zeolites generally involves a supersaturated solution of Si and Al mixed in closed reaction at high temperature (up to 180°C) for long duration to achieve nucleation and crystallization [77]. Synthesis of zeolites from commercially available Si and Al precursors is not in the scope of this chapter, but there are several reports of zeolite synthesis through industrial waste. Conversion of industrial waste to zeolites requires a thorough study of the waste material planned to be converted to zeolites and the development of a methodology to remove the non-zeolitic phase. Fly ash waste produced from different sources is prime precursors for zeolites. The main processes that are involved in zeolite production from fly ash include fusion through hydrothermal method, other multistep treatment, microwave, and ultra-sonication. Even though there are numerous reports of zeolite production from fly ash to the best of our knowledge, there are no reports on utilization of these zeolites for catalytic organic transformations. However, there are substantial reports of these zeolites as adsorbent material for CO_2, NH_3, and other flues gases [78].

Czuma and co-workers [79], synthesized sodium-zeolite type catalyst through a fusion method combining hydrothermal aging. In this process, fly ash was used as a precursor for zeolite. To carry on the zeolite synthesis, fly ash was mixed with sodium hydroxide in ratio w/w 1:0.8. The mixture was heated at 700°C for 6 h. The resulting mixture was washed with water till pH 10 and dried at 200°C under vacuum. The prepared material was then impregnated with nickel nitrate by using wet impregnation method. The impregnation was followed by drying in oven overnight at 105°C and calcination at 500°C for 5 h. Brar and co-workers [80], developed coal fly ash derived zeolite base catalyst. In this work, the coal fly ash was first passed through 200 mesh sieves, washed with water and dried at 80°C in an oven. Subsequently the dried fly ash was stirred in water for 30 min and formed suspension was kept still for 5 min so that heavier particles were settled down. The supernatant was decanted and centrifuged. The collected smaller particles were dried at 100°C and heat treated at 800°C at the rate of 5°C/min. Later, they were soaked in 10% HCl and washed with water till neutral pH, further dried in hot air oven. From the collected material, zeolite was synthesized using fusion assisted ultrasonication. NaOH and sodium aluminate (Na:Al = 1:1) were ground in mortar pestle and the mixture is heated in furnace at 750°C for 2 h. After cooling, the fused material was ground and dispersed in water using an ultrasonic probe. The obtained homogenous gel was crystallized by heating on an oil bath at 110°C for 16 h, later washed with deionized water and dried overnight in oven at 80°C to get the final catalyst.

2.6 Red Mud

Red mud, reddish colored slurry – is a bauxite residue generated during the production of alumina. It mainly contains silica, iron oxide, and other metal oxides remaining after the extraction of alumina. Red mud poses substantial environmental challenges because of its high alkalinity and possible leaching of heavy metals. Despite challenges with red mud as a waste, researchers have explored red mud for its catalytic properties due to high metal oxide contents. The most comprehensive utilization of red mud waste includes repurposing to infrastructural material, extraction of valuable components, and environmental remediation [81]. Some reports in recent years have shown the use of red mud as catalyst for both oxidations and reductions. Qiu and co-workers reported reduction of nitrobenzene by using a red mud derived catalyst [82]. The catalyst was prepared by drying the collected red mud waste in an oven at 105°C to get a powder that was sieved to a uniform particle size. The red mud powder was mixed with powder reed in minimum quantity of water to make slurry that was dried in oven (105°C) and calcined in muffle furnace (650-800°C). The prepared catalyst was used for nitro reduction. Another report of red mud-based catalyst for imine synthesis was reported by Ding and co-workers [83, 84]. The catalyst preparation protocol involves washing the red mud with water and making a suspension in excess of water and vigorously stirring for 2 h and finally separating the solid by centrifugation. The process was repeated several times for de-alkalization. In the second step, the separated red mud was again mixed with water and subjected to 1 atm CO_2 pressure while continuously stirring for 24 h. The solid was separated by centrifugation. The collected red mud sample was digested in 6 M HNO_3 at 70°C for 2 h; next, ammonium hydroxide was added dropwise into the mixture until pH 8 was achieved, finally the catalyst was separated through centrifugation. In another report by Yang and co-workers [85], the collected red mud was calcined under H_2 atmosphere and elevated temperature (from 30 to 800°C at a rate of 10°C/min and kept at 800°C for 2 h), after which the mixture was left to cool down naturally to room temperature and was subsequently used for catalysis.

2.7 Conclusion

In conclusion, this chapter provides a comprehensive overview of sustainable and ecological waste management. Particularly, this chapter focuses on transformation of solid waste into higher value products or upcycling of waste. Various possible ways to reuse waste – especially solid waste – are recycling, composting, waste to energy, construction or infrastructure, repurposing, and upcycling. This chapter predominantly discusses the upcycling of solid waste to catalyst preparation. Various methodologies have been discussed that are previously reported for catalyst preparation from solid waste. Two main sources of solid waste – agricultural/agro-industrial and industrial waste – were thoroughly discussed, with specific examples from the recent literature.

The key highlights of this chapter are:

- The environmental benefits of utilizing solid waste for catalyst preparation are clear. The upcycling of solid waste helps to mitigate unwanted environmental impacts of solid waste disposal such as soil and water contamination by landfills and incineration. In addition, repurposing or upcycling of waste to valuable products, like catalysts, can be an excellent addition to circular economy.
- The various waste treatment methodologies, sources and types of waste allow customization and optimization of the desired catalysts. Solid waste – mainly classified as agricultural/agro-industrial waste and industrial waste – can be upcycled to catalysts with several traditional methods. Nevertheless, some recent advancements in the catalyst preparation methodologies, like sol-gel processes, bi-functional modification and physical mixing, provide an interesting avenue for further development of solid waste-derived catalysts.
- Further research in the area of waste-derived catalysts is needed. In addition to continued innovation, the collaboration between researchers, industries, and policymakers is essential to maximize the benefits of this sustainable approach. This chapter advocates for a major shift towards the "waste to value" strategy to promote resource efficiency and sustainability.

AI disclosure statement: The author acknowledges the use of AI-based tools for minor language editing or grammatical correction. However, all intellectual contributions, analyses, and conclusions presented in this chapter are entirely the author's own.

References

1. Hoornweg, D., Bhada-Tata, P., Kennedy, C., Environment: Waste production must peak this century. *Nature*, 502, 615–617, 2013.

2. Chen, Z., Wei, W., Chen, H., Ni, B.-J., Eco-designed electrocatalysts for water splitting: A path toward carbon neutrality. *Int. J. Hydrogen Energy*, 48, 6288–6307, 2023.
3. Klose, F., Scholz, P., Kreisel, G., Ondruschka, B., Kneise, R., Knopf, U., Catalysts from waste materials. *Appl. Catal., B*, 28, 209–221, 2000.
4. Bennett, J.A., Wilson, K., Lee, A.F., Catalytic applications of waste derived materials. *J. Mater. Chem. A*, 4, 3617–3637, 2016.
5. Rautela, R., Sharma, A., Ranjan, V.P., Rathika, K., Pratap, V., Yadav, B.R., Kumar, S., Turning Solid Waste into Catalysts: A Path for Environmental Solutions. *ChemPlusChem*, 89, e202400246, 2024.
6. Chen, Z., Wei, W., Chen, H., Ni, B.-J., Recent advances in waste-derived functional materials for wastewater remediation. *Eco-Environ. Health*, 1, 86–104, 2022.
7. Khan, H.M., Iqbal, T., Yasin, S., Ali, C.H., Abbas, M.M., Jamil, M.A., Hussain, A., Soudagar, M.E., Rahman, M.M., Application of Agricultural Waste as Heterogeneous Catalysts for Biodiesel Production. *Catalysts*, 11, 1215, 2021.
8. Tursi, A., A review on biomass: importance, chemistry, classification, and conversion. *Biofuel Res. J.*, 6, 962–979, 2019.
9. Kumar, A. and Sharma, S., A grinding-induced catalyst- and solvent-free synthesis of highly functionalized 1,4-dihydropyridines *via* a domino multicomponent reaction. *Green Chem.*, 13, 2017–2020, 2011.
10. Mendonça, I.M., Paes, O.A.R.L., Maia, P.J.S., Souza, M.P., Almeida, R.A., Silva, C.C., Duvoisin Jr., S., de Freitas, F.A., New heterogeneous catalyst for biodiesel production from waste tucumã peels (Astrocaryum aculeatum Meyer): Parameter's optimization study. *Renewable Energy*, 130, 103–110, 2019.
11. Betiku, E., Akintunde, A.M., Ojum, T.V., Banana peels as a biobase catalyst for fatty acid methyl esters production using Napoleon's plume (Bauhinia monandra) seed oil: A process parameters optimization study. *Energy*, 103, 797–806, 2016.
12. Balajii, M. and Niju, S., Banana peduncle – A green and renewable heterogeneous base catalyst for biodiesel production from Ceiba pentandra oil. *Renewable Energy*, 146, 2255–2269, 2020.
13. de S. Barros, S., Pessoa Jr., W.A.G., Sá, I.S.C., Takeno, M.L., Nobre, F.X., Pinheiro, W., Manzato, L., Iglauer, S., de Freitas, F.A., Pineapple (Ananás comosus) leaves ash as a solid base catalyst for biodiesel synthesis. *Bioresour. Technol.*, 312, 123569, 2020.
14. Gohain, M., Laskar, K., Phukon, H., Bora, U., Kalita, D., Deka, D., Towards sustainable biodiesel and chemical production: Multifunctional use of heterogeneous catalyst from littered Tectona grandis leaves. *Waste Manage. (Oxford)*, 102, 212–221, 2020.
15. Mares, E.K.L., Gonçalves, M.A., Teresa Souza da Luz, P., Narciso da Rocha Filho, G., Zamian, J.R., Rafael Vieira da Conceição, L., Acai seed ash as a

novel basic heterogeneous catalyst for biodiesel synthesis: Optimization of the biodiesel production process. *Fuel*, 299, 120887, 2021.

16. Dai, Y.-M., Chen, K.-T., Wang, Y.-J., Chen, C.-C., Application of Peanut Husk Ash as a Low-Cost Solid Catalyst for Biodiesel Production. *Int. J. Chem. Eng. Appl.*, 5, 276–280, 2014.
17. Wang, J.-X., Chen, K.-T., Wu, J.-S., Wang, P.-H., Huang, S.-T., Chen, C.-C., Production of biodiesel through transesterification of soybean oil using lithium orthosilicate solid catalyst. *Fuel Process. Technol.*, 104, 167–173, 2012.
18. Liu, X., He, H., Wang, Y., Zhu, S., Piao, X., Transesterification of soybean oil to biodiesel using CaO as a solid base catalyst. *Fuel*, 87, 216–221, 2008.
19. Liu, H., Su, L., Liu, F., Li, C., Solomon, U.U., Cinder supported K_2CO_3 as catalyst for biodiesel production. *Appl. Catal. B: Environ. Energy*, 106, 550–558, 2011.
20. Sun, H., Duan, J., Chen, P., Lou, H., Zheng, X., Room temperature transesterification of soybean oil to biodiesel catalyzed by rod-like CaxSiOx + 2 solid base. *Catal. Commun.*, 12, 1005–1008, 2011.
21. Mguni, L.L., Mukenga, M., Jalama, K., Meijboom, R., Effect of calcination temperature and MgO crystallite size on MgO/TiO_2 catalyst system for soybean oil transesterification. *Catal. Commun.*, 34, 52–57, 2013.
22. Falowo, O.A., Oyekola, O.O., Oladipo, B., Taiwo, A.E., Ilesanmi, A., Davies, O., Synthesis of Sulfonated Magnetic Nano-catalyst Using Rice Husk Ash for Corncob Hydrolysis: Kinetic and Thermodynamic Study. *Waste Biomass Valorization*, 15, 973–987, 2024.
23. Pathak, G., Rajkumaria, K., Rokhum, S.L., Wealth from waste: M. acuminata peel waste-derived magnetic nanoparticles as a solid catalyst for the Henry reaction. *Nanoscale Adv.*, 1, 1013–1020, 2019.
24. Yusuff, A.S., Adeniyi, O.D., Azeez, S.O., Olutoye, M.A., Akpan, U.G., Synthesis and characterization of anthill-eggshell-Ni-Co mixed oxides composite catalyst for biodiesel production from waste frying oil. *Biofuels, Bioprod. Biorefin.*, 13, 37–47, 2019.
25. Huang, W.-H., Wu, R.-M., Chang, J.-S., Juang, S.-Y., Lee, D.-J., Manganese ferrite modified agricultural waste-derived biochars for copper ions adsorption. *Bioresour. Technol.*, 367, 128303, 2023.
26. Ahmad, M., Usman, A.R.A., Lee, S.S., Kim, S.-C., Joo, J.-H., Yang, J.E., Ok, Y.S., Eggshell and coral wastes as low-cost sorbents for the removal of Pb2+, Cd2+ and Cu2+ from aqueous solutions. *J. Ind. Eng. Chem.*, 18, 198–204, 2012.
27. Yeddou, N. and Bensmaili, A., Equilibrium and kinetic modelling of iron adsorption by eggshells in a batch system: effect of temperature. *Desalination*, 206, 127–134, 2007.
28. Cree, D. and Rutte, A., Sustainable Bio-Inspired Limestone Eggshell Powder for Potential Industrialized Applications. *ACS Sustainable Chem. Eng.*, 3, 941–949, 2015.

29. Rizqi, O., Brilianti, N.F., Wasito, H., Setiawan, B.D., Nanoporous of waste avian eggshell to reduce heavy metal and acidity in water. *Sustainable Chem. Pharm.*, 10, 163–167, 2018.
30. Habte, L., Shiferaw, N., Mulatu, D., Thenepalli, T., Chilakala, R., Ahn, J.W., Synthesis of Nano-Calcium Oxide from Waste Eggshell by Sol-Gel Method. *Sustainability*, 11, 3196, 2019.
31. Jai, P.H., Wook, J.S., Kyu, Y.J., Gil, K.B., Mok, L.S., Removal of heavy metals using waste eggshell. *J. Environ. Sci.*, 19, 1436–1441, 2007.
32. O'Connor, J., Hoang, S.A., Bradney, L., Rinklebe, J., Kirkham, M.B., Bolan, N.S., Value of dehydrated food waste fertiliser products in increasing soil health and crop productivity. *Environ. Res.*, 204, 111927, 2022.
33. Mignardi, S., Archilletti, L., Medeghini, L., De Vito, C., Valorization of Eggshell Biowaste for Sustainable Environmental Remediation. *Sci. Rep.*, 10, 2436, 2020.
34. Nazir, G., Rehman, A., Park, S.-J., Valorization of shrimp shell biowaste for environmental remediation: Efficient contender for CO2 adsorption and separation. *J. Environ. Manage.*, 299, 113661, 2021.
35. Hart, A., Mini-review of waste shell-derived materials' applications. *J. Environ. Manage.*, 38, 514–527, 2020.
36. Asikin-Mijan, N., Lee, H.V., Taufiq-Yap, Y.H., Synthesis and catalytic activity of hydration–dehydration treated clamshell derived CaO for biodiesel production. *Chem. Eng. Res. Des.*, 102, 368–377, 2015.
37. Risso, R., Ferraz, P., Meireles, S., Fonseca, I., Vital, J., Highly active CaO catalysts from waste shells of egg, oyster and clam for biodiesel production. *Appl. Catal., A*, 567, 56, 2018.
38. Gupta, A.R. and Rathod, V.K., Waste cooking oil and waste chicken eggshells derived solid base catalyst for the biodiesel production: Optimization and kinetics. *Waste Manage. (Oxford)*, 79, 169–178, 2018.
39. Erchamo, Y.S., Mamo, T.T., Workneh, G.A., Mekonnen, Y.S., Improved biodiesel production from waste cooking oil with mixed methanol–ethanol using enhanced eggshell-derived CaO nano-catalyst. *Sci. Rep.*, 11, 6708, 2021.
40. Munnik, P., de Jongh, P.E., de Jong, K.P., Recent Developments in the Synthesis of Supported Catalysts. *Chem. Rev.*, 115, 6687–6718, 2015.
41. Liang, Y., Xiong, C., Zhou, X., Xue, C., Ji, H., Preparation of Mo-based catalyst appreciated from agricultural waste for efficient high-value conversion of 1-hexene. *Chem. Eng. J.*, 455, 140648, 2023.
42. Awogbemi, O., Vandi Von Kallon, D., Aigbodion, V.S., Trends in the development and utilization of agricultural wastes as heterogeneous catalyst for biodiesel production. *J. Energy Inst.*, 98, 244–258, 2021.
43. Ofori-Boateng, C. and Lee, K.T., The potential of using cocoa pod husks as green solid base catalysts for the transesterification of soybean oil into biodiesel: Effects of biodiesel on engine performance. *Chem. Eng. J.*, 220, 395–401, 2013.

44. Boro, J., Konwar, L.J., Thakur, A.J., Deka, D., Ba doped CaO derived from waste shells of T striatula (TS-CaO) as heterogeneous catalyst for biodiesel production. *Fuel*, 129, 182–187, 2014.
45. Joshi, G., Rawat, D.S., Lamba, B.Y., Bisht, K.K., Kumar, P., Kumar, N., Kumar, S., Transesterification of Jatropha and Karanja oils by using waste eggshell derived calcium based mixed metal oxides. *Energy Convers. Manage.*, 96, 258–267, 2015.
46. Tomano, N., Prokaew, A., Boonyuen, S., Ummartyotin, S., Development of Sr/Cao catalyst derived from cuttlebone (Sepia officinalis) for biodiesel production. *J. Magn. Magn. Mater.*, 30, 40–47, 2020.
47. Syazwani, O.N., Rashid, U., Mastuli, M.S., Taufiq-Yap, Y.H., Esterification of palm fatty acid distillate (PFAD) to biodiesel using Bi-functional catalyst synthesized from waste angel wing shell (Cyrtopleura costata). *Renewable Energy*, 131, 187–196, 2019.
48. Mansir, N., Teo, S.H., Mijan, N.-A., Taufiq-Yap, Y.H., Efficient reaction for biodiesel manufacturing using bi-functional oxide catalyst. *Catal. Commun.*, 149, 106201, 2021.
49. Lee, S.L., Wong, Y.C., Tan, Y.P., Yew, S.Y., Transesterification of palm oil to biodiesel by using waste obtuse horn shell-derived CaO catalyst. *Energy Convers. Manage.*, 93, 282–288, 2015.
50. Gehlot, M.R. and Shrivastava, S., Solid industrial waste generation and its valorization in developing sustainable building materials- A state of the art review. *Mater. Today: Proc.*, 2023, https://doi.org/10.1016/j.matpr.2023.07.134.
51. Xu, D., Yang, L., Ding, K., Zhang, Y., Gao, W., Huang, Y., Sun, H., Hu, X., Shatir, S., S.-Hassan, A., Zhang, S., Zhang, H., Mini-Review on Char Catalysts for Tar Reforming during Biomass Gasification: The Importance of Char Structure. *Energy Fuels*, 34, 1219–1229, 2020.
52. Martín-Lara, M.A., Piñar, A., Ligero, A., Blázquez, G., Calero, M., Characterization and Use of Char Produced from Pyrolysis of Post-Consumer Mixed Plastic Waste. *Water*, 13, 1188, 2021.
53. Harussani, M.M., Sapuan, S.M., Rashid, U., Khalina, A., Ilyas, R.A., Pyrolysis of polypropylene plastic waste into carbonaceous char: Priority of plastic waste management amidst COVID-19 pandemic. *Sci. Total Environ.*, 803, 149911, 2022.
54. Lee, S., Kim, Y.T., Lin, K.-Y.A., Lee, J., Plastic-Waste-Derived Char as an Additive for Epoxy Composite. *Materials*, 16, 2602, 2023.
55. Adeniyi, A.G., Iwuozor, K.O., Emenike, E.C., Ajala, O.J., Ogunniyi, S., Muritala, K.B., Thermochemical co-conversion of biomass-plastic waste to biochar: a review. *Green Chem. Eng.*, 5, 31–49, 2024.
56. Gebre, S.H., Sendeku, M.G., Bahr, M., Recent Trends in the Pyrolysis of Non-Degradable Waste Plastics. *ChemistryOpen*, 10, 1202–1226, 2021.

57. Wang, H., Shao, Y., Mei, S., Lu, Y., Zhang, M., Sun, J.-K., Matyjaszewski, K., Antonietti, M., Yuan, J., Polymer-Derived Heteroatom-Doped Porous Carbon Materials. *Chem. Rev.*, 120, 9363–9419, 2020.
58. Gottlieb, E., Matyjaszewski, K., Kowalewsk, T., Polymer-Based Synthetic Routes to Carbon-Based Metal-Free Catalysts. *Adv. Mater.*, 31, 1804626, 2019.
59. A.-Fagúndez, N., Laserna, V., Alba-Rubio, A.C., Mengibar, M., Heras, A., Mariscal, R., López Granados, M., Poly-(styrene sulphonic acid): An acid catalyst from polystyrene waste for reactions of interest in biomass valorization. *Catal. Today*, 234, 285–294, 2014.
60. Vink, H., A new convenient method for the synthesis of poly(styrenesulfonic acid). *Makromol. Chem.*, 182, 279–281, 1981.
61. Baigl, D., Seery, T.A.P., Williams, C.E., Preparation and Characterization of Hydro soluble, Partially Charged Poly(styrenesulfonate)s of Various Controlled Charge Fractions and Chain Lengths. *Macromolecules*, 35, 2318–2326, 2002.
62. Yeung, C.W.S., Loh, W.W., Lau, H.H., Loh, X.J., Lim, J.Y.C., Catalysts developed from waste plastics: a versatile system for biomass conversion. *Mater. Today Chem.*, 21, 100524, 2021.
63. Kučera, F. and Jancář, J., Preliminary Study of Sulfonation of Polystyrene by Homogeneous and Heterogeneous Reaction. *Chem. Pap.*, 50, 224–227, 1996.
64. Teixeira, L.L.A., Araujo, R.O., Santos, J.L., Guimaraes, M.N., Ribeiro, V.M.L., Pocrifka, L.A., Tenório, J.A.S., de Araujo, J.R., de Oliveira, S.M., d. N. Batista, L., de Souza, L.K.C., Production of solid acid catalyst using waste cigarette filters for esterification. *Environ. Sci. Pollut. Res.*, 31, 8072–8081, 2024.
65. Zhao, Z., Sun, L., Li, Y., Feng, W., Polymer-derived carbon materials for energy storage devices: A mini review. *Carbon*, 210, 118066, 2023.
66. Collett, C.H. and McGregor, J., Things go better with coke: the beneficial role of carbonaceous deposits in heterogeneous catalysis. *Catal. Sci. Technol.*, 6, 363–378, 2016.
67. Gollakota, A.R.K., Volli, V., Shu, C.-M., Progressive utilization prospects of coal fly ash: A review. *Sci. Total Environ.*, 672, 951–989, 2019.
68. Alterary, S.S. and Marei, N.H., Fly ash properties, characterization, and applications: A review. *J. King Saud Univ. - Sci.*, 33, 101536, 2021.
69. Xuan, X., Yue, C., Li, S., Yao, Q., Selective catalytic reduction of NO by ammonia with fly ash catalyst. *Fuel*, 82, 575–579, 2003.
70. Khatri, C., Mishra, M.K., Rani, A., Synthesis and characterization of fly ash supported sulfated zirconia catalyst for benzylation reactions. *Fuel Process. Technol.*, 91, 1288–1295, 2010.
71. Jain, D., Khatri, C., Rani, A., Fly ash supported calcium oxide as recyclable solid base catalyst for Knoevenagel condensation reaction. *Fuel Process. Technol.*, 91, 1015–1024, 2010.

72. Zhang, A., Wang, N., Zhou, J., Jiang, P., Liu, G., Heterogeneous Fenton-like catalytic removal of p-nitrophenol in water using acid-activated fly ash. *J. Hazard. Mater.*, 201-202, 68–73, 2012.
73. Bhatti, A.H., Waris, M., Hussaine, I., Chougulec, S.S., Pasupuleti, K.S., Kariimh, I., Bhatti, U.H., Zhang, R., Renaissance of fly ash as eco-friendly catalysts for rapid CO_2 release from amines. *Carbon Capture Sci. Technol.*, 11, 100198, 2024.
74. Malpani, S.K., Goyal, D., Katara, S., Rani, A., Green, efficient and economical coal fly ash based phosphomolybdic acid catalysts: preparation, characterization and application. *Chem. Pap.*, 75, 3017–3034, 2021.
75. Katara, S., Kabra, S., Goyal, D., Hada, R., Sharma, A., Rani, A., Fly ash to solid base catalyst: Synthesis, characterization and catalytic application. *Mater. Today: Proc.*, 42, 1409–1416, 2021.
76. Yoldi, M., Fuentes-Ordoñez, E.G., Korili, S.A., Gil, A., Zeolite synthesis from industrial wastes. *Microporous Mesoporous Mater.*, 287, 183–191, 2019.
77. Buzukashvili, S., Sommerville, R., Rowson, N.A., Waters, K.E., An overview of zeolites synthesised from coal fly ash and their potential for extracting heavy metals from industrial waste water. *J. Mater. Educ.*, 33, 65–132, 2011.
78. Belviso, C., State-of-the-art applications of fly ash from coal and biomass: A focus on zeolite synthesis processes and issues. *Prog. Energy Combust. Sci.*, 65, 109–135, 2018.
79. Czumab, N., Zarębska, K., Motak, M., Gálvez, M.E., Da Costa, P., Ni/zeolite X derived from fly ash as catalysts for CO_2 methanation. *Fuel*, 267, 117139, 2020.
80. Mahajan, A., Chhibber, M., Brar, L.K., Coal fly ash derived zeolite: a solid-state base for convenient synthesis of diphenyl ethers. *New J. Chem.*, 48, 13917–13926, 2024.
81. Wang, M. and Liu, X., Applications of red mud as an environmental remediation material: A review. *J. Hazard. Mater.*, 408, 124420, 2021.
82. Wang, D., Wang, H., Luo, L., Zhou, Q., Qiu, X., Trash to treasure: Converting red mud into efficient catalysts for the hydrogenation of p-nitrobenzene compounds. *J. Environ. Chem. Eng.*, 10, 108161, 2022.
83. Chen, D., Ding, Y., Xia, C., He, L., Cao, Y., Turning hazardous red mud into useful catalysts for the carbonylation of amines to N-formamides. *Mol. Catal.*, 533, 112761, 2022.
84. Cao, Y., Chen, D., Wang, Y., Shi, H., Feng, B., Xia, C., Ding, Y., He, L., Red mud-mediated cross-coupling of alcohols and amines to imines over MnOx catalysts. *Mol. Catal.*, 535, 112883, 2023.
85. Li, H., Shi, B., Zhang, H., Fu, X., Yang, H., Preparation and application of red mud-based heterogeneous Fenton catalyst. *Appl. Catal., A*, 656, 119142, 2023.

3

Solid Waste as Catalyst Precursors: Types and Current Scenario

Niharika Shringi[1], Smriti Dwivedi[2*] and Anita Kushwaha[3]

[1]*Department of Chemistry, School of Sciences, JECRC University, Jaipur, Rajasthan, India*
[2]*Department of Applied Sciences (Chemistry), Galgotias College of Engineering and Technology, Greater Noida, U.P., India*
[3]*Department of Applied Sciences (Chemistry), IEC College of Engineering and Technology, Greater Noida, U.P., India*

Abstract
Often considered useless, solid waste now stands out as a potential source for valuable products, reducing reliance on natural resources. Driven by sustainability, ample waste represents an underutilized resource for the production of catalytic materials. This chapter presents an overview of recent developments in different types of solid waste-derived heterogeneous catalysts from incineration, industrial wastewater, e-waste, calcium-containing agricultural waste, plastic waste, and other waste materials. Herein, the diverse applications of these catalysts, including thermochemical plastic treatment, NH_3 selective catalytic reduction, biodiesel production, integrated CO_2 capture and utilization, oxidative coupling of methane, and hydrogen production from municipal solid waste, etc. are comprehensively discussed. In the end, present challenges and future aspects are also discussed.

***Keywords*:** Solid waste, sustainable heterogeneous catalysis, solid waste management, biomass, E-waste, industrial waste, carbonaceous materials

3.1 Introduction

The 21st century has witnessed unprecedented levels of industrialization and urbanization to meet global demands, which have led to the unavoidable

**Corresponding author*: s.dwivedi@galgotiacollege.edu

Deepti Goyal, Sakshi Kabra Malpani and Renuka Jain (eds.) Catalysis from Waste: Transforming Solid Residues into Value, (59–88) © 2026 Scrivener Publishing LLC

generation of solid waste. Managing this ever-growing waste menace while striving to adhere to sustainable practices has become a substantial multi-faceted challenge for the global community. In 2020, a global population of eight billion generated 2.5 billion tons of waste; this figure is predicted to increase to 5.9 billion tons by the year 2050 [1]. Solid waste is one of the most hazardous pollutants, and its inappropriate disposal causes severe adverse effects on the environment and public health [2]. Solid waste valorization has emerged as a critical approach in addressing the dual challenge of waste management and resource deficit. Among the innovative strategies for waste valorization, transforming solid waste into catalyst precursors represent an exclusively sustainable approach, offering significant environmental protection and economic benefits. This approach amalgamates with the principles of the circular economy, transforming 'waste to valuable materials' for catalytic applications across various industrial applications. The utilization of catalysts derived from waste reduces the reliance on fossil fuel-based raw materials, making them cost-effective, greener alternatives over their traditional counterparts. In recent years, several reviews have been published on the types of wastes, their applications in catalysis, and as resource materials for the preparation of heterogeneous catalysts [3, 4]. The waste materials can be used either directly as a catalyst, pre-catalyst, or catalytic support, or can undergo thermal or chemical modifications before employment. Based on the source of waste, it can be derived from industries such as red mud generated from the Bayer process and slag waste from blast furnaces during iron extraction. Biowaste is another category in this type, which includes poultry waste, such as chicken bones and eggshells, as well as agricultural waste, including rice husks, and domestic waste, including leftover food, fruit peels, and waste cooking oil. These waste materials are complex in composition, and some of them require thermal/chemical treatments to improve their catalytic activity. Nevertheless, in some cases, these additional components impart favorable properties. Interestingly, even a given single waste may vary in composition due to differences in originating sources or the exact process involved during its generation [5].

For the direct application, the modification of waste material is necessary due to the presence of additional catalytic components, which affect the catalytic performance of the prepared catalysts. Nevertheless, in some cases, these additional components also impart favorable properties. In the processing and application of these waste-derived catalysts, their physicochemical attributes play a significant role. Based on the composition, the solid waste appropriate for preparing catalyst precursors can be categorized into i) rich in transition metals or metal oxides, and ii) rich in calcium or non-transition metals. This chapter represents a comprehensive overview of the recent advancements in the utilization of catalysts derived from waste materials. Herein, various categories of catalysts are illustrated from different waste resources. For convenience, studies have been grouped into the following categories: municipal waste-derived waste catalysts, biowaste-derived catalysts, e-waste-derived catalysts, polyethylene terephthalate (PET)-based catalysts, and agricultural waste-derived catalysts. Additionally, we provide insights into the potential applications of catalysts derived from waste materials. In conclusion, the challenges and prospects of these catalysts are also discussed.

3.2 Different Types of Waste-Derived Catalysts

In the past few decades, anthropogenic waste has been transformed into valuable material with diverse applications as construction material, chemicals, and high-value functional compounds. Within the context of chemical transformations, waste materials are employed as both alternative renewable feedstock and a resource for catalyst synthesis. These waste-derived catalysts significantly contribute to making the whole process of both conventional and innovative chemical methodologies more energy and atom-efficient. In this section of the chapter, different types of solid waste-derived catalysts are outlined as follows:

3.2.1 Municipal Waste-Derived Catalysts

Freshwater is one of the vital elements for sustaining life on Earth. However, due to population growth, industrialization, and climate change, the depletion of freshwater to an alarming level has underscored the need for sustainable water management practices. Municipal wastewater-derived sludge, also known as sewage sludge, is a by-product produced by municipal wastewater treatment plants. It consists of solid materials that settle out during the treatment process [6]. The majority of municipal wastewater treatment methods use activated sludge, with an annual output of 80-90 million tons [7]. The potential reuse of this waste sludge as a catalyst for water treatment offers a sustainable solution for environmental protection. Sludge can contain a variety of organic and inorganic materials, including pathogens, heavy metals, persistent organic pollutants, and organic trace pollutants [8]. Consequently, the handling and disposal of sewage sludge is a matter of particular concern due to adverse effects on human health. The traditional approaches to sludge disposal are incineration, landfilling, and direct use in agriculture. However, the pursuit of renewable disposal alternatives and the mitigation of environmental damage related to their immediate utilization are compelling us to investigate additional potential techniques of sediment treatment. Compared to these conventional methods, pyrolysis is a potentially promising method as it reduces the volume of the solid residues. Pyrolysis also transforms sewage sludge into biochar, which has gained significant attention among policy makers and academicians [7, 9]. Sludge biochar was modified with boron carbide (B-SBC) and applied in the activation of peroxymonosulfate (PMS) for the selective decomposition of various organic pollutants. The rhodamine B dye degraded in the system using B-SBC with a pseudo-first-order rate constant was 9.8 and 3.9 times higher than in the PMS and SBC/PMS systems, respectively. Electron spin resonance (ESR), quenching experiments, and electrochemical results revealed that nonradical pathways, including $^1O_2$ generation (74.6%) and direct electron transfer (22.0%), were the major pathways for RhB oxidation in the B-SBC/PMS system. This research demonstrates the mechanism of carbocatalysis for the selective oxidation of pollutants in a non-radical pathway and provides an eco-friendly technology for sludge utilization and wastewater treatment.

Sieradzka *et al.* [10], investigated the production of high-quality syngas from sewage sludge and its hydrochars through gasification and catalytic gasification processes. The present research investigated both digested (MS1) and non-digested (MS2) municipal sludge from Silesian wastewater treatment plants (WWTPs), and two industrial sludge hydrochars (HC_IS1, HC_IS2). Hydrochars (HC_MS1, HC_MS2, HC_IS1, HC_IS2) were derived *via* hydrothermal carbonization at 200°C and 2h. The results showed that the presence of a catalyst significantly increased the hydrogen yield, from 4% to 21.7% for HC_MS1. However, using the HC_MS2, no hydrogen was found. The inclusion of SrO catalyst increased the hydrogen production to 7.6%. This research study confirmed the potential of gasification for syngas production from sewage sludge and hydrochars.

Jiang *et al.* [11], used an innovative methodology to fabricate municipal solid waste (MSW) into adsorbents through carbonation and activation with common activators such as $ZnCl_2$, KOH, H_2SO_4, etc. These activators incorporate a substantial number of pores, which mitigate tar formation and prevent tar blocking in the pyrolysis process. Mg-SB, a biochar adsorbent composed of magnesium oxide particles derived from the pyrolysis of activated sewage sludge, has been investigated for phosphate removal in aqueous environments under various conditions. Raw sewage sludge samples were air dried and sieved (0.9-2 mm) to produce Mg-SB adsorbent and then immersed in 0.5 mol L^{-1} $MgCl_2{\cdot}6H_2O$ for 12h in a mixture reactor. After removing the supernatant liquid, residual material was dried and pyrolyzed at 550°C for 2h under a nitrogen gas atmosphere. The obtained product was crushed, sieved (between 0.45 mm and 0.9 mm in size), and washed. The obtained adsorptive material was oven-dried at 80°C, sealed in a desiccator, and referred to as Mg-SB. During experiments, the Mg-SB adsorbent showed exceptional phosphate adsorption capacity at about 97,450 mg/kg in the acid region as compared to the earlier reported adsorbents. The pH in an acidic medium and higher ionic strength favored better phosphate adsorption capacity. The adsorption data were found to best fit in the pseudo-first-order model and the Langmuir model. The phosphate-loaded sludge biochar can also be used as an agricultural fertilizer to improve soil fertility and restore soil contaminated by heavy metals. Hence, the Mg-SB can serve as a promising adsorbent for phosphate removal and provide a new sustainable approach to utilizing wastewater comprehensively. In a recent study, efficient, stable, and economical sludge-based Mn-doped catalysts were utilized to treat the refractory wastewater of petrochemical industrial secondary wastewater [12]. The Mn-doped catalyst (PS-M700) was prepared *via* pyrolysis at 700°C and employed to catalyze the ozonation of petrochemical secondary wastewater. Using the PS-M700 catalyst, the total organic carbon removal efficiency of 56.6% was observed. Furthermore, the catalytic ozonation of petrochemical secondary effluent was promoted by the hydroxyl radicals. The excitation-emission matrix fluorescence spectroscopy revealed that catalytic ozonation improves the effectiveness of removing non-biodegradable chemicals as compared to ozone alone. The catalytic ozonation procedure concentrated on compounds with a molecular weight (MW) < 1 kDa in petrochemical secondary wastewater. The latest research reveals a remarkable approach to a sustainable 'waste-treating-waste' procedure for wastewater treatment.

The incineration of MSW produces residual ashes, also known as incineration bottom ash (IBA) and incineration fly ash (IFA). Depending on the leaching behavior of pollutants, landfilled incineration ashes can be categorized as hazardous or non-hazardous wastes. To resolve the disposal issue of this waste, various applications, including chemical looping combustion, carbon capture, industrial wastewater treatment, construction, and land reclamation, have been developed to valorize ash usage [13]. The employment of these ashes for catalyst development has gained the attention of researchers working on the development of sustainable catalysis. The co-pyrolysis of oily sludge (OS) and IFA is a path-breaking strategy for sustainable waste management [14]. Their previous study mere focused on the effect of IFA catalyst on the distribution and characteristics of OS pyrolysis products. Still, the role of the complex compositions of IFA involved in the catalytic mechanism remained unanswered. Other constituents, including metal oxides and inorganic counterparts, such as chlorides, hydroxides, carbonates, and sulfates, also significantly contributed to sound catalytic activity. Hu *et al.*, investigated the contribution of each component in IFA to determine the overall catalytic mechanism of IFA on OS pyrolysis. Notably, $CaCl_2$ and KCl significantly improved the pyrolysis oil upcycling, whereas IFA components altogether presented a favorable catalytic effect on the production of pyrolysis gas and coke. $CaSO_4$ and $CaCO_3$ retarded the catalytic reactions and enhanced the conversion of carbon to coke. The present research findings support waste utilization opportunities and give insights into co-processing solid wastes.

Hu *et al.*, studied the co-pyrolysis mechanism of waste textile fibers and leather with the incorporation of Ca/Fe-rich sludge ash [14]. The addition of Ca/Fe-rich sludge ash improved the co-pyrolysis thermogravimetric (TG) properties, with Ca-rich ash having the greatest catalytic effect. The weight loss rate during the slow carbonization stage at 534-900°C is 12.8%/min, and the activation energy is 100.16 kJ/mol with a decline of 30.11%. Furthermore, the activation energy was decreased by 7.38% with the use of Fe-sludge ash, and by 10.86% with a combination of Ca and Fe-rich sludge ash. The addition of Ca/Fe-rich ash inhibited the formation of nitrogen- and oxygen-containing molecules, enhancing the quality of condensable volatile products.

Ahamed A. *et al.*, utilized IBA and IFA as catalysts for plastic treatment under thermochemical conditions [13]. Ni-loaded ashes were prepared *via* hydrothermal synthesis and employed as cost-effective, waste-derived catalysts for the in-line refurbishment of volatile products from plastic pyrolysis. Both IBA and air pollution control IFA (APC) promoted the selective generation of BTEX compounds (i.e., benzene, toluene, ethylbenzene, and xylenes), alongside the formation of other gaseous and liquid species. Calcined APC (C-APC) enhanced BTEX yield, demonstrating that the pre-treatment could improve the catalytic potential of ashes. Ni was incorporated into C-APC at 15 and 30% by weight. In contrast to the Ni-loaded ZSM catalyst, Ni loading resulted in better net oil output while reducing gas yield. However, the incorporation of Ni led to the formation of heavier hydrocarbons (C_{20}-C_{35}), enabling post-treatment to extract desirable products such as BTEX.

3.2.2 Industrial Waste-Derived Catalysts

In the past decade, industrial wastes have been utilized as resource materials in environmental remediation and restoration, following the concept of the Sustainable Goals, 'treating waste with waste'. Antibiotics have made a path-breaking contribution to the treatment and prevention of disease. However, antibiotic residues produced during fermentation pose an environmental risk due to their instability, perishability, and unpleasant odor. They can also lead to bacterial drug resistance and antibiotic resistance genes, raising significant safety concerns for humans and the environment [15]. A mixture of streptomycin waste residue and layered double hydroxides (LDH)-derived $Mn_1Al_1O_x$-Str-2 catalyst was employed for NH_3 selective catalytic reduction (SCR) denitration technology. The preferred $Mn_1Al_1O_x$-Str-2 catalyst outperformed the control $Mn_1Al_1O_x$ and Mn/Al_2O_3 catalysts at temperatures ranging from 150-300°C, converting over 96% of NO_x. Moreover, the $Mn_1Al_1O_x$-Str-2 catalyst was more resistant to SO_2 and/or H_2O, with excellent regeneration ability. Analytical investigations showed that the Lewis acid sites on the surface of the $Mn_1Al_1O_x$-Str-2 catalyst played a key role in the SCR reaction [16].

Red mud (RM) is a bauxite residue and industrial solid waste generated during alumina production. With a global stockpile of 4 billion tons, 80% of this waste is accumulated around alumina plants, stretching a significant amount of land resources and posing a hazard to the natural environment in nearby places [17]. In recent years, RM has emerged as a potential raw material for iron-based catalysts due to its high iron content. In recent work, a magnetic-supported catalyst developed from red mud-based Fe-Co PBA efficiently demonstrated the successful degradation of organic pollutants by activating peroxymonosulfate under visible light [18]. Fe-Co Prussian blue analogues (RM-Co PBA-D) are used for catalyst synthesis, with RM serving as both an iron source and carrier. The Fe-Co PBA catalyst efficiently activates peroxymonosulfate (PMS) under visible light to break down tetracycline (TC) and rhodamine B (RhB) dyes. The RM-Co PBA-D's outstanding photo-generated carrier separation efficiency facilitated the RM-Co PBA-D/PMS/Vis system to entirely decompose both TC (10 mg/L) and RhB (20 mg/L) in 30 minutes, with mineralization efficiencies of 70.12% and 60.59%, respectively. Reactive oxygen species (ROS), $SO_4^{\cdot-}$ and $^1O_2$ played a governing role in the degradation process. Photo-generated carriers promoted the regeneration of active sites (Co^{2+}/Fe^{2+}), which improved organic pollutant removal.

Spent bleaching clay (SBC) is an industrial waste generated during the oil refining procedure, with an annual production of around two million tons.

It contains 20% organic components, making it a fire hazard and a major threat to soil and groundwater due to improper disposal [19]. However, SBC containing silicon and aluminum oxides can be harnessed as a precursor for silicate-based catalysts. Herein, SBC was utilized to prepare a ball-milled mixed lithium salt (Li-SBC) catalyst, predominantly consisting of lithium aluminate ($LiAlO_2$) and lithium orthosilicate (Li_4SiO_4). Cong *et al.*, reported that the merger of $LiAlO_2$ and Li_4SiO_4 demonstrated a synergistic catalytic activity and achieved an excellent yield of 95.2% of fatty acid methyl esters (FAMEs) from waste cooking oil (WCO; acid value of 1.9) at room temperature [19]. $LiAlO_2$ confirmed the synergistic impact of Li-SBC by escalating the catalyst's specific surface area by 5.8 times, whereas Li_4SiO_4 increased its basicity by a factor of 8. In contrast, $LiAlO_2$ and Li_4SiO_4 alone produced yields of 7.1% and 48.4%, respectively. The catalyst was efficiently recycled, with FAME synthesis yields exceeding 80% after four reaction cycles of soybean oil at 55°C. This study exhibits a sustainable methodology for producing FAME from waste materials by avoiding high-pressure equipment and demonstrating a more efficient and greener technology for biodiesel production.

Fly ash is a solid waste generated from a thermal power plant during electricity production. This waste is rich in silicon (Si), aluminum (Al) oxides and has been valorized through high-value synthesis of zeolites [20]. Researchers identified the potential of industrial biomass fly ash (BFA) to act as *in-situ* low-cost catalysts in biomass steam gasification in a bench-scale bubbling fluidized bed reactor, resulting in improved H_2-enriched producer gas. Granulated BFA ($G_{BFA\ fresh}$) with $CaCO_3$ and SiO_2 as main components was thermally activated to transform $CaCO_3$ to CaO ($G_{BFA\ treated}$) and its thermally modified powder form ($P_{BFA\text{-}treated}$). All BFA-based catalysts successfully improved the H_2 concentration in the pine gasification, in the series $G_{BFA\ treated} > P_{BFA\ treated} > G_{BFA\ fresh}$ > reference condition (inert material: granulated alumina). The H_2 content improved from 40.53% vol. dry and inert (N_2, Ar) free gas in the reference condition up to 54.12% vol. dry and inert (N_2, Ar) free gas using $G_{BFA\text{-}treated}$. All catalysts had a substantial effect on the H_2:CO molar ratio, with the highest increment up to 140% with $G_{BFA\text{-}treated}$. BFA-based materials exhibited significant catalytic activity towards reforming and cracking reactions.

For industrial waste valorization under environmental remediation, carbon-based catalysts derived from this waste have emerged as a promising greener alternative to traditional catalysts. Using a tyre-ash-derived activated carbon catalyst, synergistic adsorption-photocatalytic remediation of methylene blue (MB) from textile industry wastewater was studied by Munonde *et al.* [21]. The removal efficiency of the MB dye was reported to be the highest at 94.2% using NiFe LDH/AC nanocomposites among various methods such as photolysis, adsorption, and degradation on AC, NiFe LDH nanosheets. The reusability investigations revealed that the MB photo-degradation efficiency declined by 20.8%, probably due to the catalyst loss during its washing.

Chen *et al.* [22], developed a highly active sludge-derived carbon catalyst loaded with Co_3Fe_7 bimetallic alloy (DSHC(Fe)-Co) for efficient peroxymonosulfate (PMS) activation to achieve the deep oxidation of these toxic volatile organic compounds (VOCs) in the advanced oxidation process (AOP) wet scrubber. The sludge samples employed in this research were collected from the dyeing wastewater treatment plant (Jiaxing, Zhejiang Province, China). The prepared catalyst demonstrated better removal efficiency of the typical VOCs, and gaseous benzene, remained above 99% across a pH range (3-11), better than that of the single Fe or Co catalyst. The DSHC(Fe)-Co alloy outperformed other pollutants (toluene, ethyl acetate, etc.) with a degradation rate of over 98%. Analytical results showed the synergic relationship between Fe and Co, resulting in the excellent removal performance of DSHC(Fe)-Co.

Ao *et al.*, engineered a 4-benzenediazonium sulfonate (4-DBS) functionalized acidic biochar catalyst from paper waste for potential use in sustainable biodiesel production using *Jatropha curcas* oil feedstock [23]. In three separate beakers, the paper waste material was initially suspended in an aqueous $ZnCl_2$ solution in the ratios (1:1, 1:2 and 1:3). The blend was dried and underwent pyrolysis at 500°C for 2h to produce a porous, aromatic, carbonized network structure that is significantly graphitized. The resulting material was acid-activated (2 M HCl), and the resulting product (AC) was kept overnight for drying in the oven. AC was sulfonated using 4-DBS, and after certain steps, the final products were marked as PW-1, PW-2 and PW-3 concerning the weight of $ZnCl_2$ used. BET analysis showed that PW-2 with $ZnCl_2$ (1:2) possessed the highest surface area of 540 m^2g^{-1}, and pore volume of 0.084 cc g^{-1}, which significantly facilitated the mass transfer, and mechanical stress and heat were uniformly distributed throughout the transesterification reaction. The catalyst achieved low activation energy of 40.4 kJ mol^{-1} and produced 94.2 ± 0.2% of biodiesel from *Jatropha curcas* oil under optimum conditions. Additionally, the catalyst also demonstrated high reusability, with six reaction cycles and minimal activity loss due to reduced metal leaching.

3.2.3 Agricultural Waste-Derived Catalysts

Agricultural wastes are produced from various farming practices, including horticulture, seed production, animal husbandry, gardening, and forestry, which comprise at least 30% of global productivity [24]. The circular bioeconomy integrates waste recycling with the utilization of organic biomass for bioenergy and bio-based chemical synthesis. This green approach encouraged the optimum utilization of resources, reduced greenhouse gases (GHGs), and promoted sustainable economic growth. Waste-to-energy systems and food waste valorization techniques are explored to convert agricultural waste into renewable energy sources, such as bioethanol, biodiesel, and biogas, through fermentation, transesterification, and anaerobic digestion [25].

Agricultural waste serves as a potential resource for biomass, which can be utilized for renewable energy production. Biomass can be transformed into significant carbon-based catalysts through various thermochemical conversion techniques, such as pyrolysis and hydrothermal carbonization [26]. Pyrolysis involves burning in the absence of oxygen at temperatures ranging from 400-900°C, which unlocks the inherent carbon structure, resulting in a biochar with a highly porous network and bountiful oxygen-containing functional groups such as phenols and carboxyls [27]. Another technique includes hydrothermal carbonization, which operates in the subcritical or supercritical range of water (180–260°C, 2-6 MPa). This conversion is a greener alternative, yielding hydrochar rich in hydroxyl groups and providing a large surface area for catalytic interactions [28, 29]. By the chemical activation of biomass using NaOH, $AlCl_3$, $MgCl_2$, H_3PO_4, KOH [30–33], and other agents, activated carbon, the best version of carbon-based catalysts, could be achieved with striking attributes such as additional micropores and improved adsorptive and catalytic properties [32].

Carbonaceous materials derived from diversified sources, such as agricultural wastes, MSW, and industrial wastes, have played a significant role in sustainable catalysis. Waste-derived carbon materials offer advantages including cost-effective than its traditional counterparts; universal availability of biomass; ensuring sustainability by lowering waste volume; can be customized by using methodologies such as pyrolysis and hydrothermal carbonization (HTC) to improve their structural and chemical attributes; possesses high surface area and porosity [28]. Carbon nanotubes, activated carbon, biochar, and hydrochar are some notable examples of waste-derived carbonaceous materials. Activated carbon, biochar, and hydrochar are synthesized through thermochemical processes, enhancing abundant accessible surface areas and diverse functional groups. These catalysts significantly contribute to combating environmental challenges, including the adsorption of organic pollutants and heavy metals and the facilitation of contaminant degradation.

Biodiesel is mono alkyl esters of long-chain fatty acids and is known as FAME of vegetable oil or animal fats, prepared by a transesterification reaction involving methanol over an acid or base catalyst [34]. In recent years, the use of heterogeneous catalysts for biodiesel production has gained global popularity as they can be easily extracted from the product. Moreover, the recovered catalyst can be reused multiple times, thereby reducing its consumption during the reaction. The use of a catalyst provides quality products because it does not allow soap formation during the reaction [35]. To make biodiesel production more cost-effective and efficient, researchers are focusing on bio-based heterogeneous catalysts derived from waste resources. Recently, a sustainable method was demonstrated for biodiesel production from waste cooking oil (WCO) using magnetic catalyst-based biochar-derived spent coffee ground (SCG) [36]. The magnetic and basic attributes in the catalyst were enhanced through one-pot carbonization of SCG with Fe(III) and KOH as activation agents at 700°C for 2h. Biochar $Fe_2O_3/Fe_2K_6O_5$ magnetic catalyst showed good catalytic performance, producing a yield of 92.31% biodiesel (molar ratio at 6:1, 1 wt. %, and with 2h). The magnetism experiments confirmed that the prepared catalyst also possessed magnetic properties.

The residual bean shell biomass (RBBH) was utilized as a precursor for the synthesis of an efficient and regenerable catalyst for sustainable biodiesel production [37]. For the catalyst synthesis, in a muffle furnace, 25 g of RBBH powder was calcined with the temperature varying between 350 to 500°C for periods of 1h to 4h. Finally, the resulting product was referred as BHAT/t, (BHA = bean husk ash, T = calcination temperature and t = time). Characterization analysis of BHA400/3 catalyst showed K_2CO_3, CaO, MgO and $CaCO_3$ as its main components. The outstanding catalytic activity of BHA400/3 was attributed to its primary basic sites, K_2O and K_2CO_3, which yielded 97.6% ± 0.45 ester content under optimal conditions. However, after two reaction cycles, the catalyst's activity declined, but following regeneration with KOH, it continued to work well for the next three reaction cycles. The stability and regeneration ability of the BHA400/3 catalyst encourages its industrial application and can contribute to waste minimisation.

In recent years, the production of catalytically active carbon quantum dots (CQDs) from biomass has gained attention due to their unique attributes, such as ease of availability, natural abundance, renewability, cost-effectiveness, and environmental friendliness [38]. In a recent study, using okra waste, by combining carbon quantum dots (CQDs) with multi-walled carbon nanotubes (MWCNTs), a photocatalytic nanocomposite (CQDs/CNTs) catalyst was synthesized and employed for the photodegradation of methyl orange (MO) and rhodamine B (RhB) dyes under visible light. These CQDs were then distributed in a CNT solution to form CQDs/CNTs nanocomposites. The CQDs/CNTs nanocomposite achieved dye degradation rates of 98.4% for RhB and 99.1% for MO under visible light irradiations [39]. The nanocomposites also demonstrated good stability over four repeated cycles. Fluorescence quenching showed interactions between CQD composites and the lysozyme protein, indicating possible uses in biosensing.

In the last decade, the selective transformation of lignin and its model compound has emerged as an efficient method for lignin valorization to high-value biofuels and chemicals. Herein, a series of Ru-loaded biochar (BC) catalysts derived from bamboo, metasequoia, miscanthus, and poplar wood were synthesized and used to hydrolyze three lignin model compounds and alkaline-treated lignin [40]. The BC from bamboo waste with Ru loading (Ru/BABC) exhibited better catalytic activity by complete conversion with selectivities: diphenyl ether (DPE) with 80.8%, benzyl phenyl ether (BPE) with 86.0% and phenethoxybenzene (PPE) with 88.8%. Ru/BABC gave five recycling runs with the least Ru leaching, which proves its structural stability.

Z. Laldinpuii *et al.* [41], used bamboo leaf ash (BLA) for the decomposition of poly(ethylene terephthalate) (PET) bottles by methanolytic degradation, which gave dimethyl terephthalate (DMT) and ethylene glycol (EG), yielding 78% and 76%, respectively.

3.2.4 PET Waste-Derived Catalysts

PET is the fastest-growing industrial polymer obtained from fossil feedstocks with an annual worldwide production of nearly 80 Mt in 2020, making this polymeric material the 6th most-produced plastic and 3rd most commercially used packaging material [42]. PET has remarkable chemical stability, mechanical transparency, adjustable transparency, and cost-efficiency that make it the first choice for the manufacturing of plastic bottles, plastic films, food packaging, and synthetic fibers [43]. A recent report shows that PET contributes to about 12% (by volume) of worldwide solid waste [44, 45]. To combat this environmental menace, chemical depolymerization methods, including hydrolysis, glycolysis, ammonolysis and pyrolysis [46–48], can revert these polymeric forms into their stable constitutional monomers. However, these methods still have drawbacks of harsh reaction conditions, low product yields, and purification issues. To develop sustainable chemo-catalytic processes for PET recycling, many scientific endeavors have been undertaken in recent years. A. Abou-Elyazed *et al.*, synthesised metal-organic framework (MIL-101(Cr) from PET bottles by extraction of terephthalic acid and employed it as a solid catalyst for methyl oleate (biodiesel) production [49]. The produced catalytic material showed improved surface area (2618 m^2/g to waste 1673 m^2/g) and pore volume (1.42 cm^3/g; to waste 1.03 cm^3/g). Both original and waste-based MIL-101 (Cr), implementing a 1.39 molar ratio, 6 wt. % loadings, and a 4h reaction period, yielded the best biodiesel production of 86.9 and 80%, respectively. The MIL-101(Cr) PET-derived catalyst was utilized for up to three reaction cycles, while its original counterpart gave five reaction cycles.

Another green synthesis method was developed to repurpose PET bottles into terephthalic acid, which was employed to synthesize PET-MIL-101(Cr) (denoted as PM-101(Cr)) [50]. Organic solvents like HF or DMF were avoided to minimize secondary pollution. Prepared PM-101(Cr) was further modified through nitration to generate PET-NO MIL-101(Cr) (PO-101(Cr)), which was then further treated to produce amine-functionalized PET-NH 2 2 -MIL-101(Cr) (PH-101(Cr)). The catalytic activity of PH-101(Cr) demonstrated 95% MB dye removal within 4h of blue light irradiation outperforming the performance of PM-101(Cr). PH-101(Cr) successfully retained its reduction efficiency above 95% for 6h, even after being recycled up to four catalytic runs.

Karam L *et al.*, demonstrated a sustainable approach for the MIL-53(Al), an aluminum-containing metal-organic framework (MOF), which was synthesized using linkers derived from the depolymerization of PET obtained from plastic wastes. The obtained material was utilized as lamellar alumina support for the impregnation of Ni^0 particles to get Ni^0–Al_2O_3–PET catalyst and employed in the reaction of dry reforming of methane (DRM) [51]. PXRD and TPR studies revealed that the mixed oxide consisted of nickel aluminate nanoparticles within amorphous alumina. Reduction under H_2 reported that Ni^0–Al_2O_3–PET has uniformly distributed 5 nm Ni particles, which are anchored in the interlamellar porosity of the γ-alumina matrix. Ni^0–Al_2O_3–PET displayed outstanding stability, and its selectivity towards DRM, without losing its activity after 13h and the H_2:CO products ratio remained at 1. Furthermore, Ni^0–Al_2O_3–PET catalyst demonstrated equivalent performance to those of Ni^0–Al_2O_3 catalysts obtained from commercial BDC for both CH_4 and CO_2 conversions and considerably better than its conventional counterparts. This recycling can help reduce the plastic waste accumulation in the environment and provide a better substitute for industrially produced organic linkers and petroleum-based compounds.

3.2.5 Food Waste-Derived Calcium Containing Catalysts

Food waste containing calcium content has various applications [52], as shown in Table 3.1. The primary sources of waste-based calcium catalysts include seashells, eggshells, and bones, offering a sustainable solution to waste reduction [53]. Seashells, the protective exoskeletons of molluscs, primarily contain 95% $CaCO_3$. With an annual production of 18 million tons [54], this by-product can serve as a valuable raw material to fulfill the increasing $CaCO_3$ demands all over the globe. Another promising resource collected from the poultry and kitchen is eggshell waste, with the chemical composition of calcium carbonate (94%), calcium phosphate (1%), magnesium carbonate (1%), and organic matter (4%) [52, 53]. Slaughterhouse waste, mainly bone waste, is inadequately managed with a global annual production of 130 billion kg [55]. Improper disposal of these wastes without pre-treatment can produce ammonia, hydrogen sulfide, leading to the growth of pathogens and contamination of that area [56, 57]. However, utilization of this waste can serve as a reliable raw material for obtaining calcium oxide, opening a sustainable pathway to save the environment from its adverse effects.

3.2.6 E-Waste Derived Catalysts

Electronic devices have become an integral part of our daily lives. During the SARS-CoV-2 pandemic, the changes in the work environment and the transition from the official workplace to home office and remote studies led to the ever-increasing demand for electronic devices. The skyrocketing demand for electric and electronic equipment has led to an increase in e-waste generation, which reached 62 billion kg by 2022. Yet only 13.8 billion kg were gathered and recycled. In addition, by the year 2030, e-waste generation is expected to reach 182 billion kg [1, 77]. The general composition of e-waste is shown in Figure 3.1.

Along with E-waste, waste electrical and electronic equipment (WEEE) is another term which is used more often in discarded EEE appliances [79]. WEEE includes non-electronic appliances (e.g. refrigerators, air conditioning units, washing machines [80]. WEEE constitutes plastics such as polycyclic hydrocarbons (PAHs), polybrominated diphenyl ethers (PCBPs), polybrominated dibenzo-p-dioxins and dibenzofurans (PBDD/Fs); persistent organic pollutants (POPs), dioxins and glass fibers are as

Table 3.1 Various applications of food waste-derived calcium-containing catalysts.

Calcium containing wastes	Waste-derived catalyst	Synthesis methodology	Applications	Ref.
Eggshell (ES) waste	Silver nanoparticles encapsulated Calcite Fibers (ACFs)	Sol-gel method	To determine the antibacterial activity	[58]
Waste cow bone granules	Manganese-modified hydroxyapatite	Dissolution-precipitation mechanism	Removal of toluene and formaldehyde in simulated flue gas	[59]
Eggshells	Waste-derived (WD)-Ca incorporated Zn nanorods encapsulated with chitosan polymers (CZC) bandages	Sol-gel process	Control the loss of blood as well as bacterial infection	[60]
Waste eggshells and waste printed circuit boards (PCBs)	DFMs (dual-functional materials) CaO-Cu1-Oxalic (concentration of Cu at 1 wt. %)	Combining Metal extracted from waste PCBs and waste eggshells is in a single process using an organic acid	Integrated CO_2 capture and utilization (ICCU)	[61]
Sunflower seed shell waste	Sunflower seed shell matrix catalyst (SMS-750)	Catalytic alcoholysis of waste PET	Synthesis of Bis(hydroxyethyl) terephthalate (BHET)	[62]
Crab shell waste	High-calcium crab shell ash (CSA)	Chemical precipitation	CO_2 mineral sequestration	[63]
Waste eggshell (EGC), pre-served eggshell (PEC), clamshell (CLC) and crab shells (CRC)	CaO/HZSM-5, EGC, EGC and CLC	--	Pyrolysis of waste cooking oil (WCO) for enrichment of aromatics in bio-oil	[64]

(*Continued*)

Table 3.1 Various applications of food waste-derived calcium-containing catalysts. (*Continued*)

Calcium containing wastes	Waste-derived catalyst	Synthesis methodology	Applications	Ref.
Malaysia dolomite (MD) was extracted from limestone, Clamshells (CS), eggshells (ES), chicken bones (CB)	Calcium oxide (CaO)-based catalysts	-	Co-pyrolysis of empty fruit bunch (EFB) and high-density polyethylene (HDPE) to produce high-quality bio-oil	[65]
Waste oyster shells	CaO-based catalysts	Calcining at different temperatures	Oxidative coupling of methane	[66]
Eggshell waste	Calcium glycerolate catalyst	Calcination at low temperatures	Cyclopentadecanolide production *via* reactive distillation using a glycerol entrainer	[67]
Waste chicken egg shells	Chicken eggshells (ES) catalyst	Calcination at 500-900°C	One-pot synthesis of lactic acid	[68]
Waste eggshells collected from a biscuit and cake-making bakery	Eggshell derived-CaO catalyst	Drying at 105°C for 48 h followed by calcination at 900°C for ~120 minutes	Hydrogen production from MSW	[69]
Eggshell and brewed tea waste	Eggshell/tea waste-derived CaO nanoparticles.	-	For CO_2 capture	[70]
Eggshell waste	CuCa hydroxy double salt (HDS) materials	*Via* hydrothermal metal oxide-assisted treatment	Treatment of highly polluted wastewater (the initial concentration of MO is 500 ppm)	[52]
Green mussel shells, eggshells, crab shells	CaO-based sorbent	Citric acid-assisted sol-gel method	For capturing cyclic CO_2	[71]
Chicken egg shells (CES)	Nano solid base catalyst (NSBC)	Calcination of raw CES at 850 °C for 2 h	Biodiesel production	[72]

(*Continued*)

Table 3.1 Various applications of food waste-derived calcium-containing catalysts. (*Continued*)

Calcium containing wastes	Waste-derived catalyst	Synthesis methodology	Applications	Ref.
Eggshells and poplar leaves	Acid-base bifunctional CaO-SrO/biomass chars (BCs) catalyst	Calcination of eggshells and poplar leaves	Biodiesel production from sheep fat	[73]
Eggshell waste	CaO catalysts	Soft template method and thermal treatment under a controlled nitrogen atmosphere	Transesterification of pork lard with anhydrous methanol	[74]
Waste oyster shells (WOS)	WOS catalyst	Calcination	Transesterification reaction of Jatropha curcas oil (JCO)	[75]
Limestone-eggshell	Sr and K supported on limestone-eggshell-based CaO support	Impregnation and calcination	Biodiesel production	[76]

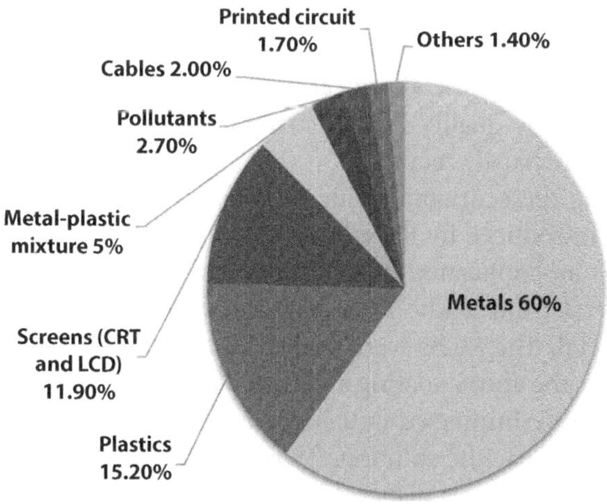

Figure 3.1 Composition of e-waste components. Reprinted with permission from [78].

organic category. Base metals (Cu, Sn, Al, Fe), rare metals (Ga, Ta), precious metals (Ag, Au, Pd, Pt) and hazardous metals (Cd, Hg, Ni, Pb, Zn, Be, Cr) are also important components of WEEE [81]. The global demand for precious metals Pt, Au, Ag, and Pd, along with the base metals Zn, Pb, Cu, and Ni, and rare earth minerals Y, La, Ce, and Nb, has recently increased significantly in electronic applications [82]. Technology critical elements (Ga, In, rare earth elements, platinum group, Nb, and Ta, and so on.) in discarded electronics can range from 0.01 up to 100 mg/kg [83]. Presently, discarded printed circuit boards (PCBs) produce approximately 3% of the total e-waste annually. PCBs comprise approximately 30% metals and 70% non-metallic substances, among which Cu is the chief metal (approx. 20% and commonly in Cu (0) [1].

A workable method for extracting base metals from low-quality PCBs with low concentrations of precious metals was devised by Ryabchuk *et al.* [84]. The Cu-iKat catalyst was prepared by leaching abandoned cellular and mobile phones and precipitating copper and other base metals from PCBs. Non-noble metals, such as Cu, Ni, Zn, Fe, Pb, Ag, and Sn, were dropped into an alkaline solution to be selectively deposited onto an alumina substrate γ-Al_2O_3 suspension after leaching various metals from PCBs with 5 M HNO_3. Following pyrolysis at 800°C, a Cu-based γ-Al_2O_3 catalyst with a 5.7 wt. % Cu concentration was obtained. The prepared Cu-iKat catalyst demonstrated significant catalytic activity for several reactions, involving the synthesis of methanol by formamide hydrogenation, deoxygenation, and hydrogenation of N-heterocycles and acetophenones, with conversion $> 99\%$ and moderate-to-high yields. Gold-enriched PCBs can be recycled using traditional hydrometallurgical methodologies. E-waste valorization may only serve low-quality PCBs, having limited precious metal content and are therefore hardly recycled.

Considering the economic value of gold, an environmentally friendly method was introduced by Bayarkhuu *et al.* [85], for creating single-atom catalytic sites on conjugated porphyrin polymer photocatalysts by recovering Au metal from acidic PCB leachate with minimum co-doping of transition metal. The PCBs were soaked in a NaOH (10 M) solution for 24 h to remove the epoxy coating and then washed with tap water. The pretreated PCBs were immersed in 0.5 L of 1 ml aqua regia at 40°C for 48 h, resulting in a blue acidic solution. Typically, conjugated porphyrin polymer, COP-180 (1.4 g) was dispersed in e-waste leachate (100 mL) at room temperature and stirred at 350 rpm for 72h under dark conditions. The powder was then collected by centrifugation, washed, and dried at 100°C overnight to yield M-COP-18 photocatalyst. The obtained M-COP-180 was further purified by oxidant treatment to produce a gold-loaded photo catalyst, named Au-COP-180. The crude M-COP-180 (1 g) while wet was redispersed in a mixture of H_2SO_4 (0.2 M) and H_2O_2 (0.02 M) solution (800 mL), and the dispersion was stirred at 300 rpm, 80°C for 18h under dark conditions. The Au-COP-180 powder was collected by centrifugation and washed with water, followed by overnight drying at 100°C. The resulting photocatalyst with Au single atoms yielded a pure H_2O_2 solution at a rate of up to 73.3 μm h^{-1} with a high H_2O_2 selectivity of up to 97.56% under white LED illumination. The produced H_2O_2 was activated to •OH radicals on the same polymer when loaded with mixed Au and Fe metal active sites. The involvement of these metal species led to a photo-Fenton reaction, completely degrading toxic microcystin LR (200 ppb) with pure H_2O_2 solution within 10 min under visible light.

Battery recycling has become a widely accepted factor due to escalating electricity demands all over the world. The estimated size of the worldwide rechargeable battery market accounted for USD 123.81 billion, and by 2034, it is estimated to reach nearly USD 219.22 billion, with a 5.88% CAGR between 2024 and 2034. [86], as shown in Figure 3.2a. The utilization of Pb-acid batteries is forecasted to be upgraded by their profitable and reliable performance (Figure 3.2b) [86].

In 2016, lead-acid batteries (LABs) and lithium-based batteries contributed 94.8% of the global battery market. According to estimation, it will reach USD 53.8 billion by 2024, with 11% CAGR during 2024-25 [87]. In a recent study, H. Chen *et al.* [88], presented a 'waste to energy' strategy to synthesize a high-performance CoO@C catalyst by using both the cathode and anode of spent lithium-ion batteries (LIBs). The CoO@C catalyst acted as peroxymonosulfate (PMS) activator and degraded the atrazine (ATZ), an agricultural herbicide frequently found in the aquatic environment. For the catalyst synthesis, the anode and cathode of the spent batteries underwent pre-treatment to get Co and graphite content. The resultant dark red metallic filtrate was filtered, collected, and subsequently used for catalyst preparation. The CoO@C/PMS system successfully degraded atrazine (ATZ) within 15 min. To improve efficiency, the CoO@C catalyst was anchored on a polyethersulfone (PES) membrane for continuous degradation of ATZ. Also, it has increased degradation efficiency of more than 90% after processing 6 L of wastewater.

Hazardous and difficult-to-process materials like spent Ni-MH batteries and aluminium foil were transformed into nanocatalysts (NCts) and efficiently catalyzed CO_2 methanation reaction [89]. The recovery of Ni from Ni-MH batteries, Al from aluminum foil, and the NCts synthesis involved the following steps: disassembly and washing of Ni-MH batteries, leaching of nickel ions, separation, crystallization, and preparation of

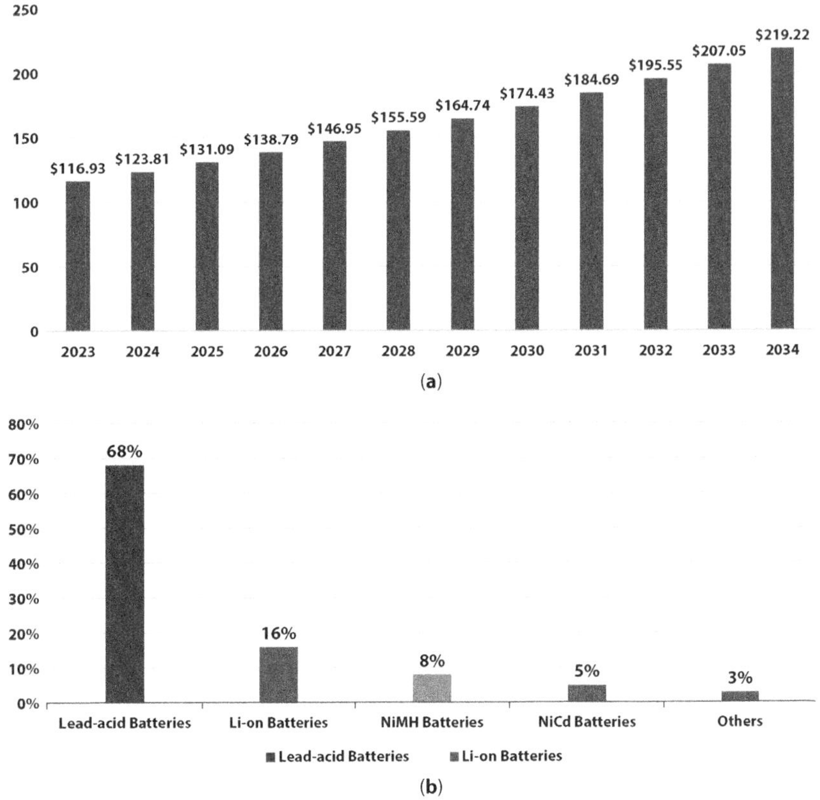

Figure 3.2 (a) Rechargeable batteries by total market value, in the year 2023–2034 (billion USD), and (b) types of rechargeable batteries utilized in the year 2023 (%). Reprinted with permission from [86].

$NiSO_4$, its reduction to $Ni(OH)_2$ hydrogel, washing and storage of $Ni(OH)_2$, dissolution of waste Al-foil, reduction to $Al(OH)_3$ pellets and their drying. $Ni(OH)_2$ and Al_2O_3 powder were mixed with Ni at 4% and 8% weight percentages, respectively, to synthesize NCts. The mixture was sonicated for 30 minutes, and the obtained suspension was dried and calcined at 550°C (4 hours) to obtain 4% or 8% (wt. %) Ni over η-Al_2O_3, designated as 4-Ni/η-Al_2O_3 and 8-Ni/η-Al_2O_3, respectively. 8-Ni/η-Al_2O_3 catalyst accomplished 99.8% selectivity, space time yield (STY) of 80.3 mmol CH_4 g_{cat}^{-1} h^{-1} of CH_4 at 400°C and Gas Hourly Space Velocity (GHSV) of 2984.16h^{-1} [89]. The spent Ni/η-Al_2O_3 nanocatalysts were recycled into Ni and Al precursors *via* acid leaching and selective NaOH precipitation, showing their sustainability and economic viability.

3.3 Challenges

The production of catalysts using solid waste as a resource faces miscellaneous challenges and knowledge voids that need to be filled for successful integration into sustainable practices. In-depth knowledge of the chemical and physical processes involved in transforming into effective catalysts is paramount. Still, the scarcity of comprehensive knowledge regarding the fundamental mechanisms of catalyst formation from waste hinders the optimization and control of these processes. In addition, elucidating the structure-activity relationships of waste-derived catalysts remains elusive, which impact the predictability of their performance in different reaction conditions [90]. The diversity in the composition of waste is the key challenge, urging the development of adaptable methodologies to ensure consistent catalyst quality across different waste types. The presence of contaminants in solid waste is another issue that needs to be addressed. This requires efficient purification methodologies to check their influence during catalyst synthesis. Moreover, optimizing the conversion process, such as pyrolysis or chemical treatments, is important for improving efficiency and cost-effectiveness. Laboratory to industrial scale-up faces challenges in maintaining catalyst quality and cost-effectiveness. Comprehensive studies on the long-term behavior of the catalyst under various operating conditions are required to ensure its performance and stability. Moreover, the lack of a comprehensive life cycle assessment indicated a critical gap, limiting an in-depth understanding of the environmental impact of waste-derived catalysts. An overall understanding of these scientific aspects is important for designing catalysts with enhanced stability, reactivity, and selectivity.

Another crucial aspect that is largely missing from published research is the profitability type analysis, which should include a comparative study of the production cost of these catalytic materials to those of their more conventional counterparts. Additionally, indicators of catalytic performance, such as space, time, yield comparisons, and catalyst lifetime, should be provided to support the catalyst's relevancy in the general application. Furthermore, the practicality of ease and cost of collection of sufficient quantities of waste viable for catalyst production is also an important point to be studied. Covering these scientific voids is important for achieving economically viable catalysts from waste resources amalgamated with sustainability.

3.4 Future Perspectives and Recommendations

In the present scenario, the over-exploitation of limited natural resources, specifically natural fossil fuels, emphasizes the need to shift towards sustainability. Anthropogenic waste emerged as a valuable resource for several applications, such as catalyst development, chemical, construction and building materials, energy generation, and value-added functional materials. Employing solid waste-derived catalysts across several applications reduces environmental concerns and encourages the global community to follow circular economy principles. Biomass-based catalysts from low-value wastes, including agricultural remnants (maize cobs or rice husks), are transformed into catalysts through processes such as pyrolysis. These catalytic materials are cost-effective, efficient, and a remarkable eco-friendly substitute for their conventional counterparts. Furthermore, carbonaceous materials derived from wastes, such as MSW waste or food waste, can serve as active catalysts or catalyst supports after appropriate modifications. The waste materials, such as FA from coal combustion, offer dual benefits. It addresses the waste disposal challenges and the requirement for an efficient catalyst by transforming waste into a value-added catalyst. Ongoing research will further explore better, simpler procedures and explore additional waste resources.

3.5 Conclusion

In summary, this chapter discusses the synthesis of catalysts from calcium containing wastes, biomass wastes, E-waste, municipal waste, incineration waste, and industrial wastes that exhibited remarkable potential as precursors. Highlighting the multifaceted contributions, the conclusion features their significance in addressing crucial challenges such as waste management, sustainable energy production, and fostering circular economy principles. This approach adequately amalgamates innovation, environmental responsibility and industrial applications. An insight into the potential impact of waste-derived catalysts, encouraging further exploration in the upcoming years, is discussed in detail.

AI disclosure statement: The author acknowledges the use of AI-based tools for minor language editing or grammatical correction. However, all intellectual contributions, analyses, and conclusions presented in this chapter are entirely the author's own.

References

1. Wang, S., Yan, W., Zhao, F., Recovery of solid waste as functional heterogeneous catalysts for organic pollutant removal and biodiesel production. *Chem. Eng. J.*, Elsevier B.V., 401, 126104, 2020.
2. Yatoo, A.M., Hamid, B., Sheikh, T.A., Ali, S., Bhat, S.A., Ramola, S., Ali, M.N., Baba, Z.A., Kumar, S., Global perspective of municipal solid waste and landfill leachate: generation, composition, eco-toxicity, and sustainable management strategies. *Environ. Sci. Pollut. Res.*, 31, 23363–23392, 2024.
3. Hart, A., Circular economy: closing the catalyst loop with metal reclamation from spent catalysts, industrial waste, waste shells and animal bones. *Biomass Convers. Biorefin.*, 13, 11483–11498, 2023.
4. Malpani, S.K. and Goyal, D., Synthesis, analysis, and multi-faceted applications of solid wastes-derived silica nanoparticles: a comprehensive review (2010–2022). *Environ. Sci. Pollut. Res.*, 30, 28321–28343, 2023.
5. Hargreaves, J.S.J., Catalysts derived from waste materials. *Catalysis*, 30, 1–20, 2018.
6. Ansari, M., Jamali, H., Ghanbari, R., Ehrampoush, M.H., Zamani, P., Hatami, B., Heterogeneous solid acid catalysts for sustainable biodiesel production from wastewater-derived sludge: A systematic and critical review. *Chem. Eng. J. Adv.*, 22, 100718, 2025.
7. Li, G., Yu, H., Guo, H., Wang, L., Cao, P., Shi, J., Activation of peroxymonosulfate by boron carbide modified biochar catalyst derived from waste sludge for the removal of organic pollutants. *Sep. Purif. Technol.*, 361, 131319, 2025.
8. Pedrazzani, R., Bertanza, G., Brnardić, I., Cetecioglu, Z., Dries, J., Dvarionienė, J., García-Fernández, A.J., Langenhoff, A., Libralato, G., Lofrano, G., Škrbić, B., Opinion paper about organic trace pollutants in wastewater: Toxicity assessment in a European perspective. *Sci. Total Environ.*, 651, 3202–3221, 2019.
9. Li, J., Li, B., Huang, H., Zhao, N., Zhang, M., Cao, L., Investigation into lanthanum-coated biochar obtained from urban dewatered sewage sludge for enhanced phosphate adsorption. *Sci. Total Environ.*, 714, 136839, 2020.
10. Sieradzka, M., Szkadłubowicz, K., Śliz, M., Kalemba-Rec, I., Kornaus, K., Kozinski, J., Wilk., M., Hydrogen-rich syngas production from sewage sludge and hydrochars *via* catalytic gasification with SrO. *Int. J. Hydrogen Energy*, 2025.
11. Jiang, M., Yang, Y., Lei, T., Ye, Z., Huang, S., Fu, X., Liu, P., Li, H., Removal of phosphate by a novel activated sewage sludge biochar: Equilibrium, kinetic and mechanism studies. *Appl. Energy Combust. Sci.*, 9, 100056, 2022.
12. Wang, J., Fu, L., Chen, X., Deng, L., Wu, C., Catalytic ozonation promoted by Mn-doped sludge-based catalyst treating refractory industrial wastewater. *Sep. Purif. Technol.*, 354, 128676, 2025.

13. Ahamed, A., Liang, L., Chan, W.P., Tan, P.C.K., Yip, N.T.X., Bobacka, J., Veksha, A., Yin, K., Lisak, G., *In situ* catalytic reforming of plastic pyrolysis vapors using MSW incineration ashes. *Environ. Pollut.*, 276, 116681, 2021.
14. Hu, Y., Xia, Z., Wang, X., Xing, B., Chen, Q., Yu, F., Zhou, N., Enhanced co-pyrolysis of textile and leather waste with Ca/Fe-enriched sludge ash: Catalytic effect on thermal behavior, volatile composition, and biochar formation. *Fuel*, 373, 132272, 2024.
15. Ben, Y., Fu, C., Hu, M., Liu, L., Wong, M.H., Zheng, C., Human health risk assessment of antibiotic resistance associated with antibiotic residues in the environment: A review. *Environ. Res.*, 169, 483–493, 2019.
16. Chen Z Chen, Z., Sun, Z., Zhao, J., Song, X., Gao, Y., Liu, G., Zhou, C., Xin, S., Xin, Y., Wang, Q., Yan, Q., Insights into the mechanism of enhancing resistance of $MnAlO_x$-Str NH_3-SCR catalyst derived from streptomycin waste residue and LDHs. *Sep. Purif. Technol.*, 356, 129974, 2025.
17. Li, G., Liu, J., Yi, L., Luo, J., Jiang, T., Bauxite residue (red mud) treatment: Current situation and promising solution. *Sci. Total Environ.*, 948, 174757, 2024.
18. Liu S Liu, S., Wang, J., Liu, Y., Yang, B., Zhao, C., Yu, S., Wang, C., Qiu, G., Liu, J., Fang, Y., Magnetic-supported catalyst derived from red mud-based Fe-Co PBA for efficient degradation of organic pollutants by activating peroxymonosulfate under visible light. *Sep. Purif. Technol.*, 354, 128954, 2025.
19. Cong, W., Rodrigues Reis, C.E., Zhang, J., Zhang, J., Luo, X., Low temperature production of FAMEs from waste cooking oil with ball-milling synthesized mixed lithium salts derived from spent bleaching clay. *Fuel*, 388, 134492, 2025.
20. Gomes, H.G.M.F., Lopes, D.V., Moura, J.M., Ribeiro, J.P., Cruz, N.C., Matos, M.A.A., Tarelho, L.A.C., Biomass fly ash granules as a promising catalyst to promote producer gas quality from residual forest biomass steam gasification. *Energy*, 319, 2025.
21. Munonde, T.S., Madima, N., Ratshiedana, R., Nosizo Nomngongo, P., Mofokeng, L.E., Dima, R.S., Synergistic adsorption-photocatalytic remediation of methylene blue dye from textile industry wastewater over NiFe LDH supported on tyre-ash derived activated carbon. *Appl. Surf. Sci.*, 679, 161205, 2025.
22. Chen, Z., Zhao, X., Liu, C., Zhao, W., Zhu, Z., Zhou, Y., Yang, Y., Qi, J., Li, J., Enhanced VOCs catalytic degradation through AOP wet scrubbing with sludge-derived carbon loaded Co-Fe alloy. *J. Environ. Chem. Eng.*, 13, 115338, 2025.
23. Ao, S., Dwivedi, P., Chowdhury, A.P., Lallianrawna, S., Dhakshinamoorthy, A., Rokhum, S.L., Paper waste-derived functionalized biochar catalyst for production of biodiesel using Jatropha curcas oil feedstock. *Bioresour. Technol. Rep.*, 29, 102031, 2025.
24. Tiwari, A.K., Jha, S., Awasthi, R.R., Singh, A.K., Ojha, R.P., Dikshit, A., Bhardwaj, A.K., Nanomaterial production using agro-waste: a current

status and future outline, in: *Sustainable Management of Agro-Food Waste: Fundamental Aspects and Practical Applications*, pp. 249–261, Academic Press, 2025.
25. Stefi, A.L. and Vorgias, K.E., Valorizing Bio-Waste and Residuals, in: *Advances in Biochemical Engineering/Biotechnology*, pp. 1–36, Springer, Berlin, Heidelberg, 2025.
26. Khandaker, T., Islam, T., Nandi, A., Anik, M.A.A.M., Hossain, M.S., Hasan, M.K., Hossain, M.S., Biomass-derived carbon materials for sustainable energy applications: a comprehensive review. *Sustainable Energy Fuels*, 9, 693–723, 2025.
27. Muhammad, S., Yahya, E.B., Abdul Khalil, H.P.S., Marwan, M., Albadn, Y.M., Recent Advances in Carbon and Activated Carbon Nanostructured Aerogels Prepared from Agricultural Wastes for Wastewater Treatment Applications. *Agriculture*, 13, 1, 208, 2023.
28. López-Fernández, E., Comendador, J., Ramírez, Muñoz-Morales, M., Llanos, J., Valorization of organic wastes as carbon sources for electrode manufacturing in hydrogen production: A critical review. *Electrochim. Acta*, 520, 145872, 2025.
29. Rathika, K., Kumar, S., Yadav, B.R., Enhanced energy and nutrient recovery *via* hydrothermal carbonisation of sewage sludge: Effect of process parameters. *Sci. Total Environ.*, 906, 2024.
30. Bhattarai, D.P., Aryal, S., Mishra, P.K., Shrestha, T., Homagai, P.L., Oli, H.B., Shrestha, R.L.S., Exploring electrochemical performance of Zanthoxylum armatum seed-derived activated carbon using phosphoric acid (H3PO4) for sustainable energy storage applications. *Carbon Trends*, 19, 100467, 2025.
31. Efimov, M.N., Muratov, D.G., Zhilyaeva, N.A., Vasilev, A.A., Legkov, S.A., Karpacheva, G.P., Effect of pretreatment temperature and atmosphere on the structure and adsorption properties of KOH-activated porous carbon for methylene blue removal. *Inorg. Chem. Commun.*, 172, 113689, 2025.
32. Fan, M., Shao, Y., Wang, Y., Sun, J., He, H., Jiang, Y., Zhang, S., Wang, Y., Hu, X., Preparation of activated carbon with recycled $ZnCl_2$ for maximizing utilization efficiency of the activating agent and minimizing generation of liquid waste. *Chem. Eng. J.*, 500, 157278, 2024.
33. Li, X., Cheng, Q., Kang, Y., Gu, Z., Bao, H., Wang, N., Liu, Y., Hu, C., Qu, J., Synergistic adsorption and removal of nitrate in an electrically polarized column packed with $AlCl_3$-modified activated carbon. *Sep. Purif. Technol.*, 364, 132570, 2025.
34. Khan, E., Ozaltin, K., Spagnuolo, D., Bernal-Ballen, A., Piskunov, M.V., Di Martino, A., Biodiesel from Rapeseed and Sunflower Oil: Effect of the Transesterification Conditions and Oxidation Stability. *Energies*, 16, 657, 2023.
35. Chutia, G.P. and Phukan, K., Biomass derived heterogeneous catalysts used for sustainable biodiesel production: a systematic review. *Braz. J. Chem. Eng.*, 41, 23–48, 2024.

36. Silvaraja, J., Yahya, N.Y., Zainol, M.M., Lee, Y.S., Preliminary investigations of sustainable magnetic catalyst-based biochar derived spent coffee ground for biodiesel production from waste cooking oil. *Cleaner Chem. Eng.*, 11, 100148, 2025.
37. de Araújo Sobrinho, I., Ribeiro, T.S., Dias e Silva, A.C., Arrais Gonçalves, M., da Rocha Filho, G.N., Vieira da Conceição, L.R., Catalytic solid derived from residual bean husk biomass applied to sustainable biodiesel production: preparation, characterization, and regeneration study. *RSC Adv.*, 15, 7050–7068, 2025.
38. Nayak, R.R., Khairun S, H., Parveen, G., Garg, A., Chumachenko, Y.A., Shu, R., Gupta, N.K., A Review on Biomass-Derived Carbon Quantum Dots: Emerging Catalysts for Hydrogenation Catalysis. *Chem. Asian J.*, 20, e202401005, 2025.
39. Meraj, U., Laiq, E., Bushra, R., Ahmad, M., Gupta, S., Mujahid, M., Hydrothermal synthesis of okra waste-derived CQDs-CNTs nanocomposites: potent catalysts for pollutant degradation of RhB and MO dyes and its interaction with lysozyme protein. *Biomass Convers. Biorefin.*, 1–21, 2025.
40. Qiu, Y., Ai, Y., Luo, W., Li, S., Chen, Y., Yi, S., Li, K., Min, Y., Biomass waste-derived biochar catalysts with Ru loading for selective and efficient transformation of lignin and its model compound. *Fuel*, 383, 133835, 2025.
41. Laldinpuii, Z.T., Khiangte, V., Lalhmangaihzuala, S., Lalmuanpuia, C., Pachuau, Z., Lalhriatpuia, C., Vanlaldinpuia, K., Methanolysis of PET Waste Using Heterogeneous Catalyst of Bio-waste Origin. *J. Polym. Environ.*, 30, 1600–1614, 2022.
42. Bohre, A., Jadhao, P.R., Tripathi, K., Pant, K.K., Likozar, B., Saha, B., Chemical Recycling Processes of Waste Polyethylene Terephthalate Using Solid Catalysts. *ChemSusChem*, 16, e202300142, 2023.
43. Wang, J., Yuan, X., Deng, S., Zeng, X., Yu, Z., Li, S., Li, K., Waste polyethylene terephthalate (PET) plastics-derived activated carbon for CO_2 capture: a route to a closed carbon loop. *Green Chem.*, 22, 6836–6845, 2020.
44. Barnard, E., Rubio Arias, J.J., Thielemans, W., Chemolytic depolymerisation of PET: a review. *Green Chem.*, 23, 3765–3789, 2021.
45. Kang, M.J., Kim, H.T., Lee, M.W., Kim, K.A., Khang, T.U., Song, H.M., Park, S.J., Joo, J.C., Cha, H.G., A chemo-microbial hybrid process for the production of 2-pyrone-4,6-dicarboxylic acid as a promising bioplastic monomer from PET waste. *Green Chem.*, 22, 3461–3469, 2020.
46. Umdagas, L., Orozco, R., Heeley, K., Thom, W., Al-Duri, B., Advances in chemical recycling of polyethylene terephthalate (PET) *via* hydrolysis: A comprehensive review. *Polym. Degrad. Stab.*, 234, 111246, 2025.
47. Ügdüler, S., Van Geem, K.M., Denolf, R., Roosen, M., Mys, N., Ragaert, K., De Meester, S., Towards closed-loop recycling of multilayer and coloured PET plastic waste by alkaline hydrolysis. *Green Chem.*, 22, 5376–5394, 2020.

48. McNeeley, A. and Liu, Y.A., Assessment of PET Depolymerization Processes for Circular Economy. 1. Thermodynamics, Chemistry, Purification, and Process Design. *Ind. Eng. Chem. Res.*, 63, 3355–3399, 2024.
49. Abou-Elyazed, A.S., Shaban, A.K., Osman, A.I., Heikal, L.A., Mohamed, H.F., Hassan, W.M., El-Nahas, A.M., Keshta, B.E., Hamouda, A.S., Comparative catalytic efficacy of cost-effective MIL-101(Cr) based PET waste for biodiesel production. *Curr. Res. Green Sustainable Chem.*, 8, 100401, 2024.
50. Liang, F., Chen, Y.H., Huang, H.W., Chen, Y.C., Huang, P.J., Waste-to-resource strategy through green synthesis of PET-derived metal-organic frameworks for efficient photocatalytic dye degradation. *MicMM*, 384, 113431, 2025.
51. Karam, L., Miglio, A., Specchia, S., El Hassan, N., Massiani, P., Reboul, J., PET waste as organic linker source for the sustainable preparation of MOF-derived methane dry reforming catalysts. *Mater. Adv.*, 2, 8, 2750–2758, 2021.
52. Han, Y., Trakulmututa, J., Amornsakchai, T., Boonyuen, S., Prigyai, N., Smith, S.M., Eggshell-Derived Copper Calcium Hydroxy Double Salts and Their Activity for Treatment of Highly Polluted Wastewater. *ACS Omega*, 8, 46663–46675, 2023.
53. Rautela, R., Sharma, A., Prakash Ranjan, V., Rathika, K., Pratap, V., Ram Yadav, B., Kumar, S., Turning Solid Waste into Catalysts: A Path for Environmental Solutions. *ChemPlusChem*, 89, e202400246, 2024.
54. Summa, D., Lanzoni, M., Castaldelli, G., Fano, E.A., Tamburini, E., Trends and Opportunities of Bivalve Shells' Waste Valorization in a Prospect of Circular Blue Bioeconomy. *Resources*, 11, 48, 2022.
55. Hussain, F., Alshahrani, S., Abbas, M.M., Khan, H.M., Jamil, A., Yaqoob, H., Soudagar, M.E., Imran, M., Ahmad, M., Munir, M., Waste Animal Bones as Catalysts for Biodiesel Production; A Mini Review. *Catalysts*, 11, 630, 2021.
56. Mignardi, S., Archilletti, L., Medeghini, L., De Vito, C., Valorization of Eggshell Biowaste for Sustainable Environmental Remediation. *Sci. Rep.*, 10, 1, 1–10, 2020.
57. Rohmadi, R., Harwijayanti, W., Ubaidillah, U., Triyono, J., Diharjo, K., Utomo, P., *In Vitro* Degradation and Cytotoxicity of Eggshell-Based Hydroxyapatite: A Systematic Review and Meta-Analysis. *Polymers (Basel)*, 13, 2021.
58. Islam, S., Alshoaibi, A., Alnaim, N., Al Naim, H.M., Silver nanoparticles functionalized calcite fibers derived from eggshell waste: a sustainable nano-composite for microbial applications. *J. Sol-Gel Sci. Technol.*, 1–11, 2025.
59. Yang, K., Li, C., Du, X., Zhao, J., Zhang, Y., Zhu, Y., Yu, Q., Huang, L., Li, S., Liu, X., Zhai, Y., Acid-mediated modification of waste hydroxyapatite to enhance Ca-O-Mn interaction for promotional simultaneous removal of toluene and formaldehyde. *Fuel*, 379, 133087, Jan. 2025.
60. Thakur, P., Anand Omar, R., Talreja, N., Chauhan, D., Ashfaq, M., Waste-derived Ca and Zn-based bimetallic (Ca/Zn) nanorods encapsulated chitosan-based haemostatic dressing bandage: A step towards waste to bandages. *J. Ind. Eng. Chem.*, 143, 327–338, 2025.

61. Xu, X., Yu, B., Hussain, M.S., Wang, Y., Li, Q., Xu, Y., Zhang, Y., Zhou, H., One-Pot Synthesis of Cost-Effective Dual Functional Material from Solid Waste for Integrated CO_2 Capture and Utilization. *Sep. Purif. Technol.*, 133309, 2025.
62. Zhao, L., Shen, G., Wen, R., Xu, T., Jiang, S., Wang, X., Wang, H., Optimization of operating conditions for the catalytic alcoholysis of waste PET for the synthesis of BHET by sunflower seed husk matrix materials. *RSC Adv.*, 15, 207–215, Jan. 2025.
63. Cao, S., Sun, J., Liu, Q., Gu, Y., Characterization of CO_2 mineral sequestration by wet carbonation of crustacean solid waste: a case study of crab shells. *Biomass Convers. Biorefin.*, 1–14, 2025.
64. Zhang, S., Xiong, J., Lu, J., Zhou, N., Li, H., Cui, X., Zhang, Q., Liu, Y., Ruan, R., Wang, Y., Synthesis of CaO from waste shells for microwave-assisted catalytic pyrolysis of waste cooking oil to produce aromatic-rich bio-oil. *Sci. Total Environ.*, 827, 154186, 2022.
65. Hakim, K.A.K.M., Tan, E.S., Habib, S.H., Hafriz, R.S.R.M., Salmiaton, A., Investigation of waste-derived and low-cost calcium oxide-based catalysts in co-pyrolysis of EFB-HDPE to produce high quality bio-oil. *J. Anal. Appl. Pyrolysis*, 177, 106375, 2024.
66. Cho, J.H., Kim, S., Park, K.T., Rhee, C.H., Park, H.W., Jung, J.C., Eco-friendly CaO-based catalysts derived from waste oyster shells for the oxidative coupling of methane. *J. Mater. Cycles Waste Manage.*, 25, 3355–3366, 2023.
67. Cheng, H., Wei, J., Liang, M., Dai, S., Liu, X., Ma, L., Wang, H., Lai, F., Calcium Glycerolate Catalyst Derived from Eggshell Waste for Cyclopentadecanolide Synthesis. *Front. Chem.*, 9, 770247, 2021.
68. Sreekantan, S., Sudhakaran, A.K., Marimuthu, B., Upcycling solid wastes: Using abundant egg shell wastes for the greener production of lactic acid under ambient conditions. *J. Environ. Chem. Eng.*, 13, 2, 115603, Apr. 2025.
69. Laghari, A.A., Jamro, I.A., Kumar, A., Chen, G., Sajnani, S., Guo, Z., Shen, Y., Zhang, J., Khoso, S., Guo, Q., Ma, W., Catalytic gasification of municipal solid waste using eggshell-derived CaO catalyst: An investigation of optimum H_2 production, production distribution, and tar compounds. *Next Sustain.*, 4, 100038, 2024.
70. Heidari, M., Borhan Mousavi, S., Rahmani, F., Akbari Sene, R., Eco-friendly, sustainable and porous eggshell/tea waste-derived CaO nanoparticles as a high-temperature CO_2 sorbent: Fabrication and textural/morphological evaluation. *J. Ind. Eng. Chem.*, 134, 169–180, 2024.
71. Isnaini, M.D., Jongsomjit, B., Yip, A.C.K., Phisalaphong, M., Waste-derived CaO from green mussel shells as a highly stabilized and superior sorbent for cyclic CO_2 capture. *Cleaner Eng. Technol.*, 26, 100956, 2025.
72. Kareem, B.A. and Fadhil, A.B., Optimized co-solvent methanolysis of mixed non-edible oils over nano solid base catalyst derived from egg shells. *J. Ecol. Eng.*, 26, 48–62, 2025.

73. Wang, S., Fu, H., Abulizi, A., Okitsu, K., Maeda, Y., Nulahong, A., Ren, T., Niu, S., Biodiesel production from sheep fat catalyzed by CaO-SrO(x)/BCs(y) acid-base bifunctional catalyst and process optimization. *Renewable Energy*, 238, 121881, 2025.
74. Wang, Y., He, C., Li, G., Liu, X., Liu, L., Jiao, Y., Sustainable hollow spherical CaO derived from waste eggshells for transesterification of lard: Catalytic characterization, performance and thermodynamics. *Renewable Energy*, 250, 123302, 2025.
75. Amesho, K.T.T., Lin, Y.C., Chen, C.E., Cheng, P.C., Shangdiar, S., Kinetics studies of sustainable biodiesel synthesis from Jatropha curcas oil by exploiting bio-waste derived CaO-based heterogeneous catalyst *via* microwave heating system as a green chemistry technique. *Fuel*, 323, 123876, 2022.
76. Olvera-Ureña, E., Rodriguez, J.A., Díaz de León, J.N., Tavizón-Pozos, J.A., Dispersion of Sr and K Species Supported on CaO Eggshell-Based Catalysts for Biodiesel Production. *Top. Catal.*, 1–14, 2025.
77. Lee, H., Coulon, F., Beriro, D.J., Wagland, S.T., Recovering metal(loids) and rare earth elements from closed landfill sites without excavation: Leachate recirculation opportunities and challenges. *Chemosphere*, 292, 133418, 2022.
78. Advantages and Disadvantages of E-Waste Recycling, RF Wireless World, Accessed: Jun. 19, 2025. [Online]. Available: https://www.rfwireless-world.com/articles/advantages-and-disadvantages-of-e-waste-recycling.
79. Robinson, B.H., E-waste: An assessment of global production and environmental impacts. *Sci. Total Environ.*, 408, 183–191, 2009.
80. Hira, M., Yadav, S., Morthekai, P., Linda, A., Kumar, S., Sharma, A., Mobile Phones—An asset or a liability: A study based on characterization and assessment of metals in waste mobile phone components using leaching tests. *J. Hazard. Mater.*, 342, 29–40, 2018.
81. Chauhan, G., Jadhao, P.R., Pant, K.K., Nigam, K.D.P., Novel technologies and conventional processes for recovery of metals from waste electrical and electronic equipment: Challenges & opportunities – A review. *J. Environ. Chem. Eng.*, 6, 1288–1304, 2018.
82. Rene, E.R., Sethurajan, M., Ponnusamy, V.K., Kumar, G., Dung, T.N.B., Brindhadevi, K., Pugazhendhi, A., Electronic waste generation, recycling and resource recovery: Technological perspectives and trends. *J. Hazard. Mater.*, 416, 125664, 2021.
83. Blengini, G.A., El Latunussa, C., Eynard, U., Torres De Matos, C., Wittmer, D., Georgitzikis, K., Pavel, C., Carrara, S., Mancini, L., Unguru, M., Blagoeva, D., Mathieux, F., Pennington, D., *Study on the EU's list of Critical Raw Materials (2020) Final Report*, 2020.
84. Ryabchuk, P., Anwar, M., Dastgir, S., Junge, K., Beller, M., From Mobile Phones to Catalysts: E-Waste-Derived Heterogeneous Copper Catalysts for Hydrogenation Reactions. *ACS Sustainable Chem. Eng.*, 9, 10062–10072, Aug. 2021.

85. Bayarkhuu, B., Cho, H., Cho, G., Hong, J., Hong, Y., Baek, S.W., Yang, S., Yavuz, C.T., Kim, H.S., Byun, J., Single-Atom Catalysts on Conjugated Porphyrin Polymer Photocatalysts *via* E-Waste for Sustainable Photocatalysis. *Adv. Funct. Mater.*, 35, 2411661, 2025.
86. Rechargeable Batteries Market Size to Hit USD 219.22 Bn by 2034. Accessed: Mar. 14, 2025. [Online]. Available: https://www.precedenceresearch.com/rechargeable-batteries-market.
87. Zhao, Y., Pohl, O., Bhatt, A.I., Collis, G.E., Mahon, P.J., Rüther, T., Hollenkamp, A.F., A Review on Battery Market Trends, Second-Life Reuse, and Recycling. *Sustainable Chem.*, 2, 167–205, 2021.
88. Chen, H., Wei, Y., Han, C., Zhong, C., Zhang, Z., Qu, X., Zhao, C., Zhao, X., Yang, L., Peroxymonosulfate activation by CoO@C catalyst from integrated recycling of spent lithium-ion battery: Performance and mechanism. *Sep. Purif. Technol.*, 355, 129658, 2025.
89. Maqbool, Q., Aharanwa, H.U., Stöger-Pollach, M., Rupprechter, G., Upcycling hazardous waste into high-performance Ni/η-Al_2O_3 catalysts for CO_2 methanation. *Green Chem.*, 27, 2706–2722, 2025.
90. Wink, K. and Hartmann, I., Recent Progress in Turning Waste into Catalysts for Green Syntheses. *Sustainable Chem.*, 5, 27–39, 2024.

4

Synthesis Techniques for Solid Waste-Derived Catalysts

Renu Hada[1*], Rajesh Kumar Meena[2], Vishwajeet Singh Yadav[3], Deepti Pal[1], Mamta Gour[1] and Anjali Soni[1]

[1]*Department of Chemistry, Medi-Caps University, Indore, M.P., India*
[2]*Department of Chemistry, Kalindi College, University of Delhi, New Delhi, Delhi, India*
[3]*Department of Chemistry, Rustamji Institute of Technology, Gwalior, M.P., India*

Abstract

The conversion of waste materials into active catalysts represents a way for waste valorization and the advancements of green chemistry. The chapter provides a detailed review of the key synthesis techniques used for the synthesis of waste-derived catalysts from waste sources such as agricultural residues, industrial, and other forms of solid waste. The chapter explores different synthesis techniques used in the literature and shows how pyrolysis can effectively convert waste into carbon rich catalytic materials. The impregnation method is assessed for its efficiency in distributing active catalytic phases onto waste-derived supports. Carbonization techniques are discussed for producing high surface area carbon materials from biomass, while the sol-gel method is also reviewed, emphasizing its value in developing specific pore architectures using templates. The hydrothermal method known for producing crystalline materials under controlled conditions is described in detail. Microwave-assisted synthesis offers rapid and energy efficient processing, and plasma-assisted synthesis enables unique chemical environments for catalyst generation. In conclusion, the chapter outlines the future perspective of the evolution of synthesis techniques of waste-derived catalysts that emphasize the importance of sustainable, efficient, and scalable catalyst preparation methods. The synthesis techniques discussed in this chapter showed significant potential in advancing waste management practices in the field of catalysis.

***Keywords*:** Solid waste, catalyst, sustainable method, pyrolysis, impregnation method, carbonization techniques, green chemistry

**Corresponding author*: Renu.hada@medicaps.ac.in

Deepti Goyal, Sakshi Kabra Malpani and Renuka Jain (eds.) Catalysis from Waste: Transforming Solid Residues into Value, (89–108) © 2026 Scrivener Publishing LLC

4.1 Introduction

In today's scenario, the concept of 'waste to wealth' through green synthesis is gaining significant attention all over the world, to address environmental challenges and promote sustainable developments. Green synthesis approach focuses on converting waste to valuable products such as catalysts, biofuels, and other value-added materials, as well as reducing environmental effects. The development of catalysts derived from solid waste such as agricultural residues, industrial by-products, and municipal solid waste, has attracted significant attention. These waste materials, often considered as liabilities but now being recognized as valuable resources to produce high performance catalysts, with the benefit of waste management as well as resource recovery [1, 2]. Solid waste includes a wide variety of materials such as straw, husks and stalks, food waste, plastics, metals, fly ash, and construction debris. They are produced in abundance and often remain unutilized. They represent a vast and largely available resource pool which can be transformed into catalysts for a variety of chemical and industrial processes. Utilization of waste is a contemporary solution towards waste management [3, 4].

By using waste to wealth approach where solid waste is converted into useful catalyst provides various environmental benefits such as reduction of solid waste volume which leads to declining environmental hazards i.e. soil contamination, water pollution, and greenhouse gas emission [5]. In addition, development of catalyst using solid waste reduces the dependence on mining and processing of resources [6–8]. The solid waste derived catalyst shows many sustainable applications in various fields including biodiesel production, hydrogen production, carbon capture, air and water treatment etc. [9–12].

A wide range of synthesis techniques can be used to convert different types of solid waste into valuable catalysts which show scalable properties for specific applications. Pyrolysis, gasification and hydrothermal carbonization are particularly effective in transforming carbon rich waste materials into valuable catalyst supports. In these methods, thermal decomposition of solid waste materials is carried out under controlled conditions to produce carbon rich solids which can be further converted to valuable catalysts through functionalization with active metals [3, 13, 14]. Pyrolysis of plastic waste can generate carbonaceous support materials to form metal catalysts for catalyzing various chemical reactions such as hydrogenation, oxidation and photocatalysis [15, 16]. Similarly, gasification of biomass and other organic waste can produce syngas, which can be utilized to produce Fischer-Tropsch catalyst [17–19].

In addition to thermochemical techniques, biotechnological methods are also gaining attraction. These approaches use the power of microorganisms and enzymes to transform organic waste into catalysts [16, 20]. Although solid waste derived catalyst shows potential applications in various fields, several challenges need to be addressed for complete utilization of these catalysts. They include optimization of conversion processes to achieve high yields, functionality and selectivity, developing tailored and efficient synthesis methods. More research is crucial for overcoming such challenges [21–23].

In conclusion, the utilization of solid waste, obtained from agricultural, industrial waste and other sources, shows a significant solution to the various environmental issues and represents advancement towards sustainability. This chapter deals with the various synthesis techniques used globally for the development of sustainable solid waste derived catalyst or catalyst support materials.

4.2 Strategies for the Synthesis of Solid Waste-Derived Catalysts

Various synthesis techniques are available for transforming solid waste into useful and scalable catalysts. They are summarized in Figure 4.1.

4.2.1 Pyrolysis

Pyrolysis is the decomposition of organic compounds at high temperature in the absence of oxygen. In this process, at elevated temperature the

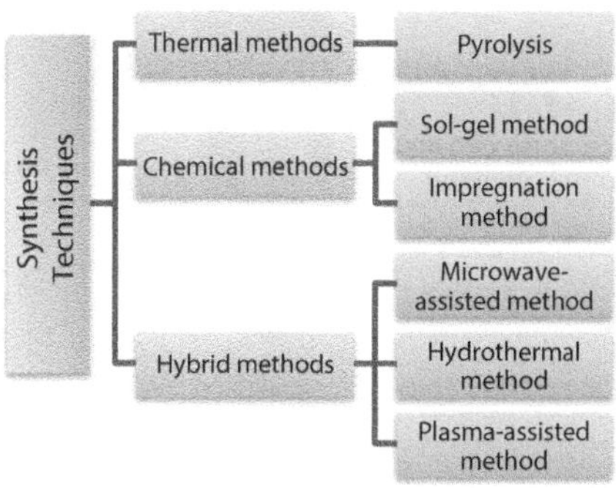

Figure 4.1 Synthesis techniques for the conversion of solid waste into catalyst or catalyst support materials.

biomass gets converted into a solid biochar - a carbon rich material which can be used as a catalytic support. Biochar catalysts can be prepared by reusing waste biomass or leftover organic materials, thus helping to control trash disposal along with the synthesis of important catalytic products. Pyrolysis can be divided into fast pyrolysis (~400 to 700°C), intermediate pyrolysis and slow pyrolysis (~300 to 500°C) depending on the reaction temperature and residence time. Figure 4.2 shows the schematic representation of pyrolysis technique. The catalysts derived from waste materials signify a feasible approach not only in reducing reliance on non-renewable resources but also in minimizing environmental and social significances. By using innovative synthesis techniques, waste materials can be converted into highly efficient catalytic materials that may help to promote sustainable resource management and circular economy. This research explores substitute synthesis methods for waste-derived catalyst and their possible impact on sustainable catalysis.

A study has been done to examine the conversion of solid waste material into a highly efficient Fenton-like catalyst using iron ore tailing and wheat straw through pyrolysis method. These Fenton-like catalysts are well known for their capacity to abolish organic pollutants in water using sophisticated oxidation mechanism. The researchers synthesized a sustainable catalyst using iron ore tailings and wheat straw which is abundant and less expensive. The catalyst shows excellent catalytic action in degrading organic contaminants and indicates its potential usage for environmental remediation application. This approach helps to open new ways for

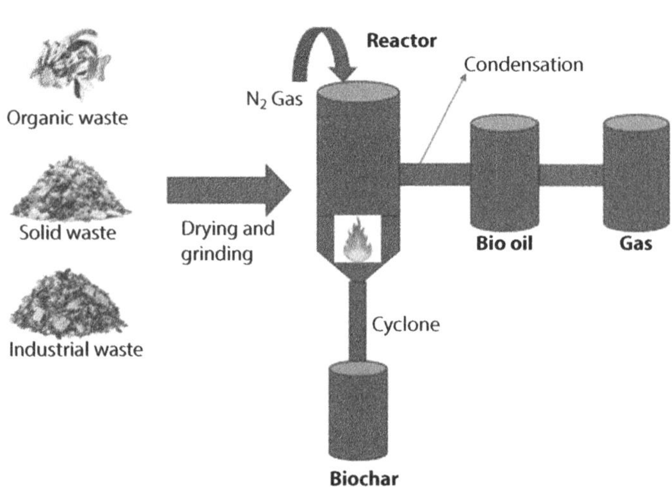

Figure 4.2 Schematic representation of pyrolysis process.

controlling pollution and reusing waste materials in catalytic applications for sustainable waste management [24].

Yang *et al.*, have converted cellulose to high-value liquid fuel through fast catalytic pyrolysis using zeolite as catalyst. Zeolite catalyst (ZSM-5 zeolite) was prepared from two solid wastes: Rice husk ash and spent fluid catalytic cracking catalyst (sFCC). The catalytic pyrolysis of cellulose to aromatics using ZSM-5 zeolite as a catalyst shows a significant enhancement in the production BTX- benzene, toluene and xylene [25].

Another work is based on the superficial synthesis of metal free Ti-Cu nano catalyst and *in-situ* catalytic pyrolysis of polystyrene-based waste reinforced in phenol to produce hydrogen and valuable liquid fuels. Three types of nano-catalysts, (Ti, $4Ti_3Cu$ and $3Ti_3Cu$) were prepared and used for the *in-situ* catalytic pyrolysis of polystyrene waste to hydrogen. It was observed that at 800°C, 97.9% hydrogen yield was achieved using $3Ti_3Cu$ nano-catalyst with complete conversion of phenol. The products achieved during catalytic pyrolysis were tert-butyl hydroperoxide, ethylamine and benzene (1,1-dimethylethoxy) [26].

Williams *et al.*, have studied the potential application of waste tires as a catalyst for the catalytic pyrolysis of high-density polyethylene to generate H_2 rich syngas. From the oxidation of processed municipal solid waste and waste tires, an ash was derived and used as a catalyst for hydrogen and syngas production through catalytic pyrolysis steam reforming of high-density polyethene. The characterization of ashes involves surface morphology, pore structure, elemental distribution and metal composition. Furthermore, the study involves the influence of process parameters, catalyst: plastic ratio and catalytic temperature on the production of syngas and hydrogen. It has been found that the ash derived from tires has pore volume and a high surface area than refused-derived fuel ash, resulting in high H_2 yield. By increasing the temperature of catalytic steam reforming process for tire ash catalysts, a significant increase in H_2 yield was observed to 83.2 $mmolg^{-1}$ plastic from 13.3 $mmolg^{-1}$ [27].

4.2.2 Sol-Gel Method

The sol-gel process is a widely used method for synthesis of solid materials from molecular precursors. In this method, a series of chemical reactions are used to convert a liquid 'sol' into a solid 'gel'. This is a common method used to make thin film coatings, ceramics, glasses, and composites. The main advantage of this method is its ability to control the composition and structure as well as the properties of the resulting materials. This can also help in creating materials with specific microstructure, high purity and consistent properties. Ultimately, this technique is an indispensable and effective tool for providing synthesis and structural control of a wide range of functional materials in engineering and materials science. The overall scheme for this procedure has been illustrated in Figure 4.3.

In recent years, the synthesis of catalysts *via* sol-gel method using waste as precursor emerged as a much-appreciated approach. In this process, metal alkoxides are hydrolyzed followed by infiltration to generate porous network structures from which inorganic compounds are obtained. Thus, produced catalysts show practical and sustainable options for waste exploitation in the field of catalysis. Because of its approach towards production of catalysts from waste, it has a significant advantage including which provides sol-gel route a favorable path in preparation of highly homogenous, well-defined catalyst with desired properties. This is fundamental to increasing the effectiveness and selectivity of catalytic processes and minimizing the harmful effects that waste can have on the environment. Moreover, sol-gel method facilitates the incorporation of various metal ions and ligands into the catalyst framework, allowing precise control over its composition and architecture. Many research works have also shown the effectiveness of the catalysts which are produced from sol-gel process with a wide range of applications, which includes the generation of silica-based products using bio-residues. In this process, such types of precursors are used which are generally derived from their bio-residue extracts also combined with conventional silica sources. The start of this process involves the production of the sols using the mixed precursor solutions, which is then continued by gelation then drying, and calcination. The materials which are produced in this approach exhibit the properties which are dependent on both the silica matrix and the bio-residue components, these characteristics make them compatible for the use in environmental remediation, sensing, and catalysis [28].

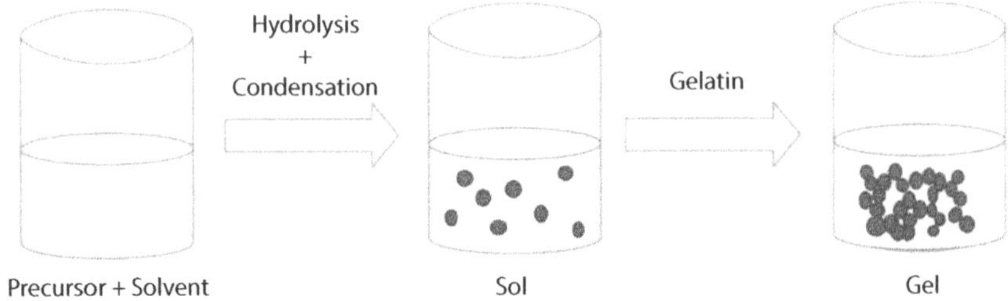

Figure 4.3 Schematic representation of sol-gel method.

Waste limescale and soybean oil residues are reused to produce recyclable catalysts for biodiesel synthesis. Through the process of gelation, the catalyst is prepared, dried and calcination is carried out. This strategy not only converts waste materials into a valuable resource but also provides a sustainable way to generate biodiesel, promoting resource efficiency and environmental conservation [29]. Ru/Al_2O_3 catalysts can be prepared by 'solvent-free preparation for CO_2 methanation', which is a new and low-cost method. The main catalyst support is ruthenium (Ru) deposited on a thin layer of aluminum oxide (Al_2O_3) which does not require a solvent in the actual deposition process, nor in the subsequent curing and activation steps. This cost-effective method greatly reduces the environmental footprint by eliminating the need for solvents and enhances catalyst production by increasing its efficiency for CO_2 methanation [30].

Mesoporous solid acid catalyst prepared from the sol-gel method is very effective for production of biodiesel using cooking oil. In this study residual cooking oil is used as a feedstock for the esterification process that takes place at room temperature. Besides, by carrying out esterification with alcohol even at room temperature, it reduces energy consumption and process cost [31]. Here, the sol-gel method is used to design the mesoporous structure of catalysts which enhances the contact between reactants and conversion efficiency. It contributes significantly to reducing environmental pollution. Using waste industrial sand as the precursor and synthesizing FAU and LTA zeolites, sustainable synthesis of materials utilizing the sol-gel process were reported [32].

This technology not only uses a precise process beyond just recycling waste, but the resulting zeolites also possess special properties. The resulting zeolites exhibit unique moisture sorption properties, making them suitable for applications such as air conditioning and industrial moisture management. These zeolites are also used as catalysts for the conversion of ethanol, allowing potentially sustainable fuel creation. Overall, this effective method focuses on waste minimization and resource efficiency goals which support the fact that how effective sol-gel synthesis is in tackling environmental challenges [33]. The synthesis, structural characterization and catalytic activity of cryptomelane type manganese oxide materials synthesized by sol-gel route have been highlighted in comparison with other synthesis methods. This comparative analysis helped in gaining valuable information about the advantages and co-benefits of the sol-gel method. This helped in improving catalytic applications and optimizing the synthesis strategy for the synthesis of tailor-made cryotolerance type manganese oxide catalysts [34].

The book 'Advances in Nanostructured Silicon Carbide Photocatalysts' highlights the importance of sol-gel processing for nanostructured SiC photocatalysts. This strategy considers finer-tuned control over the material's chemical composition and on the geometry to host energy-matched charge carriers and thereby enhancing photocatalytic efficiency. To maximize light absorption by producing sol-gel, researchers can change the characteristics of SiC materials through two essential processes in photocatalysis: photon absorption and charge separation. Because of the remarkable properties for a variety of environmental applications, nanostructured SiC photocatalysts synthesized by sol-gel methods have been studied in a lot of fields, e.g. air remediation, photocatalytic water purification. The versatility and significance of sol-gel method in photocatalytic research highlight its wide range of applications in solar energy conversion, particularly in H_2 production and CO_2 reduction [35].

4.2.3 Impregnation Method

Impregnation is a green synthesis approach for synthesizing supported or mixed catalysts. In this strategy, active precursor solutions get adsorbed on solid support material. Many researchers have been using impregnation method to produce solid waste derived catalysts. As an example, Rocha *et al.*, prepared Fe catalyst from olive mill waste for the heterogeneous Fenton and persulfate process using wet and hydrothermal impregnation method. In this study, researchers studied the ability to adsorb phenolic acid and other impurities from laboratory prepared wastewater to mimic olive mill waste effluent. Study reveals that catalyst prepared from wet impregnation method shows superior catalytic activities than the catalyst prepared from hydrothermal impregnation method [36]. Recyclable and reusable Ru based catalyst was generated using waste for hydrogenation reduction of ethyl levulinate into gamma valerolactone. The catalyst was prepared using impregnation method which showed similar catalytic activity as compared with catalyst generated from high purity ruthenium containing compounds [37].

Magnetite impregnated calcium oxide extracted from eggshell was prepared *via* impregnation method for transesterification of palm oil off-grade to produce biodiesel. Catalyst thus prepared was characterized using XRD and BET surface area. The catalyst was tested for transesterification where 10:1 molar ratio of methanol/palm oil, 6% catalyst weight, 70°C reaction temperature and 2h reaction time were found optimal conditions giving 90% yield of biodiesel [38]. Along with this, MgO-CaO/SiO_2 catalyst derived from geothermal solid waste and dolomite was successfully synthesized through impregnation method for biodiesel production. The physicochemical properties of prepared catalysts were characterized through FTIR, SEM, and BET techniques. At a ratio of 15:85 for MgO-CaO: SiO_2 catalyst, 93% biodiesel conversion was seen. The biodiesel produced through this method showed good quality in accordance with SNI 7182-2015 [39].

4.2.4 Microwave-Assisted Synthesis

Microwave-assisted synthesis works on a principle of aligning dipoles of the material in the presence of microwaves. This process may result in the production of internal heat which is responsible for the reduction of processing time and energy required. Thus, the reaction time can be reduced by adopting microwave-assisted synthesis. The microwave synthesizer has potential to react with various organic and inorganic materials depending on their permanent dipole moment. The method is green, efficient and economical for the generation of solid waste derived catalysts. An example of microwave-assisted assembly is schematically represented in Figure 4.4.

Hada *et al.*, have successfully synthesized a highly effective nickel oxide ($NiFA_k$) nanoparticles loaded fly ash catalyst using nickel nitrate as oxidizer and urea as a fuel *via* microwave-assisted solution combustion synthesis. To enhance the dispersion and bonding of nickel oxide nanoparticles over fly ash, the latter was chemically and thermally activated to be used as a supporting material. Surface characterization of prepared catalyst revealed

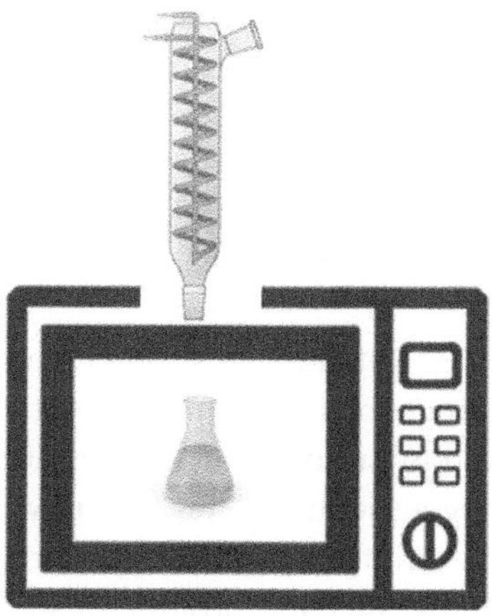

Figure 4.4 Microwave-assisted synthesis assembly.

that chemical activation of fly ash increased loading of nickel oxide due to the formation of calcium silicate hydrate phase that increase surface hydroxyl groups. The XRD study confirmed that the synthesized $NiFA_k$ nano-catalyst is nanocrystalline with 6.4 nm crystallite size. Through decomposition reaction, the catalytic activity of $NiFA_k$ on hydrogen peroxide was determined under optimized reaction conditions [40].

Recently, researchers have focused on the production of sustainable biodiesel through microwave-assisted synthesis using waste frying oil as raw material and Ca/Si/Al oxides as catalyst *via* ultrasonic transesterification reaction. A novel catalyst was synthesized by impregnating $Ca(OH)_2$-$CaCO_3$ with kaolin. The catalyst $Ca(OH)_2$-$CaCO_3$/kaolin (CSAO) has the benefits of having both acidic and basic sites of kaolin and $Ca(OH)_2$-$CaCO_3$ respectively. By utilizing transesterification of waste frying oil, they have compared the performance of biodiesel production and observed that the highest yield of biodiesel (98.7%) was obtained by 1.5 wt.% $Ca(OH)_2$-$CaCO_3$/kaolin catalyst/ wt. waste frying oil, with 15% amplitude of ultrasonic irradiation for 4 min and 1:6 oil to methanol (O/M) molar ratio [41].

Similarly, biodiesel can be synthesized from oleic acid (OA) using bio-waste catalyst through microwave irradiations. In this process, sulfonic acid functionalized banana peel waste-derived heterogenous catalyst was utilized. The yield of biodiesel was observed to be 97.9 ± 0.7% under improved reaction conditions (80°C temperature, molar ratio of methanol to OA to be 20:1, catalyst loading of 8 wt.% for 55 minutes [42]. In the same way, the usage of sandbox seed shell (SSS) waste as feedstock for the synthesis of biocatalyst for the microwave-assisted trans-esterification of sandbox oil (SBO) into biodiesel (SBOB) has been reported. Through calcination of SSS at 500°C for 3h, green solid alkali biocatalyst, calcined sandbox shell ash (CSSA) was obtained which was further characterized and found to contain 44.99% of potassium, 1.54% of calcium and 1.87% of magnesium with a surface area of $6.11 m^3/g$ and pore size of 18.142 nm. The optimized conditions for maximum yield of SBOB (98.27 wt.%) were- 8:1 methanol to SBOB ratio, 1.5 minutes reaction time, 450 W heating power and 2 wt.% CSSA loading [43].

Yadav *et al.*, have investigated the possibility of using O. Sativa husk for the fabrication of heterogenous carbon-based catalyst. By facile carbonization method followed by sulfonation, a series of heterogenous catalysts were fabricated utilizing agricultural waste as raw material. The catalysts were used in an esterification method for synthesizing fatty acid methyl ester (FAME) by changing the concentration of sulfuric acid from 2 to 10 mML^{-1}. The yield of FAME reaches 99.6 0.2% at 80°C with H_2SO_4 concentration for 60 minutes reaction time with 24:1 methanol to OA ratio and

8 wt.% catalyst dosage. It was also observed that as the chain length of carbon increases, the yield of FAME decreases [44].

4.2.5 Hydrothermal Method

The hydrothermal approach entails synthesizing materials in order to produce catalysts in an aqueous environment at high pressures and temperatures. Initially in this method, the precursor materials are dissolved in a solvent, usually water, which is then hydrothermally treated inside a reactor chamber or autoclave. Conditions such as temperature, pressure, and duration are important, which should be carefully regulated to support the necessary chemical reactions and crystallization processes, which can ultimately result in the formation of the catalyst material. During this method, the precursor materials are first dissolved, nucleated, and proliferated, leading to the production of catalyst particles with distinct surface properties, size, and shape. This method provides a good way to create catalysts with uniform particle size distribution, high crystallinity, and improved catalytic activity. Furthermore, the precise tailoring of catalyst characteristics for explicit applications is made possible by the insertion of dopants, modifiers, or template agents under controlled hydrothermal conditions. This method is a flexible and dependable way to create different catalyst materials with tunable properties, which is why it is frequently used in both industrial and scholarly settings. A flow chart for general hydrothermal synthesis is given in Figure 4.5.

Heterogeneous acid catalysts from guarana industry leftovers were synthesized using both classic carbonization/sulfonation and one-pot hydrothermal carbonization/sulfonation techniques. These catalysts had adequate physicochemical parameters, which were confirmed using characterization techniques such as X-ray diffraction, thermogravimetry, FTIR spectroscopy, scanning electron microscopy, N_2 sorption, and Boehm titration. The catalytic activity of sulfonated-based carbon catalysts synthesized *via* traditional and hydrothermal procedures was analyzed in detail. The catalyst was tested for the esterification of OA with methanol, as prepared catalysts showed results with the yields of 93 and 96% for the traditional and hydrothermal procedures respectively. Reuse studies showed that the catalyst prepared *via* hydrothermal method was stable for four cycles, but the standard catalyst's yields declined. Furthermore, the hydrothermal catalyst demonstrated higher activity in esterifying free fatty acids in cooking oil residues, with a maximum conversion of 63% under optimal circumstances. Utilization of industrial waste for catalyst synthesis is an environmentally friendly approach that minimizes waste pollution as well as

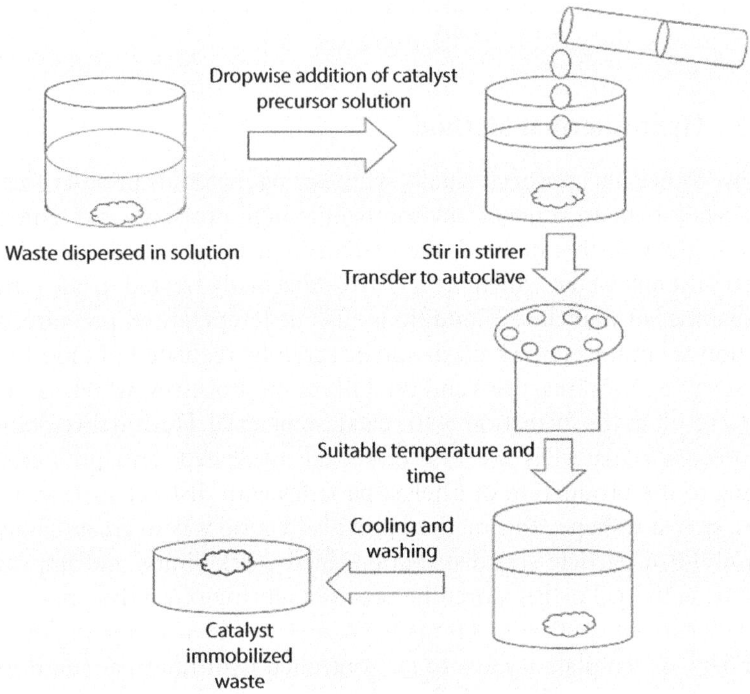

Figure 4.5 Flow chart for hydrothermal synthesis of waste-derived catalysts.

providing a cost-efficient and effective catalytic performance [45]. A study describes a new process to convert production sand waste generated from oil and gas exploration into zeolite like Linde Type A (LTA) and Faujasite (FAU), tackling the environmental problems associated with sand waste disposal.

An efficient process to extract nano-silica from waste sand and produce high-pure silica (> 98 wt.%) is investigated by means of a sol-gel hydrothermal method. A seed-assisted synthesis approach is employed to facilitate the successful hydrothermal transformation of nano-silica to FAU and LTA zeolites. Hydrothermal conditions on the formation of zeolites were elaborately studied. When hydrocarbons and air are present, the synthesized zeolites function exceptionally well as adsorbents for selective moisture adsorption, FAU in particular shows a higher water adsorption capacity. Further demonstrating the adaptability and usefulness of FAU catalyst, dehydration of ethanol has been tested, the results showed 37% conversion of ethanol and 13.8% ethylene selectivity. The catalyst has been proved to be very useful in oil and gas industries [33].

The synthesis and characterization of Ni/mesoporous silica (Ni/MS) catalysts from Lapindo mud for hydrocracking waste cooking oil into biofuels were carried out by Paramesti *et al.* MS was synthesized hydrothermally utilizing CTAB as a template. Nickel metal was loaded into the MS at 4, 6, and 8 wt.% using Ni $(NO_3)_2.6H_2O$ salt precursors *via* wet impregnation, yielding Ni(4)/MS, Ni(6)/MS, and Ni(8)/MS catalysts. XRD, FT-IR, SAA, and AAS analysis revealed specific surface areas of 63.08, 91.45, and 120.45 m^2/g for Ni (4)/MS, Ni (6)/MS, and Ni (8)/MS, respectively. The hydrocracking of waste cooking oil with these catalysts produced liquid products of 80.57, 74.63, and 75.77 wt.%, with total biofuel generation of 55.46, 50.93, and 54.05 wt.%, respectively. Notably, Ni (4)/MS showed higher catalytic performance, making it the most efficient catalyst for turning waste cooking oil into biofuel [46].

Forestry sawdust waste is used to generate biochar (SDBC-P) *via* an H_3PO_4 assisted hydrothermal carbonization process. A novel zirconium-coordinated biochar catalyst, Zr-SDBC-P is also prepared using a self-assembly method. Hydrothermal carbonization in the presence of H_3PO_4 generates oxygen-containing functional groups. Thus, synthesized Zr-SDBC-P-4 catalyzed the conversion of ethyl levulinate (EL) to biomass-derived γ-valerolactone (GVL) with 98.2% EL conversion and 85.4% GVL selectivity within 5h at 170°C, demonstrating the activity of Zr-SDBC-P-4. Zr-SDBC-P-4 shows good chemical stability over six cycles, which proves the potential application of Zr–SDBC-P-4 for ubiquitous conversion. This study emphasizes the significance of biomass waste in the development of catalysts for sustainable and environmentally friendly chemical processes. Single-phase small-pore Nap zeolites are synthesized by alkaline hydrothermal treatment (HT) of natural clinoptilolite (CLIP), which acts as a source of Si and Al, in an alternative approach compared to the conventional chemical reagents. NaP forms two isomorphs: NaP1 and NaP2, by systematic adjustments of molar concentration, crystallization time, and temperature, in contrast to NaP exhibiting high activity. This research illustrates the potential of low cost and widely available feedstocks for achieving sustainable synthesis processes of synthetic zeolites with high value as it mitigates the cost of production and environmental impact associated with standard synthetic procedures [47]. Stafin *et al.*, studied the synthesis of zeolites using diatomite, a byproduct of industrial filtration, as the silicon source *via* hydrothermal treatment, with or without pre-treatment by microwave irradiation for 3-24h. According to the characterization results, applying microwave radiation for a few minutes reduces the synthesis time to 15h when compared to the approach without pre-treatment, cutting manufacturing costs. The effectiveness

Table 4.1 Waste-derived catalysts developed using various methods.

Method	Waste	Catalyst	Applications	References
Pyrolysis	Iron ore tailings and wheat straw	Fenton-like catalysts	Removal of organic contaminants	[24]
	Polystyrene waste	Ti-Cu Nano-catalyst	Production of hydrogen and liquid fuel	[26]
	Waste tires and municipal solid waste	Ash as catalysts	Production of hydrogen-rich syngas	[27]
	Agricultural waste materials	Oxides or carbonates active catalyst components	Biodiesel production	[53]
Impregnation approach	Olive mill waste	Bio-based catalysts	Wastewater treatment	[36]
	Polyethylene terephthalate	Ni-Co/CeO_2 nano-catalyst	Generation of clean energy and chemicals	[54]
	Egg shell	Calcium-based metal oxide	Biodiesel production	[55]
	Silica oxide and dolomite	MgO-CaO/SiO_2 catalyst	Biodiesel production	[39]
Sol-gel method	Industrial sand waste	FAU and LTA zeolites	Fuel production	[33]
	Waste limescale and soybean dregs oil	Silica-based materials	Catalysis, sensing, and environmental remediation	[29]
Hydrothermal method	Lapindo mud	Ni/mesoporous silica (Ni/MS) catalysts	Biofuel generation	[46]
Microwave synthesis	Fly ash	Nickel oxide nanoparticles-loaded fly ash)	Advanced oxidation	[40]
Plasma Synthesis	Corn cob waste	Carbon nanoparticles	Biomass transformation	[52]

of microwave radiation in solubilizing the residue is proven, showing its potential for converting diatomite waste into a flexible material useful in a variety of industrial applications like catalyst production, environmental pollutant adsorption, and water treatment *via* ion-exchange resins [48]. In another study, a solid acid catalyst was synthesized from areca nut husk using a low-temperature hydrothermal carbonization method, resulting in enhanced sulfonic active sites (3.12%) and a high acid density (1.88 mmol/g). This catalyst facilitated an efficient conversion of OA to methyl oleate, achieving a 96.4% yield in 60 minutes under optimized conditions (1:25 OA: methanol ratio, 80°C, 9 wt.% catalyst dosage) with a low activation energy of 45.377 kJ mol. The biodiesel synthesis was confirmed by GC-MS and NMR, and the catalyst showed high reusability with an 85% yield after four cycles [49].

4.2.6 Plasma-Assisted Synthesis

A highly energized state of matter containing charged particles is known as plasma which can be used to synthesize catalytic materials. Plasma can be generated using various techniques such as direct current discharge (DC) and radio frequency (RF) [50]. Recently, Ekanayake *et al.*, synthesized heterogeneous 2D nanosheets of MgO catalyst using salt recovered from inexhaustible seawater *via* plasma-assisted synthesis approach. Thus, the prepared catalyst showed >99% of benzaldehyde and acetophenone conversion which was higher than the conversion obtained with regular MgO catalyst [51]. In another study, researchers have been using low temperature plasma technology for the production and modification of electrocatalytic materials for water electrolysis which proved as one of the most promising technology for hydrogen production [52]. Carbon nanoparticles were also synthesized with *in situ* surface acid activation in a one pot plasma process with a mixture of sulfuric acid and carbon precursor. As prepared carbon acid catalysts were used to convert corn cob waste to reducing sugar with the yield and selectivity 37% and 79% at 180°C for 18h respectively [52].

A summary of synthesis methods for waste derived catalyst is given in Table 4.1.

4.3 Conclusion

To tackle the issues related to different waste materials all over the world, utilization of waste has become an essential part for industries, scientific communities and society. This chapter has focused lightly on management of solid waste through production of useful catalytic materials using these waste materials. All these syntheses techniques described in this chapter can be used to produce valuable catalysts from a wide variety of solid waste materials generated from domestically or industrially including agricultural waste, food waste, which leads both environmental and economic benefits to society. By utilizing these catalytic techniques, it can be possible to develop sustainable and environment friendly materials as well as address the issue of waste management. Thus, this chapter will be useful for those who want to produce catalysts from solid waste materials.

Acknowledgement

We would like to express our sincere gratitude to Prof (Dr.) Ashu Rani, Prof (Dr.) Arpan Bhardwaj and Prof (Dr.) Vijay Devra for their guidance, valuable suggestions throughout the writing of this chapter. We are also thankful to Medicaps University, Indore, Kalindi College, Delhi, Rustamji Institute of Technology, Gwalior for providing us with the necessary facilities. Our special thanks go to the editors and the publishing team for providing us the opportunity to contribute this book.

AI disclosure statement: The author acknowledges the use of AI-based tools for minor language editing or grammatical correction. However, all intellectual contributions, analyses, and conclusions presented in this chapter are entirely the author's own.

References

1. Hossain, S.S., Bae, C.J., Roy, P.K., Recent progress of wastes derived nano-silica: Synthesis, properties, and applications. *J. Cleaner Prod.*, 377, 1–11, 2022.
2. Chen, Z., Zhao, Y., Liu, Q., Zhang, J., Waste-Derived Catalysts for Water Electrolysis: Circular Economy-Driven Sustainable Green Hydrogen Energy. *Nano Micro Lett.*, 15, 1, 1–18, 2023.
3. Fahmy, T.Y.A., Fahmy, Y., Mobarak, F., El-Sakhawy, M., Abou-Zeid, R.E., Biomass pyrolysis: past, present, and future. *Environ. Dev. Sustainability*, 22, 17–45, 2020.
4. Parrillo, F., Ardolino, F., Boccia, C., Arconati, V., Ruoppolo, G., Arena, U., Waste-derived catalysts for tar cracking in hot syngas cleaning. *Waste Manage.*, 179, 1–15, 2024.

5. Boro, J., Deka, D., Thakur, A.J., A review on solid oxide derived from waste shells as catalyst for biodiesel production. *Renewable Sustainable Energy Rev.*, 16, 904–910, 2012.
6. Wang, S., Yan, W., Zhao, F., Recovery of solid waste as functional heterogeneous catalysts for organic pollutant removal and biodiesel production. *Chem. Eng. J.*, 397, 126104–126116, 2020.
7. Adeleye, A.T., Ayeni, O.A., Ibrahim, T.H., Ibhadode, A.O., Efficient synthesis of bio-based activated carbon (AC) for catalytic systems: A green and sustainable approach. *J. Ind. Eng. Chem.*, 94, 180–195, 2021.
8. Si, X., Liu, Y., Zhao, L., Li, J., Catalytic production of low-carbon footprint sustainable natural gas. *Nat. Commun.*, 13, 1, 1–9, 2022.
9. Chen, Z., Wei, W., Chen, H., Ni, B.J., Recent advances in waste-derived functional materials for wastewater remediation. *Environ. Eng. Health Lett.*, 9, 2, 1–12, 2022.
10. Mastry, M.C., Ahmad, A.S., Gupta, R.K., Thomas, S., Processing renewable and waste-based feedstocks with fluid catalytic cracking: Impact on catalytic performance and considerations for improved catalyst design. *Front. Chem.*, 11, 1–14, 2023.
11. Qiu, B., Yang, C., Shao, Q., Liu, Y., Chu, H., Recent advances on industrial solid waste catalysts for improving the quality of bio-oil from biomass catalytic cracking: A review. *Renewable Sustainable Energy Rev.*, 154, 111822–111844, 2022.
12. Sharma, V., Tsai, M.L., Nargotra, P., Chen, C.W., Kuo, C.H., Sun, P.P., Di Dong, C., Agro-Industrial Food Waste as a Low-Cost Substrate for Sustainable Production of Industrial Enzymes: A Critical Review. *Bioresour. Technol.*, 362, 127774–127790, 2022.
13. Li, W., Wei, M., Liu, Y., Ye, Y., Li, S., Yuan, W., Wang, M., Wang, D., Catalysts evaluation for production of hydrogen gas and carbon nanotubes from the pyrolysis-catalysis of waste tyres. *Int. J. Hydrogen Energy*, 44, 19563–19574, 2019.
14. MacDermid-Watts, K., Pradhan, R., Dutta, A., Catalytic Hydrothermal Carbonization Treatment of Biomass for Enhanced Activated Carbon: A Review. *Waste Biomass Valorization*, 12, 2171–2189, 2021.
15. Gupta, J., Kumari, M., Mishra, A., Swati, Akram, M., Thakur, I.S., Agroforestry waste management – A review. *Chemosphere*, 287, 132107–132125, 2022.
16. Buzatu, A.R., Todea, A., Peter, F., Boeriu, C.G., The Role of Reactive Natural Deep Eutectic Solvents in Sustainable Biocatalysis. *J. Mol. Liq.*, 403, 123735–123752, 2024.
17. Ali, S., Zabidi, N.A.M., Subbarao, D., Correlation between Fischer-Tropsch catalytic activity and composition of catalysts. *Chem. Cent. J.*, 5, 68–76, 2011.
18. ten Have, I.C. and Weckhuysen, B.M., The active phase in cobalt-based Fischer-Tropsch synthesis. *J. Catal.*, 402, 1–14, 2021.

19. Valero-Romero, M.J., García-Mateos, F.J., Kapteijn, F., Rodríguez-Mirasol, J., Cordero, T., Fischer-Tropsch synthesis over lignin-derived cobalt-containing porous carbon fiber catalysts. *Appl. Catal., B*, 321, article no. 122078, 2023.
20. Mathipa-Mdakane, M.G. and Steenkamp, L., Aldolase: A Desirable Biocatalytic Candidate for Biotechnological Applications. *Biotechnol. Rep.*, 39, 00823–00834, 2024.
21. Gil, A., Challenges on waste-to-energy for the valorization of industrial wastes: Electricity, heat and cold, bioliquids and biofuels. *Environ. Nanotechnol. Monit. Manage.*, 18, article no. 100615, 2022.
22. Berenguer, C.V., Andrade, C., Pereira, J.A.M., Perestrelo, R., Câmara, J.S., Current Challenges in the Sustainable Valorisation of Agri-Food Wastes: A Review. *Processes*, 11, article no. 20, 2023.
23. Roy, P., Mohanty, A.K., Dick, P., Misra, M., A Review on the Challenges and Choices for Food Waste Valorization: Environmental and Economic Impacts. *ACS Environ. Au*, 3, 3–28, 2023.
24. Gao, L., Wang, L., Li, S., Cao, Y., Highly active Fenton-like catalyst derived from solid waste-iron ore tailings using wheat straw pyrolysis. *Environ. Sci. Pollut. Res.*, 29, 32155–32166, 2022.
25. Yang, K., *et al.*, Catalytic pyrolysis of cellulose using zeolite synthesized from solid wastes based on solvent-free route. *J. Anal. Appl. Pyrolysis*, 179, article no. 106498, 2024.
26. Nabgan, W., *et al.*, Facile synthesis of precious metal-free Ti-Cu nanocatalyst for enhanced hydrogen and liquid fuels production from *in-situ* pyrolysis-catalytic steam reforming reaction of polystyrene waste dissolved in phenol. *Appl. Catal., B*, 325, article no. 122279, 2023.
27. Li, Y. and Williams, P.T., Waste derived ash as catalysts for the pyrolysis-catalytic steam reforming of waste plastics for hydrogen-rich syngas production. *J. Anal. Appl. Pyrolysis*, 177, article no. 106374, 2024.
28. Zanotti, K., Igal, K., Colombo Migliorero, M.B., Gomes Zuin, V., Vázquez, P.G., Synthesis of Silica-Based Materials Using Bio-Residues through the Sol-Gel Technique. *Sustainable Chem.*, 2, 488–501, 2021.
29. Chen, C., *et al.*, Waste limescale derived recyclable catalyst and soybean dregs oil for biodiesel production: Analysis and optimization. *Process Saf. Environ. Prot.*, 149, 474–485, 2021.
30. Haddad, R., *et al.*, Solvent-free Preparation of Ru/Al_2O_3 Catalysts for CO_2 Methanation: An Example of Frugal Innovation. *Chem. Mater.*, 35, 8442–8453, 2023.
31. Cerón Ferrusca, M., Romero, R., Martínez, S.L., Ramírez-Serrano, A., Natividad, R., Biodiesel Production from Waste Cooking Oil: A Perspective on Catalytic Processes. *Processes*, 11, 7, article no. 1952, 2023.
32. Krishnaveni, M. and Kannan, C., Mesoporous solid acid catalysts esterifying waste cooking oil to biodiesel at room temperature. *Pet. Sci. Technol.*, 41, 6, 1–15, 2023.

33. Prasertsab, A., *et al.*, Transformation of Production Sand Waste to FAU and LTA Zeolites for Selective Moisture Adsorption and Ethanol Conversion. *Top. Catal.*, 66, 19–20, 1631–1648, 2023.
34. Almquist, C., Krekeler, M., Jiang, L., An investigation on the structure and catalytic activity of cryptomelane-type manganese oxide materials prepared by different synthesis routes. *Chem. Eng. J.*, 252, 249–262, 2014.
35. Kelin, H., *et al.*, Advances in Nanostructured Silicon Carbide Photocatalysts. *Acta Phys.*, 38, 11, article no. 2201021, 2022.
36. Rocha, K.O., *et al.*, Olive mill waste bio-based catalyst application in advanced oxidation processes for wastewater treatment. *Catal. Today*, 432, article no. 114618, 2024.
37. Yao, X., *et al.*, Novel Approach of Reusing the Precious Ru in the Wastewater Derived from the Depolymerization of Lignite to Prepare Ru-AC Catalysts Doped with Nb_2O_5. *Energy Fuels*, 36, 12, 6305–6316, 2022.
38. Helwani, Z., *et al.*, Impregnation of CaO from eggshell waste with magnetite as a solid catalyst (Fe_3O_4/CaO) for transesterification of palm oil off-grade. *Catalysts*, 10, 2article no. 164, 2020.
39. Widayat, W., *et al.*, Preparation of MgO-CaO/SiO_2 catalyst from dolomite and geothermal solid waste for biodiesel production. *Int. J. Renewable Energy Dev.*, 12, 3, 541–549, 2023.
40. Hada, R., Goyal, D., Singh Yadav, V., Siddiqui, N., Rani, A., Synthesis of NiO nanoparticles loaded fly ash catalyst *via* microwave assisted solution combustion method and application in hydrogen peroxide decomposition. *Mater. Today: Proc.*, 28, 119–123, 2020.
41. Ali, R.M., Salama, E., Hamad, H.A., A novel technology for microwave-assisted synthesis of new Ca/Si/Al composite oxide-based catalyst for boosting the ultrasound-assisted biodiesel production. *Process Saf. Environ. Prot.*, 194, 674–687, 2025.
42. Devasan, R., *et al.*, Microwave-assisted biodiesel production using bio-waste catalyst and process optimization using response surface methodology and kinetic study. *Sci. Rep.*, 13, article no. 2570, 2023.
43. Betiku, E., Oraegbunam, J.C., Falowo, O.A., Ojumu, T.V., Latinwo, L.M., Sustainable microwave-supported biodiesel production using sandbox oil and its waste shell as a nanoparticle green alkali heterogeneous catalyst. *Process Biochem.*, 142, 1–12, 2024.
44. Yadav, G., Yadav, N., Ahmaruzzaman, M., Microwave-assisted sustainable synthesis of biodiesel on *Oryza sativa* catalyst derived from agricultural waste by esterification reaction. *Chem. Eng. Process. Process Intensif.*, 187, article no. 109327, 2023.
45. Ribeiro, F.C.P., *et al.*, Determination of the thermal stability of sulfonic groups in heterogeneous acid catalysts derived from residue of guarana amazon biomass. *J. Therm. Anal. Calorim.*, 148, 23–35, 2023.
46. Paramesti, C., *et al.*, The influence of metal loading amount on Ni/mesoporous silica extracted from Lapindo mud templated by CTAB for conversion

of waste cooking oil into biofuel. *Bull. Chem. React. Eng. Catal.*, 16, 22–30, 2021.
47. Moreno-Torres, J.A., Espejel-Ayala, F., Ramírez-Bon, R., Coutino-Gonzalez, E., Sustainable strategies to synthesize small-pore NaP zeolites using natural minerals. *J. Mater. Sci.*, 59, 423–434, 2024.
48. Stafin, G., *et al.*, Synthesis of zeolites from residual diatomite using a microwave-assisted hydrothermal method. *Waste Manage. (Oxford)*, 126, 853–860, 2021.
49. Yadav, G., Yadav, N., Ahmaruzzaman, M., Microwave-assisted synthesis of biodiesel by a green carbon-based heterogeneous catalyst derived from areca nut husk by one-pot hydrothermal carbonization. *Sci. Rep.*, 12, article no. 21455, 2022.
50. Nguyen, T.M., Kaushik, N., Nguyen, L.T., Nguyen, G.T., Nguyen, T.H., Plasma-Assisted Decoration of Gold Nanoparticles on Bioinspired Polydopamine Nanospheres as Effective Catalyst for Organic Pollutant Removal. *Appl. Sci.*, 15, article no. 5280, 2025.
51. Ekanayake, U.G.M., *et al.*, Sustainable Claisen-Schmidt chalcone synthesis catalysed by plasma-recovered MgO nanosheets from seawater. *Sustainable Mater. Technol.*, 32, article no. e00394, 2022.
52. Qin, L., Lee, S., Ha, J., Li, O.L., Selective production of value-added chemicals from cellulosic biomass waste *via* plasma-synthesized catalysts. *Catal. Today*, 420, article no. 114030, 2023.
53. Nath, B., *et al.*, Agricultural Waste-Based Heterogeneous Catalyst for the Production of Biodiesel: A Ranking Study *via* the VIKOR Method. *Int. J. Energy Res.*, 2023, article no. 7208754, 2023.
54. Nabgan, W., *et al.*, Production of hydrogen and valuable fuels from polyethylene terephthalate waste dissolved in phenol reforming and cracking reactions *via* Ni-Co/CeO_2 nano-catalyst. *J. Anal. Appl. Pyrolysis*, 154, article no. 105018, 2021.
55. Rahman, W.U., *et al.*, Biodiesel synthesis from eucalyptus oil by utilizing waste egg shell derived calcium based metal oxide catalyst. *Process Saf. Environ. Prot.*, 122, article no. 2019.

5

Characterization Techniques for Solid Waste-Derived Catalysts

Tanvi Vats* **and Shubhanshi Sharma**

Department of Applied Chemistry, Gautam Buddha University, Greater Noida, Uttar Pradesh, India

Abstract

With the continuously growing population, along with rapid urbanization and industrialization, one of the pressing global issues is the generation of solid waste and its inadequate management. An innovative and promising approach to tackle this challenge involves converting solid waste into valuable materials like catalysts. This strategy not only helps in reducing the volume of solid waste but also facilitates the development of cost-effective and sustainable catalysts, offering viable alternatives to conventional ones. Waste materials from various sources such as agriculture, industry, and households can be repurposed to create efficient catalysts. This aligns well with the principles of green chemistry and supports the concept of a circular economy by minimizing waste and enhancing recycling practices globally. Catalysts derived from solid waste have shown significant potential in fields such as biofuel production and wastewater treatment. However, it is essential to thoroughly characterize these catalysts and evaluate the specific reactions they are best suited for. In this chapter we discuss the various techniques used in catalyst chemistry with suitable examples. It also gives insight into the specific application of a given characterization tool for a specific catalyst.

Keywords**:** Brunauer-Emmett-Teller, energy dispersive X-ray spectroscopy, Fourier transform infrared spectroscopy, inductively coupled plasma mass spectrometry, differential scanning calorimetry, inductively coupled plasma mass spectrometry, thermogravimetric analysis

**Corresponding author*: tanvi.vats@gmail.com

Deepti Goyal, Sakshi Kabra Malpani and Renuka Jain (eds.) Catalysis from Waste: Transforming Solid Residues into Value, (109–136) © 2026 Scrivener Publishing LLC

5.1 Introduction

Management of solid waste is a challenge today. The ever growing population, global development, and advancements in the industrial fields are the reasons behind a pile of solid waste generation, which has caused severe sustainability issues. This focuses on the need of identifying, formulating, accepting and utilizing these materials to repurpose them into useful materials. One such way of optimum utilization is to convert these solid wastes into catalysts. This way, the solid waste is managed better, and useful materials are obtained for catalytic applications [1].

Solid waste can be excellently used to make catalysts as the waste is present in plenty, and it is cheap. Additionally, the requirement of catalysts has increased because of an increasing demand for energy efficient and environmentally safe technologies in the fields such as environment conservation, energy regeneration, and chemical manufacture. Catalysts obtained from solid wastes help the industry meet their financial and environmental goals by providing a sustainable alternative to conventional methods [2].

An important factor that encourages the use of catalysts derived from solid waste is the inefficient waste management solutions. Traditional methods of waste disposal like landfilling and incineration can cause severe environmental issues such as degradation of soil, build-up of leachate, and emissions of greenhouse gas. Turning garbage into useful materials has its own benefits, two of which being, reduction of pollution and encouraging conservation of resources. This makes garbage look like a resource instead of a liability [3].

The basic idea of converting solid waste into catalysts is to form a circular economy and induce sustainable development and to promote waste utilization and optimization, waste reduction and being responsible towards our environment. Solid waste catalysts are sourced from various places including municipal and industrial waste and agricultural by-products. The waste is accumulated and processed to obtain different catalytic materials like silica, metals and carbon. Furthermore, one of the major issues of waste management, i.e., waste disposal can be fruitfully addressed, thereby opening the doors of technical and economic advancements [3–5]. Such catalysts, other than being environmentally friendly, also meet the principles of green chemistry [4]. The characterization of these catalysts derived from solid waste is very important for understanding their physical and chemical properties, structural qualities, and catalytic behavior. These methods help in greater understanding on how the initial structure and composition of solid wastes can cause the change in properties of the material thus obtained. For instance, the by-products of agricultural industry, such as rice husks and sugarcane bagasse consist of silica and other important minerals, which have a great potential to be converted into active components during making of catalysts. Likewise, industrial waste products such as fly ash and slag are abundantly rich in metal oxides, which are suited in the reactions involving heterogeneous catalysis [3].

This chapter discusses the important characterization techniques for solid waste-derived catalysts, focusing on their importance in tackling global waste management problems and encouraging industrial practices.

5.2 Characterization and Analytical Tools

The solid waste characterization primarily focuses on studying the physical, chemical, and catalytic properties of the prepared catalysts, which can be used to enhance their performance and make them suitable according to the usage in certain reactions. The main aim of characterization is to make these catalysts more useful and usable in a broad area of use. Some of the applications of characterization of catalysts have been depicted in Figure 5.1.

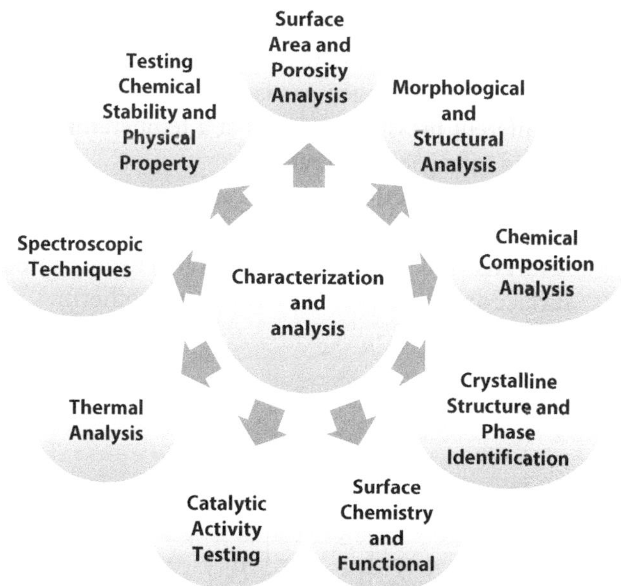

Figure 5.1 Characterization and analytical tools employed for catalysts study.

5.2.1 Surface Area and Porosity Analysis

Keeping in mind that catalysis is essentially a surface phenomenon, it becomes crucial to determine the surface area and porosity to analyze the quality of prepared catalyst and its effectiveness. Two methods are used in analyzing the surface area and porosity of a chemically synthesized catalyst. These processes have been discussed below-

(i) Brunauer-Emmett-Teller (BET) Surface Area Measurement:
BET is a very common technique used to analyze the surface area and porosity of any catalyst. It is an extrapolation of the theory of Langmuir's isotherm which is adapted for adsorption in multiple layers. This process infers that the gas molecules get adsorbed on the surface of the catalyst in several layers. On measuring the amount of gas adsorption that may have occurred, BET calculates the surface area of the material. This results in the formation of an adsorption isotherm, which is then used to calculate the surface area of the catalyst, prepared using solid waste.

I. **Principle**: It works on the theory of Langmuir's isotherm of multiple adsorptions. It calculates the amount of gas molecules adsorbed on the surface of the catalyst to form an isotherm of the data and calculates the surface area.
II. **Procedure**:
   - **Degassing**: The sample is initially degassed to eliminate any adsorbed impurities.
   - **Adsorption**: Nitrogen gas is introduced to the sample at different pressures, usually at a temperature of 77 K (liquid nitrogen temperature).
   - **Isotherm Measurement**: The amount of gas adsorbed is measured as a function of pressure.
   - **Analysis**: The BET equation is applied to determine the surface area based on the adsorption isotherm.

BET Equation
The BET equation is:

$$\frac{\frac{p}{p_0}}{v[1-\left(\frac{p}{p_0}\right)]} = \frac{c-1}{v_m c}\left(\frac{p}{p_0}\right) + \frac{1}{v_m c}$$

Where, p = equilibrium pressure of the gas
p_0 = saturation pressure of the gas
v = volume of gas adsorbed
v_m = volume of gas adsorbed at monolayer coverage
c = BET constant

The total surface area, S_{total} and the specific surface area, S_{BET} are given by:

$$S_{total} = \frac{v_m N_s}{v}$$

$$S_{BET} = \frac{S_{total}}{a}$$

Where, v_m = volume of gas adsorbed at monolayer coverage
N = Avogadro's number
s = adsorption cross section of the adsorbate
a = area occupied by one molecule (usually for nitrogen, about 0.162 nm^2)

Applications of BET to Catalysts

The BET method is widely used to evaluate the surface area and porosity of catalysts, which are crucial factors influencing their catalytic activity. BET analyzes the specific surface areas and helps in identification of active sites on the surface of the catalyst. It can also be used to check the texture of the catalyst and analyze the pore size and volume. These factors are vital in identification of the catalyst's properties. Additionally, BET helps in creating a feasible catalytic design and improving its effectiveness in a chemical reaction.

I. **Surface Area Measurement**: This method gives important information about the surface area and pore size analysis. A greater surface area means higher catalytic activity.
II. **Porosity Analysis**: BET, as discussed above, can be used to study the pore size. These pores are important in reactions which involve diffusion of reactants through porous heterogeneous catalysts.
III. **Catalyst Design**: Researchers can analyze the performance of a catalyst by studying the BET data of surface area and porosity and consequently modify the catalyst according to the requirements of applications [5].

(ii) Barrett-Joyner-Halenda Method (BJH)

BJH is also used in analysis of the pore size distribution and volume of mesoporous catalysts (2-50nm). This technique works on the capillary condensation and principles of desorption of gas. Just like BET, BJH also analyzes the adsorption-desorption isotherms, by using nitrogen, to showcase the texture and effectiveness of the catalysts in chemical reaction [6].

5.2.2 Morphological and Structural Analysis

I. **Scanning Electron Microscopy (SEM)**: It is an excellent characterization method to obtain high-resolution images of surface morphology, texture, shape, particle size, and irregularities on surface. It assists in understanding the active sites and identifying changes during catalyst preparation and reactions. SEM can be used to know the elemental composition in specific portions of the catalyst, described later. Thus, it can further be used in comprehensive understanding of catalysts [7].

II. **Transmission Electron Microscopy (TEM)**: TEM gives detailed nano-scale information about the size of a particle, shape and distribution of the catalyst. Other than just showing the crystalline nature of this catalyst, it also gives an analysis of the lattice structure [8, 9].

5.2.3 Chemical Composition Analysis

I. **Energy Dispersive X-Ray Spectroscopy**: In EDS, an incident beam of electrons strikes the catalyst sample to locate the electrons and calculate the elemental composition. The technique detects even the trace amount of impurities present in the sample and reports the uniformity in sample distribution. It is frequently employed with SEM and TEM to give full detail of the sample [15].

II. **Inductively Coupled Plasma Mass Spectrometry (ICP-MS)**: ICP-MS is a very sophisticated technique to measure the total elemental composition and quantify trace metals found in the catalyst sample. As even the traces of impurities can hamper the catalytic performance by great measures, ICP-MS is a very important tool as it can measure concentrations up to parts per trillion [16].

5.2.4 Crystalline Structure and Phase Identification

I. **X-Ray Diffraction (XRD)**: In solid state material science, XRD is a master tool as it helps to comprehend lattice arrangement, phase composition and lattice size. The diffraction patterns produced by XRD provide a detailed analysis of the various aspects of catalytic structure which in turn are crucial for the mechanism of catalytic action. It is commonly employed in the study of metal oxide catalysts and their reaction with active metals [17].

II. **Raman Spectroscopy**: The attractive feature of Raman spectroscopy is that it is a non-destructive technique. It is frequently employed to analyze the catalytic processes. Based on the study of inelastic scattering of a monochromatic beam of light falling on the sample, it is a powerful tool to measure the molecular vibrations and chemical bonds in a sample [18].

5.2.5 Surface Chemistry and Functional Groups

I. **Fourier Transform Infrared Spectroscopy (FTIR)**: FTIR is a well-known method which aids in identifying the functional groups and molecular bonds present on the catalytic surface. This information is crucial to predict the chemical interactions and surface chemistry of the products. FTIR is also usefully employed in knowing the reaction between the catalyst and its substrate [19].

II. **X-Ray Photoelectron Spectroscopy (XPS)**: XPS is another surface characterization technique which aids in determining the surface composition of a catalyst. It plays an important role in determination of the oxidation states, and hence predicting the exact species generated during the course of the catalytic reaction. The technique involves concentrating a beam of X-ray on the catalyst sample which leads to the evolution of photoelectrons. Based on the photoelectric effect, the technique gives us the calculated binding energy and kinetic energy of the emitted electrons. This information helps in identification of oxidation state and the changes in the surface of the catalyst undergoing a reaction [20].

5.2.6 Catalytic Activity Testing

I. **Activity and Selectivity Tests**: These are a variety of tests to measure the sample performance in the course of chemical reaction. These cluster tests give details about the rate of reaction, product selectivity, and percentage yield under different, varied conditions. They help us to draw a comparison between catalyzed and uncatalyzed reactions as well as different catalysts and different catalytic conditions [21].

II. **Temperature Programmed Reduction (TPR) and Oxidation (TPO)**: These methods are used to understand the oxidizing and reducing properties of a catalyst. TPR calculates the amount of hydrogen utilized in a reduction reaction of the catalyst, which reveals the extent of reducing metal oxides present in the catalyst sample. TPO keeps a check on the amount of oxygen involved in carrying out oxidation reactions to measure the extent of oxidation of the catalyst. These techniques provide insights into catalytic activity, stability, surface species, and metal-support interactions, helping in optimization of catalysts for specific applications by understanding redox mechanisms [10].

5.2.7 Thermal Analysis

I. **Thermogravimetric Analysis (TGA)**: It is a method used to analyze the change of weight of a catalyst material when it is heated in a controlled environment. TGA gives insights into the thermal stability, composition, and decomposition behavior of catalysts, helping to know the presence of highly reactive components, such as moisture, organic components, or metal precursors [11].

II. **Differential Scanning Calorimetry (DSC)**: It is a thermal analysis method used in catalysis to analyze the flow of heat associated with transition of phase, chemical reactions, and other thermal reactions in catalyst samples. By keeping a check on the temperature difference between a sample and a reference, this method provides an in-depth information into the thermal stability, melting points, and enthalpy changes of catalysts [12].

5.2.8 Spectroscopic Techniques

I. **Ultraviolet-Visible (UV-Vis) Spectroscopy**: It is an extensively used technique in catalysis to understand the electronic properties and surrounding environment of catalysts. By understanding the absorption of UV and visible light, it gives information about the electronic transitions of metal ions, ligands, and surface species. This technique is especially important for analyzing the catalysts activation state, detecting reaction intermediates, and understanding the interactions between catalyst components during catalytic processes [13].

II. **Nuclear Magnetic Resonance (NMR)**: This is a brilliant method of investigating the dynamics and molecular structure of catalysts and reaction intermediates that are generated. NMR provides detailed analysis of local environments around specific nuclei such as carbon or hydrogen; it reveals atom mobility coordination and the chemical bonding of the species of interest within the catalyst. It is especially powerful for detecting undercurrent towards the surface species, snapshots, and reaction intermediates in real time, as well as for monitoring catalyst structure changes during reactions [14].

5.2.9 Chemical Stability Testing and Physical Property Analysis

I. **Leaching Tests**: These tests help evaluate the potential for deactivation of catalyst due to metal leaching; ensuring integrity and performance of catalyst is maintained with time. Leaching studies are important for optimizing catalyst design and improving their durability in industrial applications [15].

II. **Cycling Tests**: These tests subject the catalyst to many cycles of reaction and regeneration, analyzing its performance activity retention and resistance to deactivation. Cycling tests are very important for understanding the catalyst's durability to maintain high performance over longer usage periods [16].

5.3 Characterization Techniques

Advanced characterization methods allow researchers to refine preparation techniques, locate active sites, and evaluate the catalysts' effectiveness in specific reactions. The following methods are frequently employed: FTIR, BET surface area analysis, SEM, TEM, and XRD. These techniques offer important information about the catalysts' shape, crystalline makeup, porosity, and surface functioning [17].

5.3.1 Brunauer–Emmet–Teller (BET)

Zhaozhou *et al.*, prepared solid acid carbon-based catalysts using rice straw (RS) waste and studied the effect of carbonization temperature on the catalytic activity [18]. It was found that the pore structure of solid acid catalysts greatly impacts the reactions of heterogeneous systems. The observed results of N_2-physisorption are shown in Table 5.1 which enlists the carbon precursor and catalysts used. It was observed that on increasing the temperature of carbonization, S_{BET} and V_{Total} of the carbon precursor first reduced and then saw a surge. It could be accounted to the presence of water molecule bound to the catalyst which changed surface morphology of RS240 to expand and curl, resulting in a high S_{BET} and V_{Total}. On removal of water molecules from the structure at about 300°C, a decrease was observed in the S_{BET} and V_{Total} of the catalyst. Even though carbonization occurred at temperatures above 300°C, it potentially could have led to the breaking of glycosidic bonds of cellulose and hemicellulose during

Table 5.1 Properties of carbon precursors and prepared catalysts [18].

Catalysts	S_{BET} (m^2/g)	V_{Total} (cm^3/g)	D_{pore} (nm)
RS240	306.8	2.94	3.82
RS300	260.5	2.44	3.74
RS350	312.1	2.95	3.76
RS240-80	222.1	2.22	3.98
RS300-80	420.9	4.05	3.84
RS350-80	337.7	3.30	3.90

pyrolysis at high temperature and removal of organic compounds of the carbon precursor. Therefore, the S_{BET} and V_{Total} increased.

The S_{BET} and V_{Total} of RS240-80 was observed to be lower than those of RS240, which depicted that some structural changes must have taken place in the structure of carbon after it was sulfonated. This change may involve a collapse due to dehydration reaction caused due to the presence of conc. H_2SO_4 and carbon degradation. The S_{BET} for RS300 increased from 260.5 to 420.9 m^2/g after sulfonation was done. Also, it was found that RS300-80 displayed a higher V_{Total} of 4.05 cm^3/g in comparison to RS300, which may be accounted to partial pore and surface cracking in RS300 due to breaking of glycosidic bond in cellulose and hemicellulose while carrying out sulfonation.

It was found that RS300 still had some of its crystalline structure of cellulose and hemicellulose intact after low-temperature carbonization at 300°C but on hydrolyzing in acidic medium with the decrease in degree of polymerization and mechanical strength, causing internal channel and surface cracking of RS300-80. Hence, RS300-80 showed a higher S_{BET} and V_{Total}. This was confirmed by SEM images of RS300-80.

It was found that S_{BET} and V_{Total} of RS350 showed a slight rise upon sulfonation, showcasing that 80°C, the sulfonation temperature, played little role on the effect on S_{BET} and V_{Total} of RS350-80. According to TGA, alkyl chains or oxygen containing functional groups potentially get damaged on attaining any temperature above the carbonization temperature thereby, proving the cellulose and hemicellulose structures of RS350 get damaged with higher temperatures, which cause a decrease in the levels of hydrolysis of cellulose of RS350 in concentrated sulfuric acid, making the internal channels of RS300-80 not to be broken easily. Thus, the S_{BET} and V_{Total} of RS350-80 only slightly increased.

However, excessive high carbonization temperatures, such as 350°C, likely cause RS350 to harden due to carbonization at elevated temperatures, making it difficult to modify using concentrated sulfuric acid. A comparison of the D_{pore} values of all carbon precursors and catalysts revealed that they remained nearly constant, around 3.85 nm. This consistency indicates that the resulting catalysts were predominantly mesoporous materials, and variations in carbonization temperature (below 350°C) or sulfonation temperature (below 80°C) had minimal impact on the average pore diameter of the carbon structure.

Figure 5.2 illustrates the pore size distribution and N_2 adsorption–desorption isotherms for all carbon precursors and catalysts. According to the IUPAC (2015) classification, the isotherms for both the carbon precursors and catalysts closely resembled type IV isotherms. The pore size

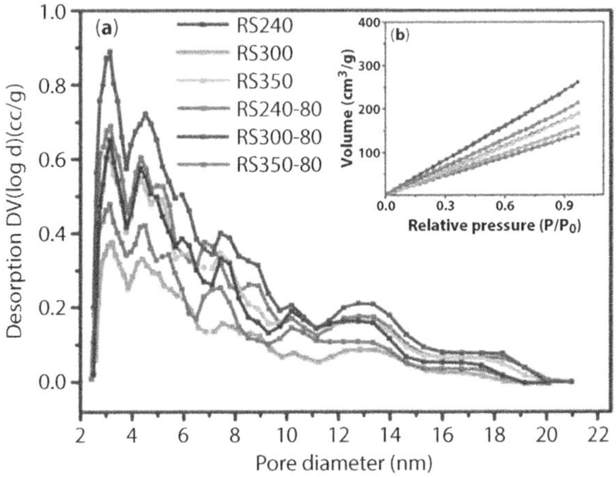

Figure 5.2 (a) Pore size distributions and (b) N_2 adsorption-desorption isotherms of all the carbon precursors and catalysts with different carbonization temperature [18].

distribution confirmed that the primary pore diameter for all samples was approximately 3.85 nm, with almost no presence of micropores. The adsorption–desorption hysteresis loops were subtle, likely due to the average pore diameters of the catalysts falling between the microporous and mesoporous ranges [18].

5.3.2 X-Ray Diffraction (XRD)

It is a widely used and powerful technique for analyzing the crystalline structure and phase composition of solid materials. When applied to catalysts derived from solid waste, XRD plays an essential role in evaluating the mineral composition, structural integrity, and phase alterations of the catalyst materials—factors that significantly influence their catalytic performance. Given that solid waste often contains a mix of minerals and non-crystalline phases, XRD is indispensable for identifying the exact existing phases, confirming the crystallinity of the catalyst structures, and assessing the extent of phase changes during the preparation or activation of the catalyst [19–23]. Key parameters that can be determined using XRD include:

a. **Phase Identification**: Since catalytic activity is highly dependent on the specific phase of the material, XRD is essential for identifying the various phases present in a catalyst sample.

b. **Crystal Structure**: Gaining an understanding of the crystal structure provides insights into the material's catalytic properties and behavior. XRD enables detailed analysis of the crystal structure of the phases within the catalyst.
c. **Particle Size**: The particle size-to-volume ratio is a crucial factor influencing catalyst efficiency, as smaller particles often exhibit unique catalytic properties compared to larger ones. XRD, through the Scherrer equation, can estimate the average size of the crystalline particles in the catalyst, aiding in the optimization of its performance.

The fundamental principle of XRD involves the interaction of X-rays with the crystalline structure of a material. When X-rays are directed at a sample, they are diffracted by the atomic planes within the crystal lattice. The resulting diffraction pattern reveals detailed information about the atomic arrangement, enabling the identification of various phases based on their distinct diffraction peaks. Bragg's law equation, $n\lambda = 2d\sin(\theta)$, connects the diffraction angle (θ), the wavelength of the incident X-rays (λ), and the interplanar spacing (d), offering a method to analyze the crystalline properties of the sample [21]. Each peak corresponds to a specific set of parallel planes within the crystal lattice. The positions of these peaks (measured in terms of 2θ) provide information about the spacing between the atomic planes. Meanwhile, the peak intensities reveal details about the arrangement and types of atoms in the unit cell, as well as the crystal's orientation. From the diffraction pattern, one can determine unit cell parameters (a, b, c, and the angles between them), which define the repeating unit of the crystal lattice. By comparing the observed diffraction pattern with standard reference patterns, the phases present in the sample can be accurately identified.

Figure 5.3(a) displays a broad band centered at $2\theta = 10–15°$, indicating the presence of amorphous silica. In contrast, Figure 5.3(b) reveals a distinct crystalline peak at $2\theta = 27.6°$, signifying the presence of the quartz crystalline phase. Upon incorporating sulfated zirconia onto the TAP surface in SZP-9, SZP-12, and SZP-15 catalysts (Figures 5.3(c–e)), the emergence of monoclinic and tetragonal zirconia phases was observed. These phases were identified at $2\theta = 25–27°$ for monoclinic zirconia and at $2\theta =$ 30, 41, 50, and 60° for tetragonal zirconia, according to the ICSD collection code 066787 [24].

Increasing the sulfated zirconia loading from 9 to 15 wt.% resulted in a corresponding increase in the intensity of both monoclinic and tetragonal phases. The SZP-15 catalyst exhibited the highest intensity and number of

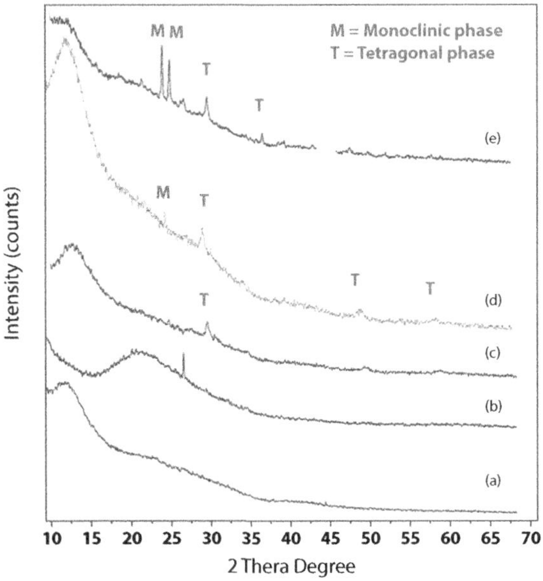

Figure 5.3 X-ray diffraction pattern of (a) perlite, (b) TAP, (c) SZP-9, (d) SZP-12, and (e) SZP-15 [27].

tetragonal phase peaks, highlighting its preferred structure. While comparing SZP-9 and SZP-12, the one which had lower sulfate content showed stronger and more numerous monoclinic phases than tetragonal ones. Additionally, the incorporation of sulfate ions into the catalysts reduced the crystallite size of both tetragonal and monoclinic phases, making the precise size determination of these peaks challenging [24, 25].

A broad hump centered at $2\theta = 10–13°$, indicative of the amorphous silica characteristic of perlite, was consistently observed in the XRD patterns of all samples (ICSD collection code 066787). With increasing sulfated zirconia loading from 9 to 15 wt.%, the intensity of both monoclinic and tetragonal phases showed a corresponding rise. The SZP-15 catalyst exhibited the highest intensity and the greatest number of tetragonal phase peaks, marking it as the preferred structure. In contrast, SZP-9 and SZP-12 with lower sulfate content, demonstrated stronger and more numerous monoclinic phases relative to tetragonal ones. The incorporation of sulfate ions reduced the crystallite size of both monoclinic and tetragonal phases, making precise size determination of these peaks impractical. The characteristic hump at $2\theta = 10–13°$ for amorphous silica was retained across all samples [24–27].

The SZP-15 catalyst demonstrated the highest intensity and the most prominent tetragonal phases, underscoring its preferred structure. In contrast, the SZP-9 and SZP-12 catalysts, with lower sulfate content, exhibited a dominance of monoclinic phases. The incorporation of sulfate ions reduced the crystallite size of both phases, making precise size determination challenging. These findings highlight the structural evolution and performance potential of the catalysts, emphasizing SZP-15 as the optimal choice for enhanced catalytic activity.

5.3.3 Scanning Electron Microscope (SEM)

A highly focused beam of electrons is bombarded on the surface of the catalyst in SEM, which results in production of a number of signals with hint at the interactions between electrons and the catalyst. These signals, which include backscattered electrons and secondary electrons, are gathered to give specific surface information about the sample. SEM captures surface morphology by analyzing electron scattering, and its wide field of view and excellent depth of focus allow for efficient examination of particle distributions across uneven catalyst surfaces [7, 28, 29].

Under traditional high-vacuum operation, SEM typically requires conductive specimens. For non-conductive materials, a conductive coating—using substances like carbon, osmium, or gold/palladium alloys—is usually applied. However, non-conductive samples can also be imaged without a coating by utilizing a low-voltage SEM mode. Additionally, environmental scanning electron microscopes (ESEM) with variable pressure capabilities enable the imaging of non-conductive materials without the need for coatings. For hydrated samples, such as biological specimens, specific preparation steps are often necessary to enhance electron optical contrast (e.g., staining), reduce sample thickness (e.g., ultrathin sectioning), and stabilize the material for imaging [30–32].

SEM is a valuable technique for analyzing heterogeneous catalysts and nanostructured materials. It provides insights into various aspects of catalysts, including the formation of active structures, phase development, texture, and factors contributing to catalyst aging. SEM is particularly effective for studying porous, non-crystalline catalyst supports such as silica and activated carbons.

SEM results can be integrated with complementary techniques such as spectroscopy, diffraction, adsorption methods, and other microscopy methods to provide a comprehensive understanding of the data obtained.

In a study by Malpani *et al.*, SZP catalysts were extensively characterized to determine their catalytic potential for vapor-phase alkylation of o-,

Figure 5.4 SEM micrographs of (a) perlite (b) SZP-15 [27].

m-, and p-cresols with tert-butyl alcohol [27]. SEM micrographs revealed that the perlite particles (Figure 5.4(a)) exhibited an irregular, layered morphology, while the SZP-15 sample (Figure 5.4(b)) showed the agglomeration of amorphous particles following sulfation with H_2SO_4. Shiny flakes observed in the SZP-15 micrograph indicated zirconia deposition on the perlite surface. Elemental analysis *via* SEM-EDX (Table 5.2) confirmed that silica and alumina were the primary constituents of perlite [33].

5.3.4 Transmission Electron Microscopy (TEM)

TEM, which provides high resolution and comprehensive structural information at the atomic scale, is a crucial tool for catalyst characterization. A high-energy electron beam is sent through an incredibly thin, electron-transparent sample—typically less than 100 nm in thickness—to create a TEM image. The electron beam is focused on the sample by a sequence of electromagnetic lenses and apertures in the microscope column, which also minimize distortions and enlarge the resulting image, which is then shown on a phosphor screen or recorded by a specialized camera [8, 9, 34].

Table 5.2 EXD analysis of perlite, SZP-9, SZP-12 and SZP-15 [27].

Samples	SiO_2 (wt%)	Al_2O_3 (wt%)	K_2O (wt%)	Na_2O (wt%)	ZnO (wt%)	FeO (wt%)	TiO_2 (wt%)	SO_3 (wt%)	ZrO_2 (wt%)
Perlite	72.74	14.79	7.02	2.10	2.04	0.91	0.40	-	-
SZP-9	77.27	16.64	-	1.40	-	-	0.40	1.31	2.98
SZP-12	74.51	16.63	-	1.67	-	-	0.32	1.24	5.63
SZP-15	69.71	14.45	-	1.83	-	-	-	2.03	11.98

TEM enables precise measurement of the particle size and shape of catalysts, providing critical insights into their morphology for optimizing catalytic activity. TEM images also reveal surface features, interfaces, and the interactions between catalyst particles and their supports—key factors in assessing catalytic performance. Additionally, TEM can detect defects and impurities within the catalytic material, which could negatively impact its stability and effectiveness [35–37].

Figure 5.5(a) presents the TEM image of perlite, showcasing its irregular morphology. In contrast, Figure 5.5(b), which depicts the dark-field image of the SZP-15 catalyst, reveals nanocrystalline, aggregated tetragonal zirconia particles dispersed across the perlite surface, visible as small dots. The particle size of the catalyst was determined to range from 9 to 25 nm. Both TEM images indicate that the catalyst exhibits a heterogeneous particle size distribution, comprising particles larger than 100 nm alongside smaller particles within the 9–25 nm range. These findings also support the

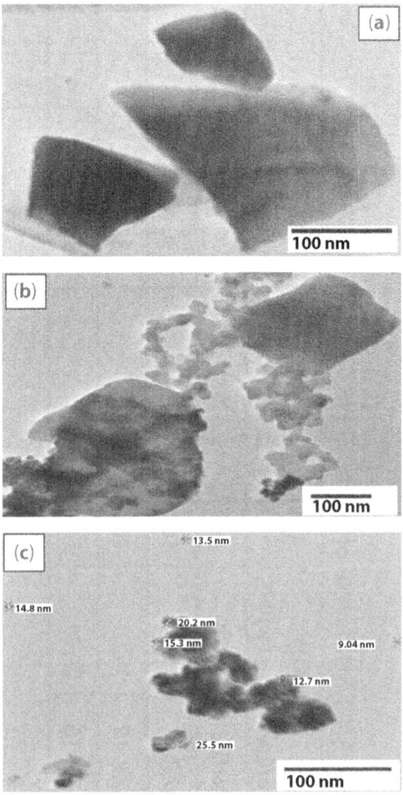

Figure 5.5 TEM micrographs of (a) perlite and (b, c) SZP-15 [27].

observation that sulfation reduces the size of the zirconia particles, effectively increasing the surface area of the final catalyst [27, 38–40]. These findings emphasize TEM's importance in understanding and optimizing catalyst properties for improved performance.

5.3.5 Fourier Transform Infrared Spectroscopy (FTIR)

FTIR is a highly effective analytical tool for identifying chemical bonds and functional groups within a substance. In the context of catalyst chemistry, FTIR plays a critical role in analyzing the composition, surface properties, and the interactions between the catalyst and reactants or products. Its significance in catalysis can be summarized in some key areas depicted in Figure 5.6.

The FT-IR spectra of perlite and TAP, presented in Figures 5.7 (a, b) along with those of SZP-9, SZP-12, and SZP-15 catalysts in Figures 5.7 (c–e), reveal notable features. A broad band observed between 3600–3300 cm^{-1} is characteristic of surface -OH groups, with its intensity progressively increasing as the sulfated zirconia loading rises. This increase is attributed to the incorporation of zirconia and sulfate. The broad nature of this band is linked to surface hydroxyl groups (–Si-OH) exhibiting strong intermolecular hydrogen bonding and the presence of physiosorbed water molecules, which diminishes in the TAP sample due to water loss during thermal activation. Additionally, a peak near 1630 cm^{-1} is identified across all samples, corresponding to the bending mode (δ O-H) of –OH groups

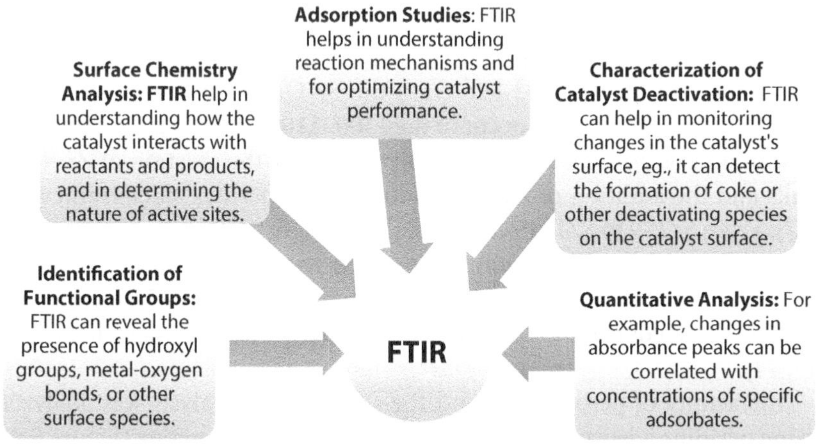

Figure 5.6 Multi-faceted role of FTIR in catalysis.

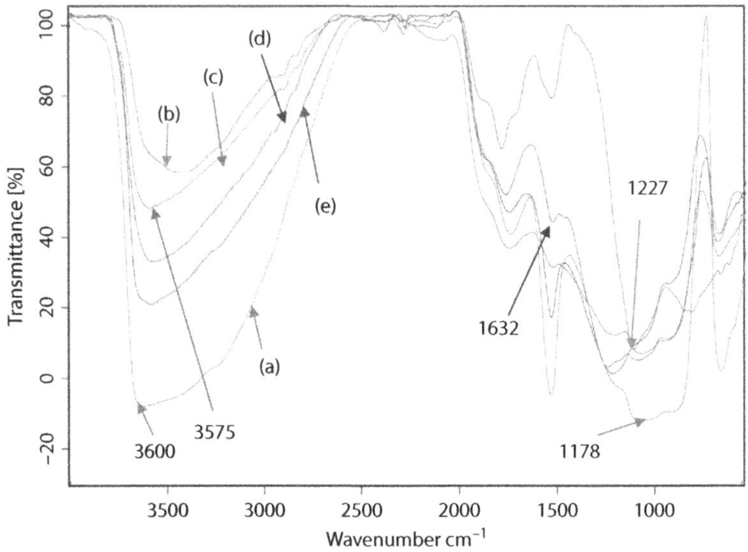

Figure 5.7 FT-IR spectra of (a) Perlite, (b) TAP, (c) SZP-9, (d) SZP-12, and (e) SZP-15 [27].

and water molecules [27]. The persistence of these bands, even after calcination at 550°C, highlights the stability of Brønsted acidity in the catalysts [33, 41].

A characteristic band corresponding to the asymmetric stretching vibration of S=O, typically observed at 1391–1382 cm^{-1}, shifts to a lower wavenumber at 1344 cm^{-1}. This shift indicates the partial ionic character of S–O bonds, which contributes to the enhanced Brønsted acidity in SZP catalysts. The absence of a band near 1400 cm^{-1} in all catalysts confirms the lack of polynuclear sulfates ($S_2O_7^{2-}$) [42].

Additionally, at 953 and 898 cm^{-1}, the Si–O–Si framework stretching vibrations, which typically occur in the 1300–1100 cm^{-1} range, shift to lower wavenumbers. This change is probably caused by the creation of Zr–O–Si links and the addition of zirconium ions to the perlite's silica framework [27]. From SZP-9 to SZP-15 catalysts, the strength of these bands gradually rises with increasing zirconium content. Furthermore, because they overlap with Si–O–Si deformation bands, the bands in the 700–550 cm^{-1} range that are ascribed to Zr–O stretching vibrations are less distinguishable [40, 43, 44].

The acidity of perlite, TAP, and the synthesized catalysts was analyzed using pyridine adsorption combined with FT-IR spectroscopy (Figure 5.8). In the spectra of perlite and TAP (Figures 5.8 a and b), broad and less

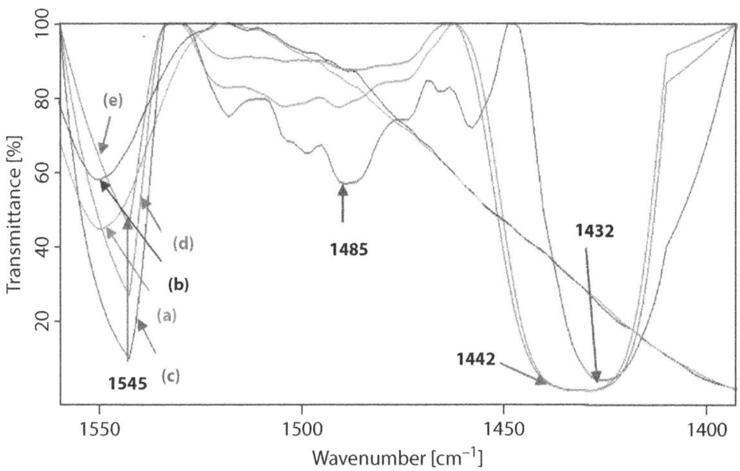

Figure 5.8 Pyridine adsorbed FT-IR spectra of (a) perlite, (b) TAP, (c) SZP-9, (d) SZP-12, and (e) SZP-15 [27].

intense bands centered at 1550 cm^{-1} indicate the coordination of pyridine with surface –OH groups. The absence of bands around 1540 and 1440 cm^{-1} confirms the lack of Brønsted and Lewis acidity in perlite and TAP.

Brønsted acidic sites were identified at 1545 cm^{-1}, while Lewis acidic sites were observed at 1445 cm^{-1}. An increase in sulfur content led to a corresponding rise in both Brønsted and Lewis acid sites. The stability of the tetragonal zirconia phase also contributed to the increase in Brønsted acid sites, as it forms the most stable structure during hydration. Consequently, Brønsted bands were broader in SZP-9 and SZP-12 (Figures 5.8 c and d) and became sharper in SZP-15 (Figure 5.8 e). Similarly, Lewis bands were more intense and broader in SZP-12 and SZP-15 but less intense in SZP-9 [27].

A band at 1485 cm^{-1}, present in all samples, was attributed to the coexistence of both Brønsted and Lewis acidic sites. Sulfate groups exhibit an inductive (-I) effect, withdrawing electron density from zirconia and generating Lewis acidic sites, while the protons associated with these sulfate groups create Brønsted acidic sites. With increasing sulfur and zirconia content from SZP-9 to SZP-15, the concentration of both acidic sites followed the same trend, consistent with previous findings [45].

5.3.6 UV-Vis Spectroscopy

This technique is employed to investigate the optical properties of catalysts by analyzing their absorption spectra. UV-Vis spectroscopy helps to detect

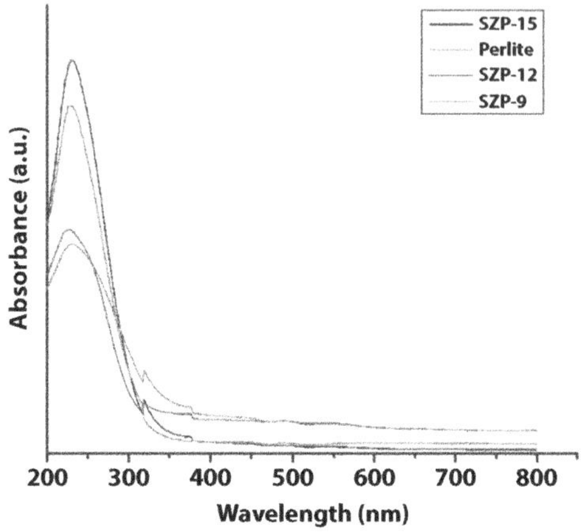

Figure 5.9 UV-Vis DRS spectra of perlite, SZP-9, 12, and 15 [27].

metal oxide species and aid in identifying electronic transitions, which are crucial for understanding how the catalyst behaves during reactions. It also offers valuable information about the oxidation states of metals within the catalyst structure, helping to optimize and design more effective catalytic systems.

A broad band in the UV-Vis curve is seen near 230 nm for perlite, showcasing the characteristic properties of silica (Figure 5.9). But the data was not significant enough to know if it is a crystalline or amorphous form of silica. The catalyst material also showed a peak very close to 210 nm, which possibly was because of charge transfer of oxygen-zirconium to the silica structure. Another band was seen at 298 nm which indicated the presence of nano sulfated zirconia, along with another band at around 375-400 nm, which showed the presence of Zr in large amounts in all these catalysts [27].

5.3.7 Thermogravimetric Analysis and Differential Scanning Calorimetry (TGA and DSC)

TGA analyzes the change in mass of the sample when it is heated, along with giving details about the stability at different temperatures, chemical composition, and decomposition patterns of the catalyst. DSC is used to check the amount of heat absorbed and released by a catalyst in a chemical reaction. It will give an idea on the melting point of the catalyst, glass transition temperature, and crystallization point. These techniques, both,

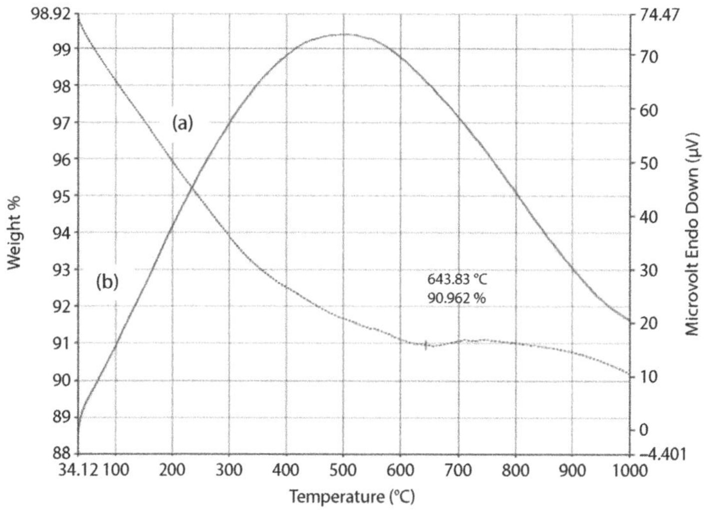

Figure 5.10 (a) TGA and (b) DSC curve of NPC-15 [33].

are significant in keeping a check on the thermal and stability aspects of composite materials.

Malpani *et al.*, saw the TGA and DSC graph of Turkish perlite supported NiO catalyst, also called as NPC-15, and observed that the graph shows two regions, one where major, 8.96% weight loss was observed at 643.83°C, and second was a minute 0.82% weight loss at greater temperatures. It was also seen that an exothermic DSC peak was observed as shown in Figure 5.10. The weight loss in the first case is mainly due to dehydration, physiosorbed water, presence of -OH groups, and decomposition of Ni salts. The second weight loss can be accounted for by the conversion of nickel hydroxide to nickel oxide [33].

5.4 Conclusion

Solid waste-derived catalysts present a promising solution to both the growing environmental challenges of waste management and the need for sustainable catalytic materials. The transformation of solid waste, such as agricultural by-products, industrial residues, and municipal waste into functional catalytic materials offers numerous benefits, including reduced environmental pollution, resource conservation, and cost-effective catalytic solutions for industrial processes. However, their full potential can only be unlocked through a deep understanding of their physicochemical properties, which requires detailed and systematic characterization. Techniques like surface area analysis, structural characterization, and catalytic testing are essential to optimize their performance across various applications such as biofuel production, wastewater treatment, and chemical processes. By advancing characterization techniques and exploring new solid waste resources, we can further enhance the efficiency, stability, and applicability of these catalysts, fostering the development of sustainable solutions for global waste management and green technology.

AI disclosure statement: The author acknowledges the use of AI-based tools for minor language editing or grammatical correction. However, all intellectual contributions, analyses, and conclusions presented in this chapter are entirely the author's own.

References

1. Das, S., Lee, S.-H., Kumar, P., Kim, K.-H., Lee, S.S., Bhattacharya, S.S., Solid waste management: Scope and the challenge of sustainability. *J. Cleaner Prod.*, 228, 658–678, Aug. 2019, doi: 10.1016/j.jclepro.2019.04.323.
2. Isahak, W.N.R.W. and Al-Amiery, A., Catalysts driving efficiency and innovation in thermal reactions: A comprehensive review. *Green Technol. Sustainability*, 2, 2, 100078, May 2024, doi: 10.1016/j.grets.2024.100078.
3. Rautela, R., *et al.*, Turning Solid Waste into Catalysts: A Path for Environmental Solutions. *ChemPlusChem*, 89, 12, Dec. 2024, doi: 10.1002/cplu.202400246.
4. Shi, Y., *et al.*, Advances and Challenges for Catalytic Recycling and Upgrading of Real-World Mixed Plastic Waste. *ACS Catal.*, 841–868, Dec. 2024, doi: 10.1021/acscatal.4c06344.
5. Brunauer, S., Emmett, P.H., Teller, E., Adsorption of Gases in Multimolecular Layers. *J. Am. Chem. Soc.*, 60, 2, 309–319, Feb. 1938, doi: 10.1021/ja01269a023.
6. Barrett, E.P., Joyner, L.G., Halenda, P.P., The Determination of Pore Volume and Area Distributions in Porous Substances. I. Computations from Nitrogen Isotherms. *J. Am. Chem. Soc.*, 73, 1, 373–380, Jan. 1951, doi: 10.1021/ja01145a126.
7. Noh, C., *et al.*, Scanning Electron Microscopy Imaging of Large DNA Molecules Using a Metal-Free Electro-Stain Composed of DNA-Binding Proteins and Synthetic Polymers. *Adv. Sci.*, 11, 28, Jul. 2024, doi: 10.1002/advs.202309702.
8. Malatesta, M., Transmission Electron Microscopy as a Powerful Tool to Investigate the Interaction of Nanoparticles with Subcellular Structures. *Int. J. Mol. Sci.*, 22, 23, 12789, Nov. 2021, doi: 10.3390/ijms222312789.

9. Franken, L.E., Grünewald, K., Boekema, E.J., Stuart, M.C.A., A Technical Introduction to Transmission Electron Microscopy for Soft-Matter: Imaging, Possibilities, Choices, and Technical Developments. *Small*, 16, 14, Apr. 2020, doi: 10.1002/smll.201906198.
10. Gervasini, A., Temperature Programmed Reduction/Oxidation (TPR/TPO) Methods, in: *Calorimetry and Thermal Methods in Catalysis*, A. Auroux (Ed.), pp. 175–195, Springer, Berlin & Heidelberg, 2013, doi:10.1007/978-3-642-11954-5_5.
11. Leal, G.F., Ramos, L.A., Barrett, D.H., Curvelo, A.A.S., Rodella, C.B., A thermogravimetric analysis (TGA) method to determine the catalytic conversion of cellulose from carbon-supported hydrogenolysis process. *Thermochim. Acta*, 616, 9–13, Sep. 2015, doi: 10.1016/j.tca.2015.07.017.
12. Yurchenko, O., Pernau, H.-F., Engel, L., Wöllenstein, J., Differential thermal analysis techniques as a tool for preliminary examination of catalyst for combustion. *Sci. Rep.*, 13, 1, 9792, Jun. 2023, doi: 10.1038/s41598-023-36878-8.
13. Bols, M.L., *et al.*, *In Situ* UV–Vis–NIR Absorption Spectroscopy and Catalysis. *Chem. Rev.*, 124, 5, 2352–2418, Mar. 2024, doi: 10.1021/acs.chemrev.3c00602.
14. Haro Mares, N., Logrado, M., Kergassner, J., Zhang, B., Gutmann, T., Buntkowsky, G., Solid-State NMR of Heterogeneous Catalysts. *ChemCatChem*, 16, 22, Nov. 2024, doi: 10.1002/cctc.202401159.
15. Newton, O.J., Takle, M.J., Richardson, J., Hellgardt, K., Hii, K.K.M., Deconvoluting Substrates, Support, and Temperature Effects on Leaching and Deactivation of Pd Catalysts: An *In Situ* Study in Flow. *ACS Catal.*, 14, 13, 9678–9686, Jul. 2024, doi: 10.1021/acscatal.4c02028.
16. Rountree, E.S., McCarthy, B.D., Eisenhart, T.T., Dempsey, J.L., Evaluation of Homogeneous Electrocatalysts by Cyclic Voltammetry. *Inorg. Chem.*, 53, 19, 9983–10002, Oct. 2014, doi: 10.1021/ic500658x.
17. Anekwe, I.M.S., Akpasi, S.O., Enemuo, E.M., Ashiegbu, D., Mustapha, S.I., Isa, Y.M., Innovations in catalytic understanding: A journey through advanced characterization. *Mater. Today Catal.*, 7, 100061, Dec. 2024, doi: 10.1016/j.mtcata.2024.100061.
18. Wei, Z., *et al.*, Preparation of Carbon-Based Solid Acid Catalysts Using Rice Straw Biomass and Their Application in Hydration of α-Pinene. *Catalysts*, 10, 2, 213, Feb. 2020, doi: 10.3390/catal10020213.
19. Perego, G., Characterization of heterogeneous catalysts by X-ray diffraction techniques. *Catal. Today*, 41, 1–3, 251–259, May 1998, doi: 10.1016/S0920-5861(98)00054-6.
20. Günzler, H. and Williams, A. (Eds.), *Handbook of Analytical Techniques*, Wiley-VCH Verlag GmbH & Co. KGaA, Weinheim & New York, 2001, doi: 10.1002/9783527618323.

21. Zaera, F. and Ma, Z., Characterization of Heterogeneous Catalysts, in: *Surface and Nanomolecular Catalysis*, pp. 1–38, CRC Press, 2006, doi: 10.1201/9781420015751.ch1.
22. Jagodzinski, H., Klug, H.P., Alexander, L.E., *X-ray Diffraction Procedures for Polycrystalline and Amorphous Materials*, 2. Auflage, p. 966, John Wiley & Sons, New York-Sydney-Toronto, 1974, Seiten, Preis: £ 18.55. Berichte der Bunsengesellschaft für physikalische Chemie, vol. 79, no. 6, pp. 553–553, Jun. 1975, doi: 10.1002/bbpc.19750790622.
23. Jellinek, M.H. and Fankuchen, I., The application of X-ray diffraction to the study of solid catalysts, in: *Advances in Catalysis*, vol. 1, pp. 257–289, Academic Press, New York, NY, USA, 1948, doi:10.1016/S0360-0564(08)60677-X.
24. Delarmelina, M., Deshmukh, G., Goguet, A., Catlow, C.R.A., Manyar, H., Role of Sulfation of Zirconia Catalysts in Vapor Phase Ketonization of Acetic Acid. *J. Phys. Chem. C*, 125, 50, 27578–27595, Dec. 2021, doi: 10.1021/acs.jpcc.1c06920.
25. Liu, E., Locke, A.J., Frost, R.L., Martens, W.N., Sulfated fibrous ZrO2/Al2O3 core and shell nanocomposites: A novel strong acid catalyst with hierarchically macro–mesoporous nanostructure. *J. Mol. Catal. A: Chem.*, 353–354, 95–105, Feb. 2012, doi: 10.1016/j.molcata.2011.11.010.
26. Marakatti, V.S., Shanbhag, G.V., Halgeri, A.B., Sulfated zirconia; an efficient and reusable acid catalyst for the selective synthesis of 4-phenyl-1,3-dioxane by Prins cyclization of styrene. *Appl. Catal., A*, 451, 71–78, Jan. 2013, doi: 10.1016/j.apcata.2012.11.016.
27. Malpani, S.K., Goyal, D., Chinnam, S., Sharma, S.K., Katara, S., Rani, A., Vapor Phase Alkylation of Isomeric Cresols with Tert-Butyl Alcohol over Perlite Supported Sulfated Zirconia Catalyst. *Sustainability*, 14, 9, 5149, Apr. 2022, doi: 10.3390/su14095149.
28. Yuan, J., Su, W., Hu, X., Li, X., Fei, C., Application of Raman imaging and scanning electron microscopy techniques for the advanced characterization of geological samples. *Microsc. Res. Tech.*, 85, 7, 2729–2739, Jul. 2022, doi: 10.1002/jemt.24093.
29. Feng, H., Xu, C., Xiao, L., Li, L., Multiscale hierarchical composite with extremely specific damping performance *via* bottom-up synergistic enhancement strategy. *Virtual Phys. Prototyping*, 20, 1, Dec. 2025, doi: 10.1080/17452759.2024.2448541.
30. Chee, S.W., Lunkenbein, T., Schlögl, R., Roldán Cuenya, B., Operando Electron Microscopy of Catalysts: The Missing Cornerstone in Heterogeneous Catalysis Research? *Chem. Rev.*, 123, 23, 13374–13418, Dec. 2023, doi: 10.1021/acs.chemrev.3c00352.
31. Lomic, G., Kis, E., Boskovic, G., Marinkovic-Neducin, R., Application of scanning electron microscopy in catalysis. *Acta Period. Technol.*, 35, 67–77, 2004, doi: 10.2298/APT0435067L.

32. Ali, A., Zhang, N., Santos, R.M., Mineral Characterization Using Scanning Electron Microscopy (SEM): A Review of the Fundamentals, Advancements, and Research Directions. *Appl. Sci.*, 13, 23, 12600, Nov. 2023, doi: 10.3390/app132312600.
33. Malpani, S.K., Goyal, D., Katara, S., Rani, A., Turkish perlite supported nickel oxide as the heterogeneous acid catalyst for a series of Claisen–Schmidt condensation reactions. *Turk. J. Chem.*, 45, 4, 1097–1114, Aug. 2021, doi: 10.3906/kim-2010-42.
34. Peters, J.J.P., *et al.*, Event-responsive scanning transmission electron microscopy. *Science (1979)*, 385, 6708, 549–553, Aug. 2024, doi: 10.1126/science.ado8579.
35. He, B., Zhang, Y., Liu, X., Chen, L., *In-situ* Transmission Electron Microscope Techniques for Heterogeneous Catalysis. *ChemCatChem*, 12, 7, 1853–1872, Apr. 2020, doi: 10.1002/cctc.201902285.
36. Chenna, S. and Crozier, P.A., Operando Transmission Electron Microscopy: A Technique for Detection of Catalysis Using Electron Energy-Loss Spectroscopy in the Transmission Electron Microscope. *ACS Catal.*, 2, 11, 2395–2402, Nov. 2012, doi: 10.1021/cs3004853.
37. Dai, J., Sun, Y., Liu, Z., Zhang, Y., Duan, S., Wang, R., Using *In situ* Transmission Electron Microscopy to Study Strong Metal-Support Interactions in Heterogeneous Catalysis. *Angew. Chem. Int. Ed.*, Sep. 2024, doi: 10.1002/anie.202409673.
38. Zhao, S., Fang, J., Wang, Y., Zhang, Y., Zhou, Y., Zhuo, S., Construction of three-dimensional mesoporous carbon nitride with high surface area for efficient visible-light-driven hydrogen evolution. *J. Colloid Interface Sci.*, 561, 601–608, Mar. 2020, doi: 10.1016/j.jcis.2019.11.035.
39. Rachmat, A., Trisunaryanti, W., Sutarno, Wijaya, K., Synthesis and characterization of sulfated zirconia mesopore and its application on lauric acid esterification. *Mater. Renewable Sustainable Energy*, 6, 3, 13, Aug. 2017, doi: 10.1007/s40243-017-0097-1.
40. Hussain, S.T., Mazhar, M., Gul, S., Increase in Concentration and Strength of Acidic Catalytic Sites on Sulfated Zirconia by Doping with Silica. *Chin. J. Catal.*, 28, 7, 622–626, Jul. 2007, doi: 10.1016/S1872-2067(07)60054-0.
41. Khatri, C., Mishra, M.K., Rani, A., Synthesis and characterization of fly ash supported sulfated zirconia catalyst for benzylation reactions. *Fuel Process. Technol.*, 91, 10, 1288–1295, Oct. 2010, doi: 10.1016/j.fuproc.2010.04.011.
42. Rodríguez-Castellón, E., *et al.*, Textural and structural properties and surface acidity characterization of mesoporous silica-zirconia molecular sieves. *J. Solid State Chem.*, 175, 2, 159–169, Nov. 2003, doi: 10.1016/S0022-4596(03)00218-4.
43. Reddy, C.R., Bhat, Y.S., Nagendrappa, G., Jai Prakash, B.S., Brønsted and Lewis acidity of modified montmorillonite clay catalysts determined by FT-IR spectroscopy. *Catal. Today*, 141, 1–2, 157–160, Mar. 2009, doi: 10.1016/j.cattod.2008.04.004.

44. Musthofa, M., Karim, A.H., Ahmad Fadzlillaah, N., Rozali Annuar, N.H., Abdul Jalil, A., Triwahyono, S., Determination of Lewis and Brönsted acid sites by gas flow-injection technique. *Mal. J. Fund. Appl. Sci.*, 6, 2, Jul. 2014, doi: 10.11113/mjfas.v6n2.195.
45. Parry, E., An infrared study of pyridine adsorbed on acidic solids. Characterization of surface acidity. *J. Catal.*, 2, 5, 371–379, Oct. 1963, doi: 10.1016/0021-9517(63)90102-7.

6

Conversion of Different Solid Waste into Heterogeneous Catalysts

Bandita Datta* **and Shreya Taran**

Department of Chemistry, Amity Institute of Applied Sciences, Amity University Kolkata, Kolkata, West Bengal, India

Abstract

The use of waste materials as a sustainable feedstock for the synthesis of platform chemicals and fuels has grown in recent years. The cost of a catalyst is an important factor for determining the overall cost of the reaction. Waste materials are useful in the context of chemical transformations because they can be used to make sustainable feedstocks and catalysts. These heterogeneous catalysts derived from waste materials help both new and established chemical processes operate more efficiently in terms of energy and atoms. This chapter is planned to offer insights focused on the identification of waste materials coming from both biological and industrial sources like red mud, fly ash, rice husk, chicken eggshells and seashells. The purpose of this chapter is to give a quick overview of the low-cost alternatives that can be successfully used as heterogeneous catalysts for various chemical reactions.

Keywords: Waste materials, chemical transformations, feedstock, heterogeneous catalysts, red mud, seashells, low-cost

6.1 Introduction

In today's world, industrialization has accelerated the speed of urbanization and has put on a significant impact on shifting the economic structure. Nonetheless, along with the increased pace of urbanization, waste generation has posed a significant challenge towards sustainable development [1].

**Corresponding author*: bdatta@kol.amity.edu

Deepti Goyal, Sakshi Kabra Malpani and Renuka Jain (eds.) Catalysis from Waste: Transforming Solid Residues into Value, (137–168) © 2026 Scrivener Publishing LLC

It has spawned an enormous quantity of waste products annually which is a serious concern for the scientists, as it is responsible for environmental contamination and unnecessary landfills. Among the various categories of waste, agricultural and industrial waste contributes majorly towards the total waste load. However, these wastes can be efficiently converted into significant resources that can be creatively recycled and repurposed with minor modifications into environmentally friendly composites with a wide range of applications. They have been converted into sustainable feedstocks and cheaper heterogeneous catalysts for manufacturing important chemicals and fuels [2].

Industrial waste like fly ash is obtained from burning coal while red mud as a byproduct of aluminum production is found to be much cheaper and readily available. Similarly, rice husk and shell ash are obtained from agricultural and animal waste respectively. Generation of industrial and agricultural waste is unavoidable, but it can be innovatively recycled to eco-friendly composites (Figure 6.1). These composites have been found to exhibit a wide range of applications in various sectors [3]. These materials are made up of important materials like alumina, silica, calcium carbonate or metal oxides that can be further fabricated to form highly valuable heterogeneous catalysts. This leads to porous structure, increased alkalinity and large surface area which is responsible for their applications not only in commercial chemical synthesis and biofuel production but also in environmental operations like wastewater treatment and pollutant degradation.

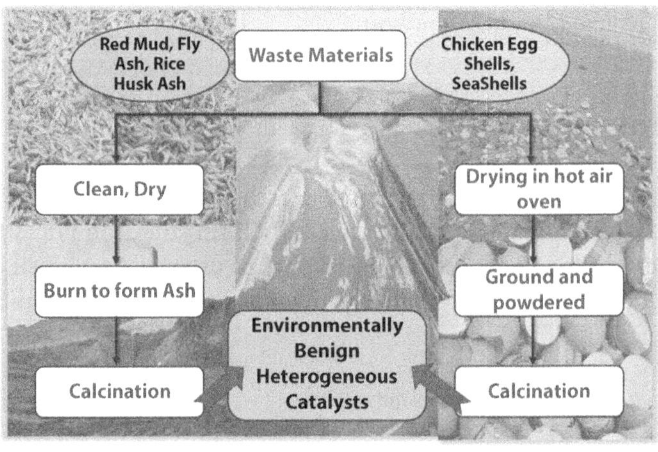

Figure 6.1 Trends in the development and utilization of waste as heterogeneous catalysts.

Overall, these innovations will help us to achieve environmental sustainability, waste management and maximizing resource efficiency [4].

In recent years, the push for sustainability has driven the exploration of waste materials as viable resources in chemical synthesis, serving as both sustainable feedstocks and cost-effective heterogeneous catalysts for producing platform chemicals and fuels. The economic feasibility of chemical reactions often hinges on catalyst costs, positioning waste-derived catalysts—sourced from biological and industrial residues such as red mud, fly ash, rice husk, chicken eggshells, and seashells—as promising alternatives [5, 6]. These materials not only reduce reliance on expensive conventional catalysts but also enhance process efficiency by improving energy and atom economy across both emerging and traditional chemical transformations.

This chapter explores the potential of these abundant byproducts, offering a detailed overview of their identification and properties. An in-depth discussion on various methods for the transformation of the waste into beneficial composites and detailed surface modifications which enhanced their catalytic activity is presented; furthermore, the various applications as heterogeneous catalysts in diverse reactions, underscoring their role in advancing greener industrial practices has been examined. Finally, inclusion of challenges and prospects gives researchers a better chance to explore more effective catalytic applications.

6.2 Red Mud

During alkaline refining of alumina, an alkaline industrial waste residue called red mud is produced. Given its complex composition, large specific surface area, and strong alkalinity, it has been challenging to find a large-scale application for red mud. Currently, 70 million tons of red mud is produced worldwide annually, and land dumping is the main method of disposal [7]. Huge attention from environmentalists has been drawn towards the plentiful resource of red mud and its rich iron content [8]. This provides a variety of intriguing catalytic behaviors in various reactions such as hydrogenation [9] and natural oil transesterification [10].

6.2.1 Properties of Red Mud

Red mud, composed of Fe_2O_3, Al_2O_3, SiO_2, TiO_2, Na_2O, CaO, etc., is also known as bauxite residue [11]. It exhibits high alkalinity due to the presence of basic oxides. It is also highly stable at high temperatures and consists of porous structure making it effective in catalysis. Due to the high surface area, it can be applied for adsorption of pollutants from wastewater and hence, is widely used in wastewater treatment.

6.2.2 Preparation and Activation of Red Mud

Red mud while calcined at high temperatures ranging 500-900°C, removes moisture and decomposes volatile impurities. This enhances its chemical stability, surface area and porosity which is responsible for its catalytic activities. Further activation of red mud ash is carried out by acid or alkali treatment which removes impurities and triggers active sites. Acid activation using HCl or H_2SO_4 makes red mud ash an acidic catalyst whereas on treatment with alkali like NaOH it gets converted into a basic catalyst. These activation processes make red mud ash highly effective for catalytic reactions and environmental applications [12].

6.2.3 Applications for Red Mud

In the research, Liu *et al.* used red mud that had been dried in air at 200°C as catalyst for soybean oil transesterification. Biodiesel yield of over 94% was recorded after three hours at 65°C when the red mud was present at a level of 4 wt.%. It was observed that the sample of red mud used was rich in calcium content when compared with typical sources from this waste stream [13]. Red mud was further used by Karimi *et al.* for the ketonization of various carboxylic acids and upgrading of pyrolysis bio-oils. 10 mole % loading of red mud and carboxylic acid were added to an autoclave that had been purged with H_2 or an inert gas before sealing and heating at 365°C for 8h. In the absence of hydrogen, a higher conversion (> 89%) as well as ketone selectivity (21–41%) were attained [10]. Other than that, hemp seed pyrolysis oil also underwent successful upgrading using 25 wt.% red mud catalyst which in turn produced less viscous and more chemically stable oil. It was also possible to make a more acidic pyrolysis oil from hardwood sawmill chips; this process resulted in increasing bio-oil heating value to 34 kJ^{-1} due to deoxygenation [14]. Sushil *et al.*, employed waste from the tannery, electroplating, and aluminum (red mud) industries as catalyst precursors. The composition and activity of red mud varied depending on the site from which it was sourced, but at 350°C, calcined red mud was active for the complete oxidation of propane under continuous flow of 1 bar air and a modest gas hourly space velocity of 5000 h^{-1}. Red mud delivered 60–90% propane conversion by 500°C. Red mud catalysts' activity was enhanced by the addition of tannery shavings, a source of Cr_2O_3, which allowed for full conversion at lower temperatures. The most active material was pure tannery shavings, but high gas flow rates couldn't be used because of their tiny particle size [15]. Evidence is available that red mud may be used as a catalyst to transform and store petroleum gas, which is a blend of methane, and other hydrocarbons discharged during offshore oil drilling. Red mud's metal oxides reacted with CH_4 to make water, CO/CO_2, and reduced metals, which then reacted to produce hydrogen. After that, the coked metal and steam interacted to produce syngas. Significant conversion was seen above 650°C during methane breakdown, which was carried out in a continuous flow reactor in a CH_4/N_2 stream between 25 and 900°C. When metal/carbon composites were oxidized under steam, more than 650°C was reached, releasing H_2 and producing CO over 800°C due to the oxidation of carbon deposits. Red mud has demonstrated potential as a catalyst for the pyrolysis of waste plastics, using as feed 40 weight% polyethylene, 35 weight% polypropylene, 18 weight% polystyrene, 4 weight% poly(ethylene terephthalate), and 3 weight% poly(vinyl chloride). Red mud produced a 98% yield of volatile hydrocarbon products at 500°C, outperforming a ZSM-5 zeolite catalyst in terms of gas yields and viscosity bio-oils. However, red mud was inert below 440°C. The additional hydrogenation and hydrocracking processes caused by the transition metal oxides (Fe_2O_3 and TiO_2) in red mud leads to different product selectivity in comparison to ZSM-5. Red mud has a comparatively smaller surface area with respect to other heterogeneous catalysts; still it seems to be a good solid acid waste catalyst [16].

Red mud is used in transesterification, ketonization, and oxidation processes. It is also used as a catalyst for methane breakdown, producing H_2 and Fe_3C as byproducts. Using 0.4 g of red mud in a continuous flow microreactor between 600 and 700°C with a 1:4 flow of N_2:CH_4, favorable hydrogen production rates of 3.80 x 10^{-5} $molg^{-1}s^{-1}$ were obtained. These rates are like those reported in the literature for high loading Fe_2O_3/Al_2O_3 catalysts [5]. A modified mesoporous red mud was employed to produce hydrogen from ammonia. Acid digestion and base re-precipitation eliminated most of the Na and Ca from the waste material, changing the red mud's crystallinity and surface characteristics. Before catalytic testing, the processed red mud was crushed, calcined, and impregnated with Ni at loadings ranging from 0 to 15 weight%. The pore volume grew from 0.08 to 0.39 cm^3g^{-1} and the surface area expanded from 8 to 225 m^2g^{-1} with calcination at 600°C, converting metal hydroxides into oxides. The interaction between Ni and Fe species in the red mud support is responsible for the high activity of ammonia decomposition. It progressed with 96% conversion over a loading of 12 weight% Ni/red mud catalyst in a continuous flow microreactor operating between 350 to 700°C [13]. In hydrodechlorination (HDC) processes, hydrogen reacts with organic molecules that have a C–Cl bond to produce HCl and a C–H bond (Equation 6.1). Red mud has been investigated as a catalyst for the catalytic detoxification of chloro hydrocarbons [17] and the hydrodechlorination of tetrachloroethylene [18].

$$R - Cl + H_2 \rightarrow R - H + HCl \tag{6.1}$$

The catalyst's viability depends on factors such as its availability; expenses associated with conversion to active product and its performance in transesterification reactions. Red mud is less active than platinum group metal catalysts like Ru/C or Pt/Al_2O_3, according to scientists [19]. Nevertheless, red mud's low cost provides value to an otherwise troublesome, alkaline waste stream, at least partially offsetting this drawback [20]. Owing to its large-scale production, red mud has been used for catalytic trans-esterification, hydro-dechlorination, and hydrogenation. It also has significant potential as an economical and abundant catalyst/reagent on Earth [21].

Although the aforementioned examples produced active oxidation catalysts, the active phase was produced using commercial chemicals in many instances, and only the inert support came from waste. Despite this, as support makes up most of the mass of heterogeneous catalysts, creating support from waste materials for oxidation catalysis it would unavoidably lower the cost of catalysts and the amount of waste that needs to be disposed of or burned. This is crucial for waste materials because their extensive storage can have negative environmental effects, as demonstrated by the Ajka red mud discharge incident in Hungary [22]. However, since the active phase of a catalyst is frequently the most expensive and rare component, it is still difficult to produce active components from waste. The ideal waste catalyst is one that is made entirely of recycled materials and requires little in the way of additives or expensive processing. This application generates low-carbon fuel while reducing emissions of harmful greenhouse gas and valorizing waste from the Bayer process.

6.2.4 Challenges and Future Prospects

The following are the challenges of red mud:

a) Red mud contains toxic metals like arsenic, lead, and chromium, making its disposal problematic.
b) Its alkaline nature (high pH) makes land disposal difficult, leading to land degradation and long-term ecological

damage. Despite its potential as a construction material, in catalysis, and other industries, large-scale applications of red mud are still limited.

c) Developing economically viable solutions to repurpose red mud remains a challenge. Due to its bulky nature, transportation and handling costs are high, which limits the feasibility of using red mud in distant industries.

Advancing technologies to reuse red mud in cement, ceramics, and metallurgical industries can turn waste into valuable products, promoting a circular economy. Red mud's metal oxide content and alkalinity offer promising potential in heterogeneous catalysis for environmental remediation, such as wastewater treatment and CO_2 capture. Ongoing research of converting red mud into high-value materials like pigments, refractory materials, and geopolymers could open new avenues for its utilization. Innovations in red mud processing to neutralize its toxic components, coupled with regulatory incentives for waste repurposing, could increase its safe, large-scale application. In the future, a combination of technological advancements, regulatory support, and increased awareness can help mitigate the challenges associated with red mud and unlock its full potential for sustainable industrial applications [23].

6.3 Fly Ash

Fly ash, a byproduct of coal combustion in power plants, has diverse and significant applications across various industries. Some of the key applications include construction industry, road construction, geotechnical applications, agriculture, manufacturing and environmental applications. Fly ash is widely used as partial replacement of Portland cement as well as eco-friendly and lightweight bricks and blocks [24]. Fly ash is further utilized in the construction of road subbase [25] and used in construction of embankments and roadways by stabilizing soil [26]. Fly ash can be used as a soil conditioner, [27] treatment of wastewater [28] production of ceramics [29] glass and synthetic gypsum [30]. Fly ash is considered now as a crucial component of sustainable development plans since it reduces greenhouse gas emissions, manages industrial waste, and protects natural resources. Interest has been generated by the possible application of fly ash as a catalyst in numerous industrial and environmental processes. Fly ash is a versatile and effective catalyst due to its unique physical and chemical properties.

6.3.1 Properties of Fly Ash

Fly ash, generated from the combustion of coal, is a fine powder composite made up of silica, alumina, iron oxide and calcium oxide. Further probe into its structure shows its spherical particle shape with high thermal stability and low permeability. The fine particles generated are found to have a large surface area making it an effective catalyst.

6.3.2 Preparations and Applications of Fly Ash in Catalysis

This comprehensive evaluation examines the various uses of fly ash as a catalyst. Fly ash is composed of silica (SiO_2), alumina (Al_2O_3), and iron oxide (Fe_2O_3), which can act as catalytic sites. These properties are exploited for the development of heterogeneous catalysts in chemical synthesis processes.

One of the utmost important catalytic applications is seen in the synthesis of zeolites. Zeolites are microporous, aluminosilicate minerals widely used as catalysts in petrochemical refining and in producing detergents. Fly ash serves as a cost-effective raw material for synthesizing zeolites. According to a study by Rattanasak *et al.*, fly ash and NaOH caused Si, Al, and other minor ions to leach and begin the process of geopolymerization. The leaching period and concentration of NaOH both affected the amount of leaching [31]. By using an alkaline hydrothermal process, fly ash was converted into zeolites with varying surface areas. Furthermore, an assessment was conducted on their ability to adsorb heavy metal ions, specifically Hg^{2+} and Pb^{2+} [32]. The newly formed zeolites display surface area and pore volume which enables them to function as efficient catalysts in cracking reactions, such as the transformation of heavy hydrocarbons into gasoline and other useful products.

Moreover, fly ash can be utilized in environmental applications to degrade pollutants through catalytic processes. Degradation of organic contaminants by photocatalysis is one of the most significant applications. Fly ash, especially when doped with titanium dioxide (TiO_2), can act as a photocatalyst for the degradation of organic pollutants in wastewater. Under UV light, TiO_2 generates electron-hole pairs that produce reactive oxygen species (ROS). These ROS oxidize and break down organic contaminants into less harmful substances. Fly ash-TiO_2 composites have been successfully used to degrade dyes like methylene blue and rhodamine B, as well as pharmaceuticals like antibiotics in wastewater treatment plants [33]. Fly ash is also found in the process to enhance carbon capture processes, contributing to the reduction of CO_2 emissions. For instance, CO_2 gets captured by sorbents based on fly ash. It is modified with amines or alkali metals which serves as an effective sorbent for CO_2 capture [34]. The porous structure of fly ash provides a large surface area for CO_2 adsorption. When modified with amines, fly ash can chemically bind CO_2, forming carbamates, which can be regenerated by heating. One of the most important applications of fly ash is transesterification of vegetable oils to form biodiesel. The process of converting vegetable oils into biodiesel involves the reaction of triglycerides with methanol in the presence of a catalyst. Fly ash is modified with calcium oxide (CaO) to enhance its basicity, which is crucial for the transesterification reaction. The CaO-fly ash catalyst facilitates the breaking of triglyceride bonds and the formation of methyl esters (biodiesel) and glycerol. Studies have shown that fly ash-based catalysts can achieve high conversion rates of oils to biodiesel, with advantages such as easy separation from the product and reusability [35]. Another application of fly ash involves production of hydrogen as a clean energy carrier. Steam reforming of hydrocarbons is a common method for hydrogen production, where catalysts play a crucial role. Fly ash is impregnated with nickel (Ni), which is an effective catalyst for steam reforming. The Ni-fly ash catalyst facilitates the reaction between hydrocarbons and steam to produce hydrogen and carbon monoxide. The Ni-fly ash catalyst has been tested for the steam reforming of methane and biomass-derived syngas, showing good catalytic activity and stability [29]. Fly ash can be employed in various oxidation reactions due to its catalytic properties like Catalytic Wet Air Oxidation (CWAO). CWAO is a process used to treat industrial wastewater containing refractory organic pollutants [36]. Fly ash is modified with transition metals such as copper (Cu) or manganese (Mn) to enhance its oxidative capabilities. The metal-modified fly ash catalyst promotes the oxidation of organic pollutants in the presence of air (oxygen) at elevated temperatures and pressures, converting them into CO_2, water, and harmless inorganic salts [37]. Fly ash-based catalysts have been used in CWAO processes to treat effluents from the textile, pharmaceutical, and petrochemical industries, achieving significant reduction in chemical oxygen demand (COD) and total organic carbon (TOC) [38]. Fly ash can be used to catalyze gasification processes, converting carbonaceous materials into syngas (a mixture of CO and H_2). Biomass gasification is a promising technology for producing renewable syngas. Fly ash is enriched with alkali metals such as potassium (K) to enhance its catalytic activity. The alkali metal-modified fly ash catalyst promotes the breakdown of biomass components (cellulose, hemicellulose, and lignin) into syngas during gasification. Gao *et al.*, reported syngas generation from biomass feedstocks like agricultural residues, wood chips and municipal solid wastes in presence of fly ash-based catalysts with high yield and improved process efficiency [39].

The Flue Gas Desulfurization (FGD) process involves removal of sulfur dioxide from flue gases released by power plants and other industrial processes. Fly ash aided FGD process has been utilized where the calcium content of fly ash gets converted to calcium sulfite, a solid by-product which precipitates out and can be easily removed. Modified fly ash has been proved to be maximum effective in removing sulfur content from the gases ejected in coal-fired power plants improving the air quality [40]. Through various catalytic reactions, another important application of fly ash is to synthesize fine chemicals like glycerol carbonate for polymer and pharmaceutical industries. Fly ash is functionalized with metal oxides like zinc oxide (ZnO) to enhance its catalytic properties. The ZnO-fly ash catalyst promotes the reaction between glycerol and urea to produce glycerol carbonate and ammonia. The use of fly ash-based catalysts in the synthesis of glycerol carbonate has been demonstrated to be efficient, offering a sustainable route for converting glycerol, a byproduct of biodiesel production, into a high-value chemical [41]. Fly ash has also proved to be beneficial in the field of agriculture. The chemical, physical and biological properties of the soil have been modified enhancing the quality of the crops. This has also taken care of the disposal problem to some accounts [42].

6.3.3 Challenges and Future Prospects

The following are the challenges of fly ash:

a) Fly ash contains heavy metals like arsenic, lead, and mercury, which can leach into groundwater or enter the atmosphere as fine particulates, posing serious environmental and health risks.
b) Large quantities of fly ash are generated globally, and only a portion is utilized in beneficial applications. Disposing of unutilized fly ash in landfills or ponds can cause land degradation and water contamination.
c) The coal-fired power plants that produce fly ash are significant contributors to CO_2 emissions, raising concerns about the carbon footprint of utilizing this byproduct.
d) The chemical and physical properties of fly ash can vary depending on the source of the coal and combustion

conditions, which may affect its suitability for specific applications.

e) As many countries transition to renewable energy, the availability of fly ash may decline. The reduction in coal-based power generation poses challenges for industries dependent on a steady supply of fly ash.

The prospects of fly ash include advances in material science. These new applications for fly ash in sustainable construction are likely to emerge. There is growing interest in using fly ash in combination with technologies that capture and store CO_2, helping to reduce the carbon footprint of concrete production. Innovative processing techniques, such as selective beneficiation, can improve the quality of fly ash, making it suitable for higher-value applications like high-performance concrete and geopolymers. Researchers are exploring the potential of incorporating nanomaterials into fly ash-based products to enhance their strength, durability, and other performance characteristics [34]. Further studies are required to explore the adsorption studies of heavy metals and other contaminants from industrial wastewater. This will extend its application in environmental remediation.

In summary, fly ash has been applied in various disciplines. The activity of fly ash can be enhanced by functionalization and has been used in chemical synthesis, environmental remediation, desulfurization, biodiesel production and many more. The modified fly ash has shown versatility and potential for heterogeneous catalysts adhering to the principles of green chemistry.

6.4 Rice Husk

Rice being one of the most important staple foods for India, produces large quantities of rice husk as a byproduct. Rice husks make up to 20% of the total amount of rice produced, and they don't decompose easily [43]. As a result, they cannot be used for animal feed and their disposal becomes a problem. It is made up of cellulose and lignin as well as silica, alkalis etc. [44]. Rice husk ash (RHA) is produced on the combustion of rice husks. RHA was found to possess unique properties like porosity, high surface area and chemical stability [45]. These properties have not only helped the researchers to utilize RHA as heterogeneous catalysts aligning with green chemistry but also resolved the disposal problem and promoted sustainable chemical processes. The silica content present on RHA has been widely used in drug administration [46], diagnosis [47], and other biological applications [48]. This part of the chapter demonstrates the synthesis, activation and potential of catalytic processes of RHA.

6.4.1 Properties of Rice Husk Ash

The reported composition of RHA states that it consists of 85-90% silica by mass and the remaining mass contains calcium oxides, potassium oxides, carbon, and other oxides [49]. Since RHA is rich in silica content, it has proved to be valuable for catalysis. Surface modification has led to an advanced form of silica which has exhibited excellent activity. Moreover, RHA is formed on combustion with high surface area which are prerequisites for a good heterogeneous catalyst [50]. Various synthetic methods have been developed to produce RHA with varying surface area, 50 to 300 m^2 per gram. RHA also shows good porosity which gives an opportunity for the reactants to bind to the catalytic sites, making it a strong catalytic species [51]. Furthermore, it has been found that RHA is inert, contains a stable structure. It can maintain its structure even in harsh conditions like high temperatures and corrosive environments. The presence of high silica content in RHA is responsible for thermal stability and resistance to sintering. Overall, RHA has proved to be a potential catalyst due to its low-cost, thermal stability, large surface area, reducing waste and promoting recycling of agricultural wastes. These properties make RHA versatile and effective support material for a wide range of catalytic applications.

6.4.2 Preparation and Activation of Rice Husk Ash

The RHA is washed and dried to remove impurities like dirt, sand, and organic matter. Pretreatment steps involve size reduction and sieving [44]. The most common method of producing RHA is through the controlled combustion of rice husks. The combustion process can be carried out in open-air pits, industrial furnaces, or fluidized bed reactors. The conditions under which combustion occurs, such as temperature, time, and atmosphere—significantly influence the properties of the resulting ash (Figure 6.2). The three categories of formation of RHA by combustion are: i) Open-Air combustion which is the simplest and least costly but often results in a heterogeneous product with varying degrees of carbon content; ii) Furnace combustion carried out in furnaces allows better control over temperature and oxygen levels, leading to a more consistent product with higher silica purity and iii) Fluidized bed combustion, an advanced technique which

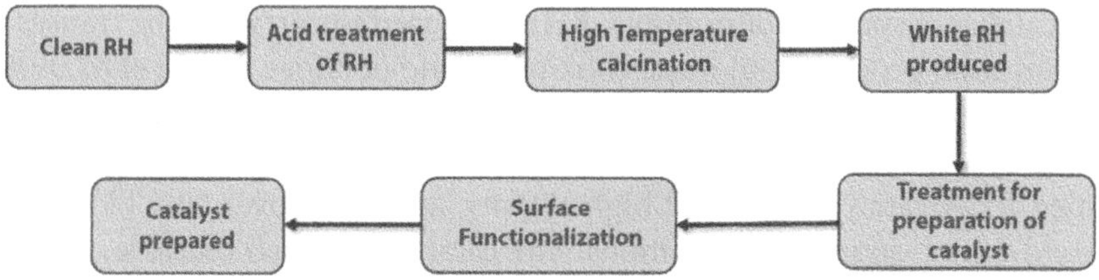

Figure 6.2 Preparation and activation of RHA for application as catalyst.

offers superior control over combustion conditions, resulting in highly uniform and pure RHA with optimized physical properties [52].

To enhance the catalytic properties of RHA, various activation techniques can be employed like chemical, thermal and impregnation with catalytic species. Chemical activation involves treatment with acids (e.g., HCl, H_2SO_4) or bases (e.g., NaOH) which removes impurities, increases surface area, and introduces active sites on the RHA surface [53]. Calcination at high temperatures is carried out for thermal activation. This increases the surface area and modifies the porosity of RHA, making it more suitable for specific catalytic applications. Another method is impregnating active metals such as platinum, palladium, or nickel onto the RHA surface, transforming it into an active catalyst for specific reactions. These activation techniques are critical for tailoring RHA to meet the demands of various catalytic processes.

6.4.3 Applications of Rice Husk Ash in Catalysis

RHA can be functionalized to act as either an acidic or basic catalyst. For acidic catalysis, RHA can be treated with sulfuric acid to introduce sulfonic acid groups on its surface, making it effective in catalyzing esterification and transesterification reactions [54]. Similarly, treatment with alkali metals can render RHA a basic catalyst, suitable for reactions like Friedel Craft's reactions, benzoylation etc. The catalytic activity of iron-incorporated rice husk ash silica for the benzylation of toluene was reported by Adam *et al.* [55]. One of the most significant applications of RHA is in the production of biodiesel, a renewable and environmentally friendly fuel. RHA-based catalysts, especially those modified with potassium hydroxide (KOH) or sodium hydroxide (NaOH), have been shown to be highly effective in catalyzing the transesterification of vegetable oils into biodiesel [56]. The use of RHA in this context not only reduces the cost of biodiesel production but also minimizes the environmental impact of catalyst disposal. For example, RHA was used to carry out the transesterification process from waste cooking oil and obtained a yield of 83.09% [57]. In another study, RHA was modified with potassium hydroxide and used as a catalyst for the transesterification of jatropha oil [58]. The modified RHA catalyst exhibited high activity, achieving over 90% biodiesel yield under optimized conditions. The catalyst was also reusable, maintaining its activity over multiple cycles, which demonstrates its potential for commercial biodiesel production.

RHA has been further employed as a catalyst in various organic synthesis reactions. For instance, it has been used directly to catalyze the synthesis of 12-Aryl-8, 9, 10, 12-tetrahydro [a] xanthene-11-ones and quinoxaline derivatives [59]. At the same time $FeCl_2{\cdot}2H_2O$ can be impregnated on RHA and used as a catalyst for synthesis of quinolines [60]. The ability to tailor the surface properties of RHA through chemical modification makes it a versatile catalyst for a wide range of organic reactions. Additionally, the photocatalytic applications were extensively investigated. On doping RHA with metal oxides, for example titanium dioxide, the modified RHA can be utilized for photocatalytic degradation of organic pollutants under UV light [61]. This has proved to be crucial for industrial wastewater treatment and the abstraction of persistent organic pollutants.

6.4.4 Challenges and Future Prospects

Although RHA has proved to be quite efficient as heterogeneous catalysis, the following are the challenges involved:

a) **Inconsistency in Ash Quality**: The catalytic activity of RHA varies depending on the composition and characteristics. The properties keep varying based on source and conditions of combustion.
b) **Scale-Up Issues**: Due to different compositions and issues with reproducibility, they cannot be productively scaled-up for commercial uses. Moreover, the modification methods need to be optimized at a larger scale.
c) **Limited Commercialization**: Inconsistency in the catalyst regeneration and their stability is a bottleneck for its use on an industrial scale.

There is still a long way to go before RHA-based catalysts are widely used in industry. This is partly because large-scale production facilities are scarce, and catalytic properties still need to be further optimized.

Future research on using RHA as a heterogeneous catalyst could focus on developing validated methodologies for the production of high-quality RHA with consistent properties. Surface modifications can lead to enhancement of catalytic activity and selectivity of RHA. Nanomaterial composites can be synthesized to produce hybrid catalysts with superior performance. Rice husk ash, a readily available and low-cost byproduct of rice production, has emerged as a promising material for heterogeneous catalysis. Its high silica content, large surface area, and chemical stability make it an ideal candidate for various catalytic applications, from biodiesel production to environmental remediation. While challenges remain in the standardization and commercialization of RHA-based catalysts, ongoing research continues to unlock new possibilities for this sustainable material [62]. As industries increasingly prioritize green and cost-effective solutions, RHA's role as a heterogeneous catalyst is likely to expand, contributing to more sustainable and efficient chemical processes.

6.5 Chicken Eggshell Ash

The disposal of chicken eggshells, a common byproduct of the poultry and food industries, presents significant environmental challenges due to their non-biodegradable nature. However, these shells, primarily composed of calcium carbonate ($CaCO_3$), offer valuable opportunities for sustainable applications. One promising use is as a heterogeneous catalyst in various chemical processes [63]. Utilizing chicken eggshell ash (CESA) as a catalyst aligns with the principles of green chemistry by converting waste into a valuable resource, thus promoting sustainability. We discuss the properties, preparation, and applications of CESA as a heterogeneous catalyst, highlighting its potential in various industrial processes.

6.5.1 Properties of Chicken Eggshell Ash

Chicken eggshells are majorly (94-97%) made up of calcium carbonate. The remaining consists of other inorganic components like magnesium carbonate, calcium phosphate, and some organic proteins. On calcination of chicken eggshells, we obtain calcium oxide forming the chicken eggshell ash (CESA) [64].

Like RHA, CESA shows excellent catalytic applications due to significant surface area and high porous structure. The calcination process of conversion of calcium carbonate into calcium oxide is performed at temperature 800-900°C and behaves as alkaline material. Laca *et al.*, demonstrated the usage of this alkalinity in industrial processes like transesterification and aldol condensation [65]. The ash also exhibits high thermal stability which has helped the researchers to use it as a catalyst under high temperature conditions [66]. Moreover, the activity of CESA can be elevated through chemical modifications by doping with transition metal ions or other metal ions.

The following are the advantages involved in the conversion of chicken eggshell (CES) waste into a valuable material:

a) Huge production of CES as waste makes it cheaper to be used as a raw material.
b) On calcination, CESA consists mainly of CaO which acts as a basic catalyst required for many chemical reactions.
c) High thermal stability makes CESA a suitable choice of catalyst for high-temperature catalytic processes.
d) Reduction in environmental pollution by decreasing the poultry industry by-product, eggshell waste and converting them into useful materials.

6.5.2 Preparation and Activation of Chicken Eggshell Ash

The preparation of CESA begins with the collection of eggshells, typically from food processing plants, restaurants, or households. The shells are first cleaned to remove any residual egg white or membrane, which could interfere with the calcination process. This is usually done by washing the shells with water, followed by drying. The production of CESA involves the calcination of cleaned and dried eggshells. The dried eggshells are calcined in a furnace with temperature maintained between 800–900°C decomposing calcium carbonate which produces CaO and carbon dioxide [67]. As described earlier the change in temperature and the duration of calcination can yield the ash with varying properties. For example, Sethupathi *et al.*, carried out calcination at different temperatures (800–1100°C) and studied the adsorption of hydrogen sulfide and sulfur dioxide. At 900°C, the adsorption was maximum [68]. Similarly, Navajas *et al.*, carried out modifications in the calcination environment by altering the atmosphere (e.g., using a CO_2-rich environment) and adding catalytic promoters. These modifications resulted in modified surface properties which enhanced the catalytic activities [69].

Further innovations were carried out by introducing several activation techniques to improve the catalytic performance of CESA. Reports are available where the CESA is treated with acids like HCl or with bases like NaOH which helps in removal of impurities and increasing the active sites. Consequently, enhancing the catalytic activity of the catalyst [70]. Moreover, varying the calcinating temperature enhances the porosity and surface area making it more suitable for specific catalytic applications. Awogbemi *et al.*, reported incorporation of catalytic promoters like iron, copper or nickel creates additional active sites showcasing better catalytic activity, especially in oxidation and reduction reactions [67]. Such activation techniques are critical steps for optimizing CESA's performance in different catalytic processes.

6.5.3 Applications of Chicken Eggshell Ash in Catalysis

As discussed earlier, the presence of CaO in CESA makes it an excellent basic catalyst for various catalytic processes. Faridi *et al.*, was able to successfully carry out transesterification of triglycerides to produce biodiesel by utilizing basic CESA. This showcased its application as a renewable and eco-friendly alternative to fossil fuels [71]. Additionally, CESA can be modified to introduce acidic sites, expanding its utility to acid-catalyzed reactions [72]. For example, doping CESA with sulfur or other acidic promoters can enable it to catalyze esterification reactions efficiently [73]. Another well-known application of CESA is the industrial process of transesterification of vegetable oils or animal fats with methanol for production of biodiesel. Catalytic sites help triglycerides to bind to it and undergo the formation of fatty acid methyl esters (FAME) which is the basic component for biodiesel. Surface modifications of CESA were carried out by treating with potassium hydroxide. K^+ ions impregnated CESA was used for production of FAME with a conversion factor of 98.36% from waste soyabean oil [74]. From the above examples we can conclude that CESA has proved to be an important alternative for costly catalyst. Furthermore, it has also taken care of disposal of poultry waste and utilized the waste oils for biodiesel production. It has also been proved that CESA shows chemical stability, reusability and can be reused for at least 8 cycles for production of biodiesel.

The applications of CESA as basic catalysts have also been extended into various organic synthesis. Chayana *et al.*, have carried out aldol condensation forming a new C-C bond with the help of CESA as a basic catalyst [75]. It has also been used for the synthesis of chalcones which are important precursors for crucial organic materials and pharmaceutics [76]. Furthermore, Kargar *et al.*, utilized CESA while carrying out synthesis of pyridine derivatives which were carried forward for the total synthesis of other drugs [77]. They proved that the above synthesis was carried out under mild conditions producing high yields with excellent selectivity. These applications of CESA stated the versatility of CESA in various complex organic reactions as cost-effective and efficient catalyst.

CESA can be functionalized by deposition of nanoparticles like CuO making CESA a bifunctional catalyst. Now it contains both acidic and basic active sites thereby increasing its applications as heterogeneous catalysts. For instance, it has been efficiently used for carrying out Knoevenagel condensations, epoxidation reactions and Michael addition reactions [78]. In summary, CESA's versatility in organic synthesis demonstrates its potential as a catalyst that can perform multiple functions. Another most important application of CESA is in the environmental field, especially for wastewater treatment. The ash shows good porosity which can be modified by the calcination process. This porosity can be used for trapping different toxic materials like metal ions or gases. Meng *et al.*, explained in their review how CESA was able to remove lead, cadmium and other toxic metals from contaminated samples of water [79]. Additionally, this ash can be employed for degradation of organic pollutants in wastewater *via* advanced oxidation processes. CESA was modified by doping with transition metal ions like iron or copper. The catalyst developed took part in Fenton's reaction or other oxidation processes to break down many harmful organic compounds into biodegradable substances [80].

6.5.4 Challenges and Future Prospects

Even though CESA has many promising applications, still there stand some obstacles that need to be dealt with:

a) The catalytic activity depends primarily on composition which may not be consistent due to different sources.
b) Another challenge is scaling up. At the laboratory scale, it is very successful but in industrial processes its consistency is yet to be maintained.
c) Although this catalyst can be used multiple times still, its activity gradually decreases. Further research is needed to increase its reusability.

Consequently, more research is desirable to develop standard protocols for production of high quality CESA with consistent properties. More surface modifications and doping with other metal ions must be carried out to increase the application horizon for chicken eggshell waste. Further investigation needs to be carried out to combine CESA with nanomaterials to create novel hybrid catalysts with improved performance.

Chicken eggshell ash, a byproduct of the poultry industry, has emerged as a promising material for heterogeneous catalysis. The availability, low cost, and high calcium oxide content of CESA is a great option for a variety of catalytic applications, such as environmental cleanup and the manufacture of biodiesel. Even though CESA-based catalyst standardization and commercialization still face obstacles, continuous research is opening new opportunities for this sustainable material [81]. CESA's function as a heterogeneous catalyst is anticipated to grow as sectors prioritize environmentally friendly and economically viable solutions increasingly, leading to development of more sustainable and effective chemical processes.

6.6 Seashells as Heterogeneous Catalysts

The global food industry generates significant amounts of shellfish waste, including seashells from mollusks like oysters, clams, and mussels. Traditionally, these seashells have been disposed of in landfills, contributing to environmental pollution and waste management challenges [82]. However, seashells, primarily composed of calcium carbonate ($CaCO_3$), present a valuable opportunity for repurposing as heterogeneous catalysts in various industrial processes [83].

Heterogeneous catalysis is essential in many industrial applications, ranging from petrochemical refining to environmental remediation [84]. Since they are in different phases they interact with the reactants at the interface and the products once formed are released from the active sites. Seashells like previously discussed solid waste can be modified into cheaper catalysts following green chemistry protocols. In this chapter we will discuss the properties, modification protocols and applications as catalysts highlighting their potential in various chemical and environmental processes.

6.6.1 Properties of Seashells

Like other shells, seashells are also primarily composed of calcium carbonate with 95-98% of whole mass [85]. The remaining mass consists of organic matrixes like various proteins and polysaccharides. These materials contribute towards the structural integrity of a shell. Seashells have been found to exist in two major polymorphic forms, i.e. aragonite and calcite. Among them calcite being the most stable form whereas aragonite is less stable and often gets transformed into calcite if stored for long-time [86]. Detailed studies of the structure of seashell revealed that seashell possesses a layered structure containing organic and inorganic layers arranged alternatively. This layered structure seems to be responsible for its strength and toughness [87].

Seashells are made up of calcium carbonate, hence on calcination it decomposes to calcium oxide which is basic in nature. On calcination, the basic sites are activated which increases the potential of seashells as heterogeneous catalysts [88]. The calcination processes enhance thermal stability and high surface area, making seashell ash (SSA) suitable for application in high temperature catalytic processes. Along with CaO, SSA contains oxides of other elements like magnesium, strontium and silica which further influences the catalytic properties further [89].

The following are the advantages of SSA over other commercial catalysts.

a) SSA is cheaper compared to other available catalysts since it is produced from seashells, which is a byproduct of the seafood industry.
b) SSA contains basic sites as it is mainly made up of calcium oxide. This makes it a sustainable basic catalyst.
c) SSA, due to its thermal stability, can withstand high temperature conditions in reactions.
d) Moreover, recycling of seashell waste helps in reducing environmental pollution by practicing waste management.

The above properties make SSA an attractive and sustainable option for heterogeneous catalysis.

6.6.2 Preparation and Activation of Seashell Ash

The seashells are collected from seafood processing plants and coastal areas. The seashells collected are washed and cleaned with water to remove any contaminants and residual organic materials present in them. These contaminants may interfere in the calcination process. The shells are then dried at low temperatures to remove moisture. The dried seashells are calcined at high temperatures between 800°C and 900°C where calcium carbonate decomposes into calcium oxide and carbon dioxide. The seashell ash (SSA) hence formed is primarily composed of calcium oxide which serves as the catalytic material [90].

The catalytic activity of SSA can be improved by varying the calcination process and activation techniques. The most common method is chemical activation. SSA is treated with acids like HCl or bases like NaOH, which removes impurities and activates active sites. As a result, the catalytic activity gets enhanced. Secondly, in some cases, thermal activation is applied by further continuing the calcination process to increase the surface area and porosity of SSA. This methodology improves the catalytic behavior of SSA. Furthermore, SSA forms can be doped with metal ions like iron, copper and nickel into the catalyst matrix. This can create additional active sites for better performance of the catalyst, especially as oxidant and reductant [91]. The above methods are crucial for optimizing catalytic performance in various industrial processes.

6.6.3 Applications of Seashell Ash in Catalysis

SSA is a well-known heterogeneous catalyst behaving as basic in nature due to its composition. In all those reactions where a basic environment is required, SSA can be efficiently employed. Jindapon *et al.*, applied this basic catalyst for carrying out transesterification of triglycerides for production of biodiesel [92]. Moreover, the calcium oxide in SSA efficiently catalyzes the synthesis of FAME from oils and fats. Nahas *et al.*, were able to convert sunflower oil and waste cooking oils into biodiesel using scallop SSA under calcination at different temperatures ranging from 600–1000°C [93]. The aim of utilizing SSA for catalysis in industrial processes has led to the generation of biodiesel from renewable sources and sustainable use of waste materials. Several reports are available in literature stating high conversion factors in transesterification reactions along with reusability of SSA [94]. Like CESA, SSA can also be converted into a bifunctional catalyst by modification of active sites. The bifunctional catalyst can be used as both acidic and basic catalyst [95]. This bifunctional catalyst has been employed for carrying out organic reactions like Aldol condensation, esterification, Michael addition reactions etc. For instance, sulfur doped SSA has proved to be an efficient catalyst for esterification processes. The treatment of SSA with other acidic promoters also enhances its efficiency for various organic synthesis [96].

Another important application of modified SSA as an environment-friendly catalyst is in the treatment of wastewater. The porosity of this catalyst helps in adsorption of toxic metal ions and organic pollutants. Popović *et al.*, have reported the application of SSA in environmental remediation for removal of lead, cadmium and arsenic from industrial wastewater [82]. Additionally, SSA finds its applications in advanced oxidation processes for the degradation of organic pollutants. Toxic and harmful organic compounds can be efficiently reduced to less toxic or biodegradable materials by utilizing iron or copper doped SSA through Fenton's oxidation process [97]. There are various applications of SSA in different organic transformations. Xeng *et al.*, demonstrated the formation of C-C bond formation using SSA *via* Knoevenagel condensation [98]. It can also be used as a basic catalyst for synthesis of bioactive compounds and pharmaceutical intermediates [99]. Surface modifications and doping methods have proved to be advantageous for SSA by increasing its catalytic activities towards complex organic synthesis reactions, making it a valuable tool for researchers and industrial chemists.

6.6.4 Challenges and Future Prospects

Below are some of the challenges faced by the science community for applications of SSA as catalyst:

a) Seashells obtained from different species have different compositions which affect the consistency in SSA catalytic activity.
b) Although SSA has been proved to be chemist-friendly in laboratories, scaling up in industrial level with similar consistencies remains a challenge.
c) The reusability of SSA has been proved to be applicable for a few cycles. In larger scales multiple cycles led to reduced catalytic activity. More research is needed to enhance durability and reusability.

Standardized procedures are required to be formulated for consistent efficiency of SSA as heterogeneous catalyst. Novel chemical modifications and doping strategies are required to increase the selectivity and specificity of the catalyst. Association with nanomaterials to create hybrid nanocomposites needs to be developed for improved catalytic performance.

In summary, we can state that SSA can be produced by thermal or calcination methods with highly active sites. The catalyst due to the presence of calcium oxide acts as a basic catalyst and being in powder form can be used as a heterogeneous catalyst. However, its consistency has been a bottleneck due to different compositions of seashells. Various other doping and surface modifications need to be carried out to convert SSA into a sustainable material. More research must be undertaken into the commercialization of SSA as a user-friendly catalyst [73]. The usage of SSA as a heterogeneous catalyst is anticipated to grow as companies place a greater emphasis on economic and environmentally friendly solutions, resulting in more sustainable and effective chemical processes.

6.7 Conclusion

Red mud, fly ash, rice husk ash, seashell and chicken eggshell ash have proved to be an efficient heterogeneous catalyst. Applications of these catalysts demonstrate an innovative approach towards sustainable chemistry. They are generally composed of metal oxides and serve as cost-effective and eco-friendly catalysts. They are derived from industrial and agricultural waste, which are abundantly available. Surface modifications and doping with various other substances lead to the formation of a new generation of highly activated catalysts which complement conventional catalysts in various chemical reactions. One of the major applications for catalysts in these categories is in transesterification reactions. Also, various complex organic molecules and pharmaceutical precursors are efficiently synthesized in the presence of the catalysts by forming new C-C and C-N bonds. Red mud is rich in iron oxide; fly ash and rice husk ash are rich in silica content whereas chicken eggshell and seashell majorly consist of calcium oxides. Hence, all these catalysts behave as basic catalysts. With the surface functionalization these basic catalysts can be further converted into acidic catalysts making them a user-friendly catalyst. These materials have high porosity and numerous active sites on the surface with enhanced surface area. The applications of these catalysts in industrial processes have reduced reliance on expensive catalysts and addressed waste management issues. Overall, abundantly available waste materials can be converted into potential materials for different chemical processes, supporting green technologies. Further research is needed to reduce variations in size, shape, moisture content, and surface characteristics along with increased selectivity, reusability, efficiencies, and consistencies to increase the potential for sustainable industrial practices.

AI disclosure statement: The author acknowledges the use of AI-based tools for minor language editing or grammatical correction. However, all intellectual contributions, analyses, and conclusions presented in this chapter are entirely the author's own.

References

1. Voumik, L.C. and Sultana, T., Impact of urbanization, industrialization, electrification and renewable energy on the environment in BRICS: fresh evidence from novel CS-ARDL model. *Heliyon*, 8, 11, e11457, 2022.
2. Nath, B., Basumatary, B., Wary, N., Basumatary, U.R., Basumatary, J., Rokhum, S.L., Azam, M., Min, K., Basumatary, S., Agricultural Waste-Based Heterogeneous Catalyst for the Production of Biodiesel: A Ranking Study *via* the VIKOR Method. *Int. J. Energy Res.*, 7208754, 23, 2023.
3. Delzell, E., Wood dust and formaldehyde, in: *IARC Working Group on the Evaluation of Carcinogenic Risks to Humans*, vol. 52, IARC Publishers, France, 1995.
4. Khan, H.M., Iqbal, T., Yasin, S., Ali, C.H., Abbas, M.M., Jamil, M.A., Hussain, A., Soudagar, M.E.M., Rahman, M.M., Application of Agricultural Waste as Heterogeneous Catalysts for Biodiesel Production. *Catalysts*, 11, 1215, 2021.
5. Balakrishnan, M., Batra, V.S., Hargreaves, J.S.J., Monaghan, A., Pulford, I.D., Rico, J.L., Sushil, S., Hydrogen production from methane in the presence of red mud -making mud magnetic. *Green Chem.*, 11, 42–47, 2009.
6. Balakrishnan, M., Batra, V.S., Hargreaves, J.S.J., Pulford, I.D., Waste materials - Catalytic opportunities: An overview of the application of large scale waste materials as resources for catalytic applications. *Green Chem.*, 13, 16–24, 2011.
7. Sushil, S. and Batra, V.S., Catalytic applications of red mud, an aluminium industry waste: A review. *Appl. Catal., B*, 81, 64–77, 2008.
8. Pontikes, Y., Rathossi, C., Nikolopulos, P., Angelopoulos, G.N., Jayaseelan, D., Lee, W.E., Effect of firing temperature and atmosphere on sintering of ceramics made from Bayer process bauxite residue. *Ceram. Int.*, 35, 401–407, 2009.
9. Wang, S.B., Ang, H.M., Tade, M.O., Novel applications of red mud as coagulant, adsorbent and catalyst for environmentally benign processes. *Chemosphere*, 72, 1621–1635, 2008.
10. Karimi, E., Teixeira, I.F., Ribeiro, L.P., Gomez, A., Lago, R.M., Penner, G., Kycia, S.W., Schlaf, M., Ketonization and deoxygenation of alkanoic acids and conversion of levulinic acid to hydrocarbons using a Red Mud bauxite mining waste as the catalyst. *Catal. Today*, 190, 73–88, 2012.
11. Wahyudi, A., Kurniawan, W., Hinode, H., Utilization of Modified Red Mud as a Heterogeneous Base Catalyst for Transesterification of Canola Oil. *J. Chem. Eng. Jpn.*, 50, 561–567, 2017.
12. Pratt, K.C. and Christoverson, V., Hydrogenation of a model hydrogen-donor system using activated red mud catalyst. *Fuel*, 61, 460–462, 1982.
13. Liu, Q., Xin, R., Li, C., Xu, C., Yang, J., Application of red mud as a basic catalyst for biodiesel production. *J. Environ. Sci.*, 25, 823–829, 2013.

14. Karimi, E., Briens, C., Berruti, F., Moloodi, S., Tzanetakis, T., Thomson, M.J., Schlaf, M., Red mud as a catalyst for the upgrading of hemp-seed pyrolysis bio-oil. *Energy Fuels*, 24, 12, 6586–6600, 2010.
15. Sushil, S., Scholz, P., Pollok, K., Ondruschka, B., Batra, V.S., Application of industrial waste based catalysts for total oxidation of propane. *Chem. Eng. J.*, 166, 568–578, 2011.
16. López, A., de Marco, I., Caballero, B.M., Laresgoiti, M.F., Adrados, A., Aranzabal, A., Catalytic pyrolysis of plastic wastes with two different types of catalysts: ZSM-5 zeolite and Red Mud. *Appl. Catal., B*, 104, 211–219, 2011.
17. Cao, J.-L., Yan, Z.-L., Deng, Q.-F., Wang, Y., Yuan, Z.-Y., Sun, G., Jia, T.-K., Wang, X.-D., Bala, H., Zhang, Z.-Y., Mesoporous modified-red-mud supported Ni catalysts for ammonia decomposition to hydrogen. *Int. J. Hydrogen Energy*, 39, 5747–5755, 2014.
18. Halász, J., Hodos, M., Hannus, I., Tasi, G., Kiricsi, I., Catalytic detoxification of C2-chlorohydrocarbons over iron-containing oxide and zeolite catalysts. *Colloids Surf., A*, 265, 171–177, 2005.
19. Ho, P.H., Woo, J.-W., Ilmasani, R.F., Han, J., Olsson, L., The role of Pd–Pt Interactions in the Oxidation and Sulfur Resistance of Bimetallic Pd–Pt/γ-Al2O3 Diesel Oxidation Catalysts. *Ind. Eng. Chem. Res.*, 60, 18, 6596–6612, 2021.
20. Jollet, V., Gissane, C., Schlaf, M., Optimization of the neutralization of Red Mud by pyrolysis bio-oil using a design of experiments approach. *Energy Environ. Sci.*, 7, 1125–1133, 2014.
21. Ordóñez, S., Sastre, H., Díez, F.V., Catalytic hydrodechlorination of tetrachloroethylene over red mud. *J. Hazard. Mater.*, 81, 103–114, 2001.
22. Anton, A.D., Klebercz, O., Magyar, A., Burke, I.T., Jarvis, A.P., Gruiz, K., Mayes, W.M., Geochemical recovery of the Torna–Marcal river system after the Ajka red mud spill, Hungary. *Environ. Sci. Processes Impacts*, 16, 2677–2685, 2014.
23. Wang, S., Jin, H., Denga, Y., Xiao, Y., Comprehensive utilization status of red mud in China: A critical review. *J. Cleaner Prod.*, 289, 125136, 2021.
24. Pranav, S., Aggarwal, S., Yang, E.-H., Sarkar, A.K., Singh, A.P., Lahoti, M., Alternative materials for wearing course of concrete pavements: A critical review. *Constr. Build. Mater.*, 236, 117609, 2020.
25. Sahu, V., Misra, A.K., Gayathri, V., Fly ash in the construction industry, in: *Handbook of Fly Ash*, K.K. Kar (Ed.), pp. 657–680, Elsevier Inc., 2022.
26. Behnood, A., Soil and clay stabilization with calcium- and non-calcium-based additives: A state-of-the-art review of challenges, approaches and techniques. *Transp. Geotech.*, 17, 14–32, 2018.
27. Kalra, N., Jain, M.C., Joshi, H.C., Choudhary, R., Harit, R.C., Vatsa, B.K., Sharma, S.K., Kumar, V., Flyash as a soil conditioner and fertilizer. *Bioresour. Technol.*, 64, 3, 163–167, 1998.

28. Gadore, V. and Ahmaruzzaman, M., Tailored fly ash materials: A recent progress of their properties and applications for remediation of organic and inorganic contaminants from water. *J. Water Process Eng.*, 41, 101910, 2021.
29. Zimmer, A. and Bergmann, C.P., Fly ash of mineral coal as ceramic tiles raw material. *Waste Manage. (Oxford)*, 27, 1, 59–68, 2007.
30. Mareya, M., Tchadjie, L., Sithole, T., Turning fly ash and waste gypsum into a resource for backfilling applications. *Case Stud. Constr. Mater.*, 20, e02703, 2024.
31. Rattanasak, U. and Chindaprasirt, P., Influence of NaOH solution on the synthesis of fly ash geopolymer. *Miner. Eng.*, 22, 12, 1073–1078, 2009.
32. Kobayashi, Y., Ogata, F., Nakamura, T., Kawasaki, N., Synthesis of novel zeolites produced from fly ash by hydrothermal treatment in alkaline solution and its evaluation as an adsorbent for heavy metal removal. *J. Environ. Chem. Eng.*, 8, 2, 103687, 2020.
33. Visa, M., Cosnita, M., Moldovan, M., Marin, C.A., Mihaly, M., Fly Ash Waste Recycling by Pt/TiO_2 Incorporation for Industrial Dye Removal. *Int. J. Environ. Res. Public Health*, 18, 8, 3887, 2021.
34. Dindi, A., Quang, D.V., Vega, L.F., Nashef, E., Abu-Zahra, M.R.M., Applications of fly ash for CO_2 capture, utilization, and storage. *J. CO2 Util.*, 29, 82–102, 2019.
35. Ho, W.W.S., Ng, H.K., Gan, S., Tan, S.H., Evaluation of palm oil mill fly ash supported calcium oxide as a heterogeneous base catalyst in biodiesel synthesis from crude palm oil. *Energy Convers. Manage.*, 88, 1167–1178, 2014.
36. Gao, K., Sahraei, O.A., Iliuta, M.C., Development of residue coal fly ash supported nickel catalyst for H_2 production *via* glycerol steam reforming. *Appl. Catal., B*, 291, 119958, 2021.
37. Jing, G., Luan, M., Chen, T., Progress of catalytic wet air oxidation technology. *Arabian J. Chem.*, 9, S1208–S1213, 2016.
38. Wang, N., Zhao, Q., Zhang, A., Catalytic oxidation of organic pollutants in wastewater *via* a Fenton-like process under the catalysis of HNO3-modified coal fly ash. *RSC Adv.*, 7, 27619–27628, 2017.
39. Nachiappan, S. and Gopinath, K.P., Treatment of pharmaceutical effluent using novel heterogeneous fly ash activated persulfate system. *J. Environ. Chem. Eng.*, 3, 3, 2229–2235, 2015.
40. Navarrete, I., Vargas, F., Martinez, P., Paul, A., Lopez, M., Flue gas desulfurization (FGD) fly ash as a sustainable, safe alternative for cement-based materials. *J. Cleaner Prod.*, 283, 124646, 2021.
41. Arora, S., Gosu, V., Subbaramaiah, V., Hameed, B.H., Lithium loaded coal fly ash as sustainable and effective catalyst for the synthesis of glycerol carbonate from glycerol. *J. Environ. Chem. Eng.*, 9, 5, 105999, 2021.
42. Jala, S. and Goyal, D., Fly ash as a soil ameliorant for improving crop production—a review. *Bioresour. Technol.*, 97, 9, 1136–1147, 2006.

43. Amutha, K., Ravibaskar, R., Sivakumar, G., Extraction, Synthesis and Characterization of Nanosilica from Rice Husk Ash. *Int. J. Nanotechnol. Appl.*, 4, 61–66, 2010.
44. Nzereogu, P.U., Omah, A.D., Ezema, F.I., Iwuoha, E.I., Nwanya, A.C., Silica extraction from rice husk: Comprehensive review and applications. *Hybrid Adv.*, 4, 100111, 2023.
45. Okoro, H.K., Alao, S.M., Pandey, S., Jimoh, I., Basheeru, K.A., Caliphs, Z., Ngila, J.C., Recent potential application of rice husk as an eco-friendly adsorbent for removal of heavy metals. *Appl. Water Sci.*, 12, 259, 2022.
46. Salavati-Niasari, M., Javidi, J., Dadkhah, M., Ball Milling Synthesis of Silica Nanoparticle from Rice Husk Ash for Drug Delivery Application. *Comb. Chem. High Throughput Screening*, 16, 458–462, 2013.
47. Prasad, R. and Pandey, M., Rice husk ash as a renewable source for the production of value added silica gel and its application: an overview. *Bull. Chem. React. Eng. Catal.*, 7, 1–25, 2012.
48. Alshatwi, A.A., Athinarayanan, J., Periasamy, V.S., Biocompatibility assessment of rice husk-derived biogenic silica nanoparticles for biomedical applications. *Mater. Sci. Eng., C*, 47, 8–16, 2015.
49. Carmona, V.B., Oliveira, R.M., Silva, W.T.L., Mattoso, L.H.C., Marconcini, J.M., Nanosilica from rice husk: Extraction and characterization. *Ind. Crops Prod.*, 43, 291–296, 2013.
50. Mosaberpanah, M.A. and Umar, S.A., Utilizing rice husk ash as supplement to cementitious materials on performance of ultra high performance concrete: – A review. *Mater. Today Sustainability*, 7–8, 100030, 2020.
51. Saoudi, M.H. and Hello, K.M., Recent use of rice husk ash silica as an efficient solid catalyst: a review. *Asian J. Green Chem.*, 6, 3, 370–378, 2022.
52. Das, S.K., Adediran, A., Kaze, C.R., Mustakim, S.M., Leklou, N., Production, characteristics, and utilization of rice husk ash in alkali activated materials: An overview of fresh and hardened state properties. *Constr. Build. Mater.*, 345, 128341, 2022.
53. Heidarinejad, Z., Dehghani, M.H., Heidari, M., Methods for preparation and activation of activated carbon: a review. *Environ. Chem. Lett.*, 18, 393–415, 2020.
54. Galhardo, T.S., Simone, N., Gonçalves, M., Figueiredo, F.C.A., Mandelli, D., Carvalho, W.A., Preparation of Sulfonated Carbons from Rice Husk and Their Application in Catalytic Conversion of Glycerol. *ACS Sustainable Chem. Eng.*, 1, 11, 1381–1389, 2013.
55. Adam, F., Appaturi, J.N., Iqbal, A., The utilization of rice husk silica as a catalyst: Review and recent progress. *Catal. Today*, 190, 2–14, 2012.
56. Sulaiman, N.F., Leong, Y.W., Lee, S.L., Goh, Z.W., Yahya, S.S.M., Sofiah, A.G.N., Process optimization of rice husk ash supported catalyst in biodiesel synthesis using response surface methodology approach. *Fuel*, 358, 130165, 2024.

57. Roschat, W., Siritanon, T., Yoosuk, B., Promarak, V., Rice husk-derived sodium silicate as a highly efficient and low-cost basic heterogeneous catalyst for biodiesel production. *Energy Convers. Manage.*, 119, 453–462, 2016.
58. Shirini, F., Akbari-Dadamahaleh, S., Mohammad-Khah, A., Aliakbar, A.-R., Rice husk: A mild, efficient, green and recyclable catalyst for the synthesis of 12-Aryl-8, 9, 10, 12-tetrahydro [a] xanthene-11-ones and quinoxaline derivatives. *C.R. Chim.*, 16, 3, 207–216, 2013.
59. Shirini, F., Akbari-Dadamahaleh, S., Mohammad-Khah, A., Rice husk ash supported FeCl2·2H2O: A mild and highly efficient heterogeneous catalyst for the synthesis of polysubstituted quinolines by Friedländer heteroannulation. *Chin. J. Catal.*, 34, 12, 2200–2208, 2013.
60. Li, Y., Xirong, L., Zhanpeng, L., Jinyun, L., Highly-Efficient and Visible Light Photocatalytical Degradation of Organic Pollutants Using TiO2-Loaded on Low-Cost Biomass Husk. *Materials*, 15, 8671, 2022.
61. Awogbemi, O., Kallon, D.V.V., Aigbodion, V.S., Pathways for Sustainable Utilization of Waste Chicken Eggshell. *J. Renewable Mater.*, 10, 8, 2217–2246, 2022.
62. Pan, Y., Li, N., Wu, C., Zhou, Q., Li, K., Unlocking the potential of metal-doping Fe_2O_3/rice husk ash catalysts for low-temperature CO-SCR enhancement. *ACS Omega*, 9, 14, 16621–16630, 2024.
63. Waheed, M., Yousaf, M., Shehzad, A., Inam-Ur-Raheem, M., Khan, M.K.I., Khan, M.R., Ahmad, N., Abdullah, Aadil, R.M., Channelling eggshell waste to valuable and utilizable products: A comprehensive review. *Trends Food Sci. Technol.*, 106, 78–90, 2020.
64. McClelland, S.C., Cassey, P., Maurer, G., Hauber, M.E., Portugal, S.J., How much calcium to shell out? Eggshell calcium carbonate content is greater in birds with thinner shells, larger clutches and longer lifespans. *J. R. Soc. Interface*, 18, 182, 20210502, 2021.
65. Laca, A., Laca, A., Díaz, M., Eggshell waste as catalyst: A review. *J. Environ. Manage.*, 197, 351–359, 2017.
66. Ngayakamo, B. and Onwualu, A.P., Recent advances in green processing technologies for valorisation of eggshell waste for sustainable construction materials. *Heliyon*, 8, 6, e09649, 2022.
67. Awogbemi, O., Inambao, F., Onuh, E.I., Modification and characterization of chicken eggshell for possible catalytic applications. *Heliyon*, 6, 10, e05283, 2020.
68. Ahmad, W., Sethupathi, S., Munusamy, Y., Kanthasamy, R., Valorization of Raw and Calcined Chicken Eggshell for Sulfur Dioxide and Hydrogen Sulfide Removal at Low Temperature. *Catalysts*, 11, 295, 2021.
69. Navajas, A., Issariyakul, T., Arzamendi, G., Gandía, L.M., Dalai, A.K., Development of eggshell derived catalyst for transesterification of used cooking oil for biodiesel production. *J. Chem. Eng.*, 8, 742–748, 2013.

70. Muliwa, A.M., Leswifi, T.Y., Onyango, M.S., Performance evaluation of eggshell waste material for remediation of acid mine drainage from coal dump leachate. *Miner. Eng.*, 122, 241–250, 2018.
71. Faridi, H. and Arabhosseini, A., Application of eggshell wastes as valuable and utilizable products: A review. *Bioresour. Technol. Res. Agr. Eng.*, 64, 2, 104–114, 2018.
72. Rahman, W.U., Yahya, S.M., Khan, Z.A., Khan, N.A., Halder, G., Dhawane, S.H., Valorization of waste chicken egg shells towards synthesis of heterogeneous catalyst for biodiesel production: Optimization and statistical analysis. *Environ. Technol. Innovation*, 22, 101460, 2021.
73. Basumatary, S.F., Brahma, S., Hoque, M., Das, B.K., Selvaraj, M., Brahma, S., Basumatary, S., Advances in CaO-based catalysts for sustainable biodiesel synthesis. *Green Energy Resour.*, 1, 3, 100032, 2023.
74. Hossain, M., Muntaha, N., Osman Goni, L.K.M., Jamal, M.S., Gafur, M.A., Islam, D., Fakhruddin, A.N.M., Triglyceride conversion of waste frying oil up to 98.46% using low concentration k+/cao catalysts derived from eggshells. *ACS Omega*, 6, 51, 35679–35691, 2021.
75. Cahyana, A.H., Ardiansah, B., Maloka, M.B., Magnetite - activated chicken eggshell (Fe3O4-ACE) composite for aldol condensation reaction. *AIP Conf. Proc.*, 2023, 020078, 2018.
76. Mardianaa, L., Ardiansaha, B., Bakri, R., Cahyana, A.H., Anita, Y., Aziza, N.P., Utilization of eggshell-derived material as a solid base catalyst for efficient synthesis of substituted chalcones. *J. Teknol. (Sci. Eng.)*, 79, 5, 175–182, 2017.
77. Kargar, P.G., Bakhshi, F., Bagherzade, G., Value-added synthesized acidic polymer nanocomposite with waste chicken eggshell: A novel metal-free and heterogeneous catalyst for Mannich and hantzsch cascade reactions from alcohols. *Arabian J. Chem.*, 16, 4, 104564, 2023.
78. Baláž, M., Boldyreva, E.V., Rybin, D., Pavlović, S., Rodríguez-Padrón, D., Mudrinić, T., Luque, R., State-of-the-Art of Eggshell Waste in Materials Science: Recent Advances in Catalysis, Pharmaceutical Applications, and Mechanochemistry. *Front. Bioeng. Biotechnol.*, 8, 612567, 2021.
79. Meng, D., Gong, X., Peng, Z., Wang, Z., Ren, D., Removal of Cd2+ in water by FeSO4 magnetic modified eggshell adsorbent. *Desalin. Water Treat.*, 247, 161–172, 2022.
80. Karoshi, G., Kolar, P., Shah, S.B., *et al.*, Recycled eggshells as precursors for iron-impregnated calcium oxide catalysts for partial oxidation of methane. *Bioresour. Bioprocess.*, 7, 48, 2020.
81. Gaide, I., Makareviciene, V., Sendzikiene, E., Effectiveness of Eggshells as Natural Heterogeneous Catalysts for Transesterification of Rapeseed Oil with Methanol. *Catalysts*, 12, 246, 2022.
82. Topić Popović, N., Lorencin, V., Strunjak-Perović, I., Čož-Rakovac, R., Shell Waste Management and Utilization: Mitigating Organic Pollution and Enhancing Sustainability. *Appl. Sci.*, 13, 623, 2023.

83. Azarian, M.H. and Sutapun, W., Biogenic calcium carbonate derived from waste shells for advanced material applications: A review. *Front. Mater.*, 9, 1024977, 2022.
84. Hou, Y., Shavandi, A., Carne, A., Bekhit, A.A., Ng, T.B., Cheung, R.C.F., Bekhit, A.E.A., Marine shells: Potential opportunities for extraction of functional and health-promoting materials. *Crit. Rev. Env. Sci. Technol.*, 46, 11–12, 1047–1116, 2016.
85. Owuamanam, S. and Cree, D., Progress of Bio-Calcium Carbonate Waste Eggshell and Seashell Fillers in Polymer Composites: A Review. *J. Compos. Sci.*, 4, 2, 70, 2020.
86. Yamaguchi, K. and Hashimoto, S., Cold sintering of calcium carbonate derived from seashells. *Open Ceram.*, 12, 100302, 2022.
87. Suwannasingha, N., Kantavong, A., Tunkijjanukij, S., Aenglong, C., Liu, H.-B., Klaypradit, W., Effect of calcination temperature on structure and characteristics of calcium oxide powder derived from marine shell waste. *J. Saudi Chem. Soc.*, 26, 2, 101441, 2022.
88. Lyubas, A.A., Kuznetsova, I.A., Bovykina, G.V., Eliseeva, T.A., Gofarov, M.Y., Khrebtova, I.S., Kondakov, A.V., Malkov, A.V., Mavromatis, V., Shevchenko, A.R., Trace Element Patterns in Shells of Mussels (Bivalvia) Allow to Distinguish between Fresh- and Brackish-Water Coastal Environments of the Subarctic and Boreal Zone. *Water*, 15, 3625, 2023.
89. Peceño, B., Alonso-Fariñas, B., Vilches, L.F., Leiva, C., Study of seashell waste recycling in fireproofing material: Technical, environmental, and economic assessment. *Sci. Total Environ.*, 790, 148102, 2021.
90. Nasir, N.F. and Hazri, M.M., Calcium Oxide from Waste Shells as Potential Green Catalyst for Biodiesel Production. *Res. Prog. Mech. Manufac. Eng.*, 1, 1, 44–55, 2020.
91. Goli, J. and Sahu, O., Development of heterogeneous alkali catalyst from waste chicken eggshell for biodiesel production. *Renewable Energy*, 128, 142–154, 2018.
92. Jindapon, W., Jaiyen, S., Ngamcharussrivichai, C., Seashell-derived mixed compounds of Ca, Zn and Al as active and stable catalysts for the transesterification of palm oil with methanol to biodiesel. *Energy Convers. Manage.*, 122, 535–543, 2016.
93. Nahas, L., Dahdah, E., Aouad, S., El Khoury, B., Gennequin, C., Abi Aad, E., Estephane, J., Highly efficient scallop seashell-derived catalyst for biodiesel production from sunflower and waste cooking oils: Reaction kinetics and effect of calcination temperature studies. *Renewable Energy*, 202, 1086–1095, 2023.
94. Hangun-Balkir, Y., Green Biodiesel Synthesis Using Waste Shells as Sustainable Catalysts with Camelina sativa Oil. *J. Chem.*, 6715232, 2016.
95. Fattahi, A.H., Dekamin, M.G., Clark, J.H., Optimization of green and environmentally-benign synthesis of isoamyl acetate in the presence of ball-milled seashells by response surface methodology. *Sci. Rep.*, 13, 2803, 2023.

96. Khan, S.G., Hassan, M., Anwar, M., ZeshanKhan, U.M., Zhao, C., Mussel shell based CaO nano-catalyst doped with praseodymium to enhance biodiesel production from castor oil. *Fuel*, 330, 125480, 2022.
97. Suteu, D., Bilba, D., Aflori, M., Doroftei, F., Lisa, G., Badeanu, M., Malutan, T., The Seashell Wastes as Biosorbent for Reactive Dye Removal from Textile Effluents. *CLEAN - Soil Air Water*, 40, 2, 198–205, 2014.
98. Sheng, X., Xu, Q., Wang, X., Li, N., Jia, H., Shi, H., Niu, M., Zhang, J., Ping, Q., Waste Seashells as a Highly Active Catalyst for Cyclopentanone Self-Aldol Condensation. *Catalysts*, 9, 661, 2019.
99. Benzekri, Z., Sibous, S., El hajri, F., Hsissou, R., Boukhris, S., Souizi, A., Investigation of snail shell waste as potential and eco-friendly heterogeneous catalyst for synthesis of 1-(benzothiazolylamino) methyl-2-naphthols derivatives. *Chem. Data Collect.*, 31, 100599, 2021.

7

Sustainable Catalysts Derived from Agricultural and Biomass Waste

Sivamani Sivalingam*, Dheedshith M. U., Kaviya J. and Lakshmi Aishwarya J. K.

Department of Chemical Engineering, Rajalakshmi Engineering College, Thandalam, Chennai, India

Abstract

In this chapter, agricultural and biomass waste materials derived cost-effective catalysts are presented. Particularly, three prominent catalysts are explored using the wastes namely, carbon-based catalysts from sawdust, silica-based catalysts from rice husk ash, and calcium oxide (CaO) catalysts from eggshells and bones. This waste conversion approach offers promising alternatives to traditional catalysts, providing numerous advantages including reduced environmental pollution, cost-effectiveness, and utilization of abundant resources. Carbon-based catalysts are highly capable of water purification and diverse catalytic processes due to their high surface area and adjustable structural properties. Silica-based catalysts demonstrate versatility in catalytic applications and water treatment. CaO catalysts exhibit remarkable efficacy in various reactions including biodiesel production. Therefore, by utilizing agricultural waste materials and biomass, researchers are addressing Industry 5.0 on environmental concerns, developing sustainable and eco-friendly catalytic technologies.

Keywords: Agricultural waste, biomass, rice husk ash, eggshell, sawdust, sustainable catalysts, silica, activated carbon

7.1 Introduction

The accumulation of agricultural and biomass waste is increasing every day and has become a growing concern with extensive environmental,

**Corresponding author*: sivamchem@gmail.com; sivamani.s@rajalakshmi.edu.in

Deepti Goyal, Sakshi Kabra Malpani and Renuka Jain (eds.) Catalysis from Waste: Transforming Solid Residues into Value, (169–192) © 2026 Scrivener Publishing LLC

economic, and health implications. Failure in disposal of these wastes and lack of proper waste management techniques create critical issues in terms of greenhouse gas emission, groundwater contamination, and loss of arable land. The agricultural residues obtained from crops such as corn stover, rice husks, sugarcane bagasse etc. are generally open-burnt or thrown into an open area, emitting noxious air pollutants like carbon monoxide, methane, and particulate matter into the atmosphere. Improper waste disposal also causes release of hazardous elements into water bodies, interfering with aquatic life and damaging human health [1, 2].

Ever-increasing waste has now proved to be one of the greatest challenges for current researchers to reduce their environmental impacts and convert them into sustainable resources. In this perspective, green chemistry has settled with a promising solution for the development of sustainable catalysts that propose an environmentally friendly alternative to conventional catalysts. So, many researchers and industries have been reported using renewable or recyclable materials, for reducing reliance on harmful chemicals. Interestingly, most of the sustainable catalysts are comparatively more efficient than conventional catalysts as they require lower energy consumption and generate less waste [3, 4].

Agricultural and biomass waste, rich in renewable carbon and other interesting elements, tend to be well explored in many catalyst developments nowadays. For example, carbon materials derived from renewable sawdust represent an important class of materials with broad applications. Waste eggshells have been used to prepare CaO catalysts through calcination. Similarly, rice husk ash (RHA) possesses the major silica constituent which is suitable for production of Si-based materials like zeolite, and metal-organic frameworks (MOFs). The techniques like calcination and pyrolysis have been reported for extraction and synthesis of appropriate metal-based catalysts, including ruthenium and copper [5, 6].

The most recent innovations are ensuring that waste conversion technologies should focus on both minimization and environmental protection in an integrated manner. Apart from addressing the challenge of increasing waste, catalysts that enable accuracy, effectiveness, and reduced reliance on fossil fuels were obtained by the researchers through the valorization of waste products like RHA and eggshells. As shown by research activity trends in this area, it has drawn massive attention within the chemical sector. The publications focussed on sawdust carbon-based catalysts, the development of various CaO catalysts from eggshells and silica catalysts from RHA have shown constant increase every year which denotes key significance of this study. Figure 7.1 (a-c) shows the increasing sustainable catalyst demand every year in the last decade. These three catalysts

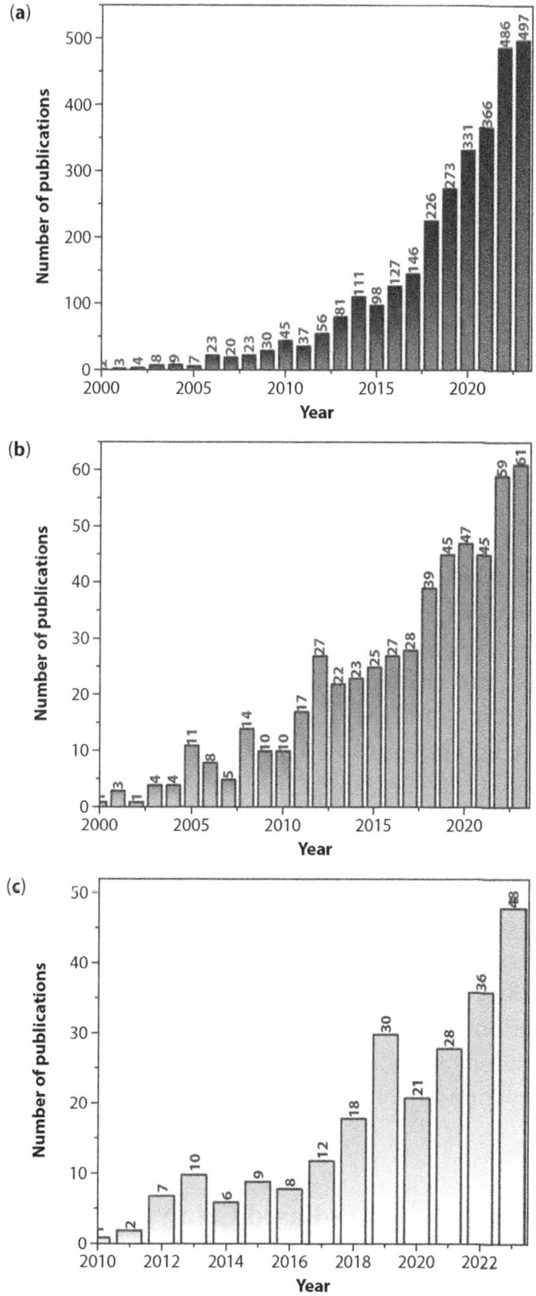

Figure 7.1 Publication trend of sustainable catalysts (a) carbon-based catalysts from sawdust, (b) CaO catalysts from eggshell, (c) silica catalysts from RHA until June 2024.

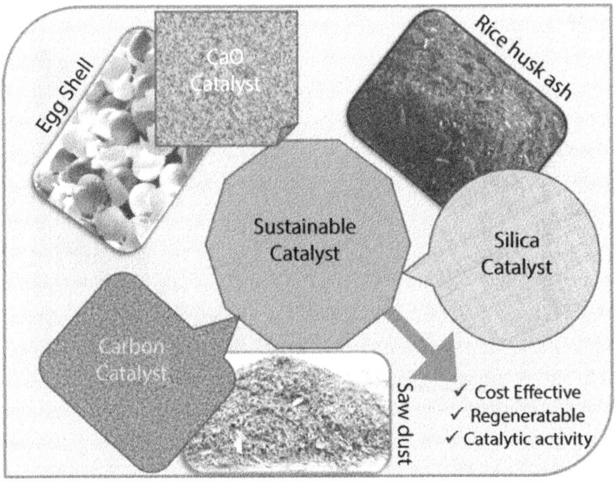

Figure 7.2 Sustainable catalyst prepared from agricultural and biomass waste materials.

show the highest and most consistent growth, highlighting their prominence and potential in sustainable catalyst research. Figure 7.2 displays all three sustainable catalysts derived from agricultural and biomass waste sources. The present chapter aims to explore and develop sustainable catalysts derived from renewable and environmentally abundant sources like eggshells, RHA, and biomass.

7.2 Carbon-Based Catalysts from Sawdust

Carbon-based catalysts are versatile, stable, efficient, and find applications in numerous catalytic processes particularly in wastewater treatment. Their high surface area and tunable surface chemistry make them attractive for adsorption and catalytic reactions, emphasizing sustainable chemical transformations [7]. Carbon is used as a catalyst for many reactions, especially in reactions involving organic compounds like alcohols, ketones, etc.

7.2.1 Synthesis of Carbon-Based Catalysts from Sawdust

The synthesis of carbon-based catalyst from Chinese fir sawdust is well reported in the literature. First and the foremost step was dispersing raw material in deionized water which helps in the removal of dust and debris, after that raw material was heated at 80°C to eliminate the moisture content.

The sawdust undergoes mechanical grinding and screening to achieve an optimal particle size, afterwards it was impregnated with 85 wt.% of H_3PO_4 solution which enhanced the catalytic property of the product. The impregnated material was further appealed to carbonization, dehydration, and pyrolysis processes at 550°C temperature. The obtained sample was cooled to room temperature and washed with deionized water to remove the acid residual. Later, researchers dried the sample for 12h at 105°C to ensure a moisture-free product which was further used in various fields such as materials science and environmental engineering [8, 9]. Figure 7.3 depicts the synthesis of carbon-based catalysts from sawdust waste.

Furthermore, sawdust waste generated from wood industries has been used to make graphitic carbon nanotubes (CNTs). The choice of sawdust addresses environmental concerns regarding waste management and the economic advantage of utilizing a readily available resource. Moreover, the wood structure contains functional groups linked to carbon chains, which naturally lend themselves into CNTs production. The general experimental procedure for synthesis of CNTs from saw dust begins with mixing the sawdust with other substances in a column reactor, including a reducing agent like commercial zinc, bed material such as calcite, and catalyst, either ferrocene or a combination of iron (Fe), molybdenum (Mo), and magnesium oxide (MgO). This mixture was then heated to a temperature of 750°C for 3 h. Notably, this heating process occurs without blowing air, likely to prevent unwanted oxidation reactions. These specific conditions were determined through optimization using thermodynamic simulation. Simulation is considered for analyzing reaction kinetics, thermodynamic equilibrium, and energy consumption to identify the optimal conditions

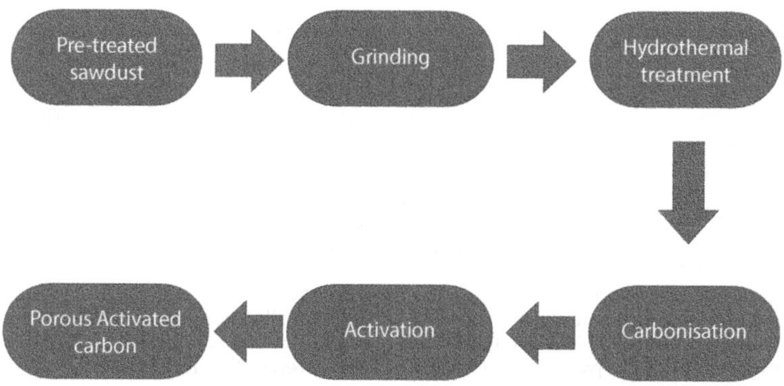

Figure 7.3 A simple schematic diagram of synthesis of carbon-based catalysts from sawdust waste.

for the desired outcome. For each batch test, the following materials were used: 10 g of sawdust, 2 g of ferrocene (or the Fe/Mo/MgO catalyst), 2 g of zinc, 5 g of calcite, and 0.6 g of clay. These clear-cut proportions ensure consistency and reproducibility in the experiments, allowing for accurate comparison of results and reliable conclusions from the data [7].

CNTs can be used as a support material which possesses superior catalytic properties as compared to traditional catalysts. CNTs are supported with metals, metallic salts or complexes where the active phase resides on the external surface of CNTs. They can also be used in the selective deposition of metallic nanoparticles within the inner layers of nanotubes; this technology is yet to be explored. In one such case, CNTs are used for the hydrogenation of biomass-derived furfural into tetrahydro furfuryl alcohol which has been achieved by using Ni-C-CNT composites to increase the efficiency of conversion [10]. These composites were formed by doping of CNTs over Ni-MOFs. It has been found that the utilization of lignin has enabled the continuous production of high-performance CNT fibers which have superior tensile strength and electrical conductivity. Furthermore, sawdust-based adsorbents can also be used in the removal of various pollutants from aqueous solution. These studies reveal the versatility and sustainability of biomass-derived materials and their diverse applications [3]. Figure 7.4 shows the schematic sketch of preparation of CNT fibers from lignin and the detailed processing.

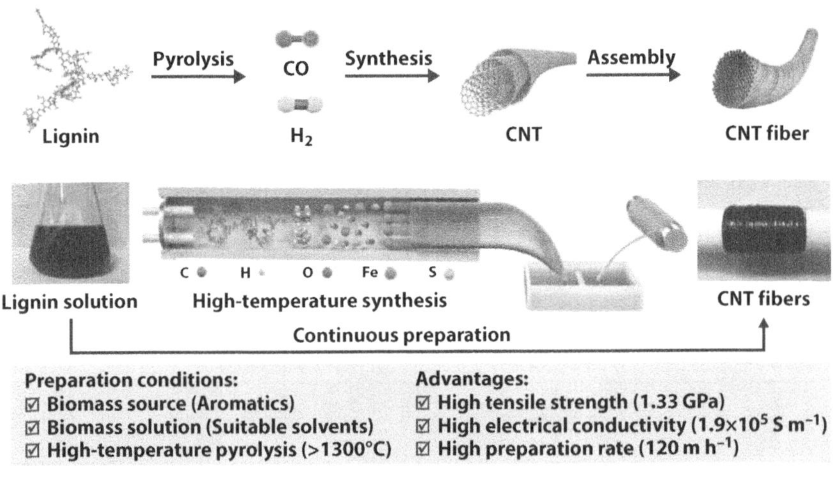

Figure 7.4 Schematic diagram of synthesis of CNT fibres from lignin. (Copyright: licensed under a Creative Commons Attribution 4.0) [11].

7.2.2 Characterization of Carbon-Based Catalysts from Sawdust

Various characterization techniques such as CHNS elemental analysis, scanning electron microscopy (SEM) and transmission electron microscopy (TEM) are used to characterize raw sawdust and sawdust-derived carbon catalysts. CHNS elemental analysis is used to find the carbon and other material content of the product. SEM and TEM are used to study the internal structure, surface morphology, macroscopic features, and microscopic overview of sawdust-derived carbon catalysts. The specific surface area and pore distribution can be determined with the help of Brunauer-Emmett-Teller (BET) analyzer. The crystalline structure and phase composition of the product can be analyzed by X-ray powder diffraction (XRD) method. Raman spectroscopy is used to analyze the molecular structure. X-ray photoelectron spectroscopy analyzes the chemical composition and their internal structure. In a study, the elemental composition of sawdust-derived activated carbon (SD-AC) was determined through elemental analysis, with percentages of carbon, oxygen, hydrogen, and nitrogen being 67.46, 26.83, 3.02, and 0.22%, respectively [12].

The size of the pores can be estimated by using density functional theory (DFT), through which the pore size was approximately found to be 0.55 nm for micropores and 2.58 nm for mesopores [13]. Through BET analysis, the surface area and the pore volume were calculated as 621 m^2 /g and 0.35 cm^3/g respectively, which help in understanding the material-efficient ion penetration [3, 14]. SD-AC electrode exhibits outstanding electrochemical performance, achieving a specific capacitance of 244.1 F/g at 1.0 A/g and demonstrating remarkable cyclic stability, retaining 87.03% of its capacitance after 10,000 cycles at 10 A/g. These properties enable the symmetric supercapacitor based on SD-AC to deliver an energy density of 19.9 Wh/kg, a power density of 650 W/kg, and an extended operational lifespan. The superior performance of the SD-AC electrode is attributed to its large surface area and hierarchical pores, which facilitate efficient ion penetration and transport, thereby enhancing both capacitance and rate capability. Additionally, the presence of abundant functional groups contributes extra faradaic pseudo-capacitance. This study presents a sustainable approach to transforming waste biomass into high value AC, offering great potential for efficient energy storage solutions [8, 9].

7.2.3 Applications of Carbon-Based Catalysts from Sawdust

- **Hydrogenation and Dehydrogenation**

Carbon is used as a catalyst in hydrogenation reaction or in conversion of unsaturated hydrocarbons to saturated ones [15]. Carbon-based catalysts act crucial in removing hydrogen atoms from organic compounds in dehydrogenation reactions [16]. For instance, carbon-supported metal catalysts, like nickel or platinum, are often used to dehydrate alkanes to produce alkenes. This process is essential in the petrochemical industry for producing various intermediates and synthesizing plastics, polymers, and other organic compounds [16].

- **Oxidation and Decomposition**

Carbon is used as a catalyst in oxidation reactions for conversion of carbon monoxide to carbon dioxide [17]. Carbon breaks down large, high molecular-weight hydrocarbons into smaller molecules to produce fuels and other chemical compounds [9].

- **Water Purification**

Granulated AC has been used in water filtration for inducing oxidation or reduction process with the organic components present in water and is also used to remove rotten odor due to sulfide and chloride. AC is a highly effective material for purifying water. The strength of carbon is its vast surface area as created by a network of tiny pores [18]. As water passes through the carbon, all organic and inorganic impurities and other compounds that have unpleasant smell and taste will get trapped in the pores of the AC. Later, the water leaving the system is cleaner and purer, all these effects are due to the extensive surface area of AC which increases the adsorption capacity.

Carbon based catalysts are predominantly being used in the sewage water treatment process. The treatment of sewage water is a fundamental process for protecting the environment and ecosystems from municipal and industrial liquid waste. This treatment process involves the collection of effluent or municipal wastewater followed by a three-stage treatment process including primary, secondary, and tertiary treatment processes to mitigate its impacts on the environment [19, 20]. This process is essential for maintaining water quality and prevents the spread of waterborne diseases. Modified AC is a developing technology used in the adsorption of contaminants from wastewater. AC has high adsorption capacity due to its highly porous structure and large surface to volume ratio. The porous structure allows the adsorption of micropollutants such as chlorine, methane, and organic compounds and helps to remove odor, as the water passes through the AC filter depending on the pore size of particles which get stuck in filters. The wide pores of the carbon trap impurities, gradually purify and improve the water quality. The AC filtration is influenced by factors such as surface area, impurity concentration, treatment duration, and pH etc. [21].

- **Ammonia synthesis**

Synthesis of ammonia is typically done by the Haber-Bosch process while using carbon as a catalyst. In the production of ammonia, carbon is often used as a supporting agent to enhance the action of iron or other metal catalysts. The main role of carbon is to stabilize and support the activated iron radical which thereby enhances the efficiency of the reaction. In this reaction, carbon increases the chances of interaction of the catalyst metal with the reactants [5, 22].

7.3 Silica-Based Catalysts from Rice Husk Ash

India holds the second most prominent place in rice production, next to China, contributing to 21.5% of global rice production (Food and Agriculture Organization). During the crop year 2022/2023, 135 million metric tons of milled rice was produced in India out of which approximately 90 million tonnes of rice husk is produced during processing rice, as a by-product. The produced rice husk is often burned to produce RHA. The composition of the rice husk varies with the variety of rice, soil of different regions and climate. Generally, it comprises cellulose (50%), hemicellulose (15-20%), lignin (25%–30%), and silica (15%–20%), the rest is moisture content with some metal content. Smaller amounts of potassium, sodium, magnesium, calcium, iron, and phosphorus can also be found in the RHA. Since the calorific value of rice husk is 50% of that of coal, it is inefficient to use as a direct fuel. Many researchers have found that because of proper burning of rice husk, RHA is produced and possesses around 90% SiO_2 [23]. This silica extracted from RHA has proved to be an effective catalyst in many processes. In addition, due to this unique property and high silica content of RHA, it can convert into pozzolanic admixtures, high-performance concrete and renewable silica sources.

7.3.1 Methods of Silica Extraction from Rice Husk Ash

Various methods like sol-gel, precipitation, and heat treatment are reported to extract silica from RHA. The heat treatment method stands out for its simplicity, effectiveness, and environmental friendliness. High-purity ultra-fine silica can be obtained by pre-treating RHA under an oxygen environment to remove impurities. The purity and properties of the extracted silica is influenced by various factors like heating rates, pyrolysis temperature, and rice husk particle size [23]. In processes like acetic acid steam reforming and hydrogen production, the potential of silica as a catalyst can be witnessed [9].

- **Hydrothermal Method**

This method focuses on the rice husk's solubility with a suitable leaching agent, typically an alkaline solution like sodium hydroxide (NaOH) or potassium hydroxide (KOH). The alkaline solution reacts with silica in the RHA, forming alkaline silicate. The separation of minor metal oxides and residues like cellulose and hemicellulose, makes it easier to separate the silica from RHA. Further acid treatment helps to neutralize the solution, then the precipitate is collected followed by drying to obtain silica of high purity [24]. The primary chemical process involved the dissolution of the silica in the RHA, followed by the leaching procedure; the dissolved silicate-containing liquid phase undergoes vacuum-filtration to keep away the solid residues. The silicate solution is separated from the residual solid particles, which is then used to create spherical silica particles. The silicate solution is mixed with acetic acid and polyethylene glycol (PEG), which helps in silica precipitation. The precipitated silica is subsequently filtered, cleaned, and calcined to create pure, spherical silica particles. The silica obtained from this process is highly pure with a purity of 99.1%. Compared to the original RHA, the calcium, magnesium, and potassium content in the silica particles is significantly reduced, demonstrating the effectiveness of the process in removing these impurities. Despite the sodium level increased slightly due to removal of other metallic impurities, the overall purity in this method is higher than that of many other methods [17].

- **Calcination Method**

Calcination is another heat treatment method in which a solid substance has been treated thermally in the presence of limited oxygen, or a controlled oxygen environment. It is carried out in a furnace to remove the volatile matter or to incur thermal decomposition of the impurities by heating the substance below its melting point. Rice husk is exposed to air or oxygen during the thermal calcination process, which raises the material's temperature to a maximum of 500–700°C, producing RHA [25, 26]. During this process, the cellulose and lignin (a part of the organic material) in the rice husk burn off and undergo thermal breakdown [26].

- **Pyrolysis method**

In pyrolysis, rice husk is heated at high temperatures, typically between 400 and 650°C, in a controlled environment inside a reactor or furnace, as a part of the pyrolysis process. The organic matter in rice husk composed primarily of cellulose and lignin, breaks down thermally, releasing volatile compounds and leaving behind an inorganic material residue, primarily containing silica. After that, silica-rich ash is gathered, processed to improve its quality and removed contaminants for a range of uses. The calcination occurs in the presence of air or oxygen while pyrolysis occurs in the absence of oxygen. The characteristics and composition of the final products are impacted by these variations in circumstances. Although both calcination and pyrolysis are thermal breakdown processes, they differ in their characteristics and end product properties [27]. The lack of oxygen during pyrolysis enables the creation of bio-oil and gas, which are not generated during calcination. The silica-rich ash produced during pyrolysis also differs from that produced during calcination, as it is more porous and has a higher surface area. Figure 7.5 shows the different factors affecting the high purity of silica content.

Table 7.1 summarizes the different methods of silica extraction, detailing the process conditions, chemical treatments, and their outcomes. The methods vary in terms of temperatures, durations, and types of acids or bases, resulting in silica with different characteristics suitable for various industrial applications.

The pores of mesoporous silica range from 2 to 50 nm diameter. It is unique due to its mesoporous nature. According to the IUPAC terminology, this type of silica falls in between macroporous (>50 nm) and microporous (<2 nm) types. Because of their unique qualities, mesoporous silica

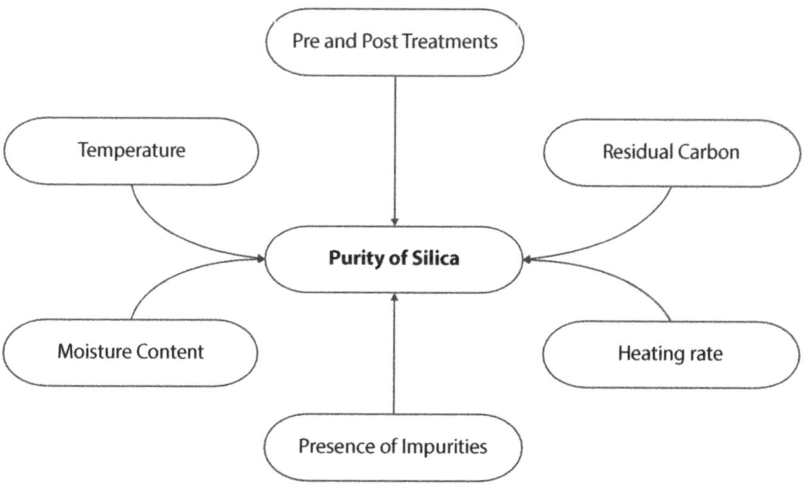

Figure 7.5 The different factors affecting the high purity of silica content.

Table 7.1 Different methods to extract silica from RHA.

Process	Temperature and duration	Chemical treatment	Outcomes	References
Thermal treatment	500-700 °C for 2-6 h	HCl and NaOH	Amorphous silica	[14]
Thermal treatment	Varied	Not specified	High-purity, chemically reactive silica (white color)	[28]
Pre-treatment followed by calcination	600 °C	HNO_3	Pure white amorphous silica with no organic or inorganic impurities	[29]
Calcination	500°C and 700°C for 2-3 h	Not specified	Amorphous silica	[30]
Leaching with various acids	Not specified	Acetic acid, oxalic acid, HCl, HNO_3	Improved properties of ashes (comparable to mineral acid leaching)	[12]
Pre- and post-treatment, followed by incineration	500-800°C under air, argon, and oxygen for 4 h	3% HCl, 10% H_2SO_4, 3% NaOH	Pure activated silica (1N HCl most effective for metal removal)	[31]
Roasting	500-600°C (amorphous), 600-750°C (crystalline)	Not specified	Amorphous and crystalline silica	[20]
Calcination	High temperature (800°C)	Not specified	Surface melting, carbon fixation, amorphous state below 800°C	[32]
Controlled burning	Below 700°C	Not specified	Amorphous silica (transition to crystalline above 850°C)	[33]

(*Continued*)

Table 7.1 Different methods to extract silica from RHA. (*Continued*)

Process	Temperature and duration	Chemical treatment	Outcomes	References
Low-temperature alkali extraction	Not specified	Not specified	Amorphous silica	[34]
Alkali extraction followed by acidification	Around 100°C (open system)	NaOH, acidification	Mesoporous silica	[33]
Burning followed by alkaline extraction	700°C for 6 h	2.5 N NaOH	90.3% silica, nanosilica with 656 m^2/g surface area	[35]
Purification followed by acid precipitation	Not specified	NaOH, acid precipitation	High-purity 25 nm silica powder with 274 m^2/g surface area	[36]
Acid treatment followed by controlled thermal combustion and nano-silica extraction	Not specified	NaOH, H_2SO_4	90.94% silica ash, 98.8% pure, nanosized amorphous silica (high surface area)	[37]
Heat treatment	700°C for 6 h	Not specified	95% silica powder, increased surface area (54 to 81 m^2/g after wet milling)	[1]
Controlled pyrolysis	700°C for 2 h	Not specified	20-30 nm silica nanoparticles (avoiding higher temperatures to prevent melting)	[38]
Precipitation with acid washing, carbonization, and activating agents	550-700°C, optimum at 600°C for 1 h	Sodium hydroxide, zinc chloride, phosphoric acid	Amorphous and crystalline silica at different conditions	[39]

nanoparticles like MCM-41 and SBA-15 are frequently used in applications like drug delivery, imaging, catalysis etc.

7.3.2 Synthesis of Silica Catalysts from Rice Husk Ash

The production of silica catalysts from RHA involves a multi-step process that begins with silica extraction from the ash. Quality silica is obtained through the alkaline or sol-gel method. The ash reacts with an alkaline solution or solvent to dissolve the silica component. Extracted silica is purified and the impurities are removed to enhance the quality for catalytic applications. The physical processes such as filtration, washing, and drying techniques are used to remove the impurities and improve the silica catalyst surface area and reactivity. An alternate way of extracting silica can involve the impregnation method using functionalization with a specific organic group or with a metal precursor to enhance their catalytic activity [23]. For example, metal ions like copper, nickel, or iron can be impregnated onto a silica surface to create a heterogeneous catalyst with specific properties. Organic groups like carboxyl or amino can also be functionalized onto a silica surface to enhance its selectivity and reactivity.

RHA ruthenium-based catalyst (RHA-Ru) is reported to have a notable pore volume of 0.5 cm^3/g, a high surface area of 120 m^2/g, and a ruthenium loading of 10 wt.%. These results supported silica catalyst efficacy because of its well-defined mesoporous structure. The prepared catalyst exhibits excellent catalytic activity and selectivity, resulting in effective mass transfer and improved catalytic performance in hydrogenation reactions [40]. To find effective catalyst [40], Adam *et al.* attempted to synthesize RHA-supported copper-based catalyst (RHA-Cu) that has a notable pore volume of 0.7 cm^3/g, a substantial surface area of 150 m^2/g, and a copper loading of 15 wt.%. Due to the well-dispersed copper species on the silica support, these catalysts have high catalytic activity and selectivity, which makes it easier for reactants to adsorb and activate during ethanol dehydrogenation reactions.

7.3.3 Applications of Silica Catalysts from Rice Husk Ash

Silica catalysts obtained from RHA are possible candidates for wide industrial usage, especially in the areas of environmental management and energy recovery. Silica from RHA could be used to make heterogeneous catalysts for the oxidation of some pollutants found in water, leading to better removal of some persistent materials [27]. Silica supported manganese oxide as a catalyst, accomplishes the transformation of volatile organic compounds (VOCs) like acetone with great effectiveness, solving both issues of waste burning and air pollution [17]. The applicability of silica from RHA increases with the fabrication of more sophisticated silicon carbide materials for structural purposes [41]. Such adaptability of rice husk silica emphasizes the necessity to establish proper procedures for its extraction and subsequent applications. Despite the promising advantages of RHA silica, there are still hindrances regarding the optimization of its processing technology and achieving the needed level of production economy.

7.4 CaO Catalysts from Eggshells

Due to increasing consumption of eggs every day to make essential protein, eggshells waste simultaneously producing in large quantities worldwide. This waste can be converted into a CaO catalyst when heated at high temperatures. Since the eggshell contains 95% calcium carbonate, 3% magnesium, and traces of sodium, potassium, zinc, manganese, iron, and copper, it is more suitable to prepare a CaO catalyst. Calcium carbonate can be decomposed during heating to form calcium oxide and carbon dioxide. Eggshell-derived CaO and metal doped catalyst was synthesized from waste chicken eggshell and later used in transesterification reaction [42].

7.4.1 Synthesis of CaO Catalysts from Eggshell

CaO catalyst can be derived from $CaCO_3$ present in the eggshell by calcination process. The different raw materials required slightly varying conditions to process unique CaO production, but the calcination process takes place all the way [43]. CaO catalyst is generally obtained from various sources such as limestone, calcite on marble, quicklime, calcium hydroxide, seashell, dolomite, etc. The eggshell, a biomass, is generated abundantly and used as a raw material in producing CaO.

Calcination is a heat treatment process that involves heating a substance (eggshell) to a high temperature (800°C) without air or a limited supply of oxygen. During this process, the substance decomposes and the volatile components such as carbon dioxide (CO_2) are driven off. In the calcination of eggshell, the calcium carbonate ($CaCO_3$) undergoes thermal decomposition and produces CO_2 as a volatile component and calcium oxide as a desired product. The chemical reaction involved in this process is as follows:

$$CaCO_3 \rightarrow CaO + CO_2 \tag{7.1}$$

The production of CaO catalyst from eggshell involves various steps from the collection of eggshells to production of CaO catalyst [44], which are given below:

- The eggshells are collected and cleaned to remove the impurities.
- Then, they are dried to remove moisture content.
- Later heated at a high temperature in the furnace.
- The eggshells are collected from the furnace and finely powdered using mechanical grinders. This process increase the surface area of the CaO catalyst to improve reactivity.
- CaO powder is analyzed to find its composition purity using FTIR, XRD, and SEM techniques.
- To enhance the catalytic activity or selectivity, the CaO catalyst may undergo additional treatments.
- Finally, the synthesized CaO catalyst can be used for various applications.

Figure 7.6 outlines the step-by-step procedure, from collecting and cleaning the eggshells to producing calcium oxide nanoparticles. Chicken eggshells were collected and soaked in boiled water for 10 minutes to solidify the gelatinous materials adhering to the inner wall of the eggshell, by which the removal is more straightforward. It was washed, rinsed with distilled water to remove impurities, and dried in the oven at 105°C for 24 h. Then, it was grounded and sieved to the micrometer sieve range of 63 micrometers. The grounded eggshell powder was taken into a crucible and calcined in a furnace at 900°C for 3 h. Later, it was refluxed in water at 60°C for 6 h and dried up to 105°C for 7-8h. The powder was then calcined at 800°C for 3 h, which converts the hydroxide form into highly porous calcium oxide nanoparticles [19]. Eggshell waste was also used in the production of Multimodel CaO (MM-CaO) catalyst. For this, waste was washed with excess water to remove the impurities/dust particles from the surface of the eggshell and dried at room temperature. Then, the samples were placed in a ball mill to make fine powder. The samples were washed, dried and calcined for 2 h at 800°C to convert $CaCO_3$ in the eggshells into MM-CaO.

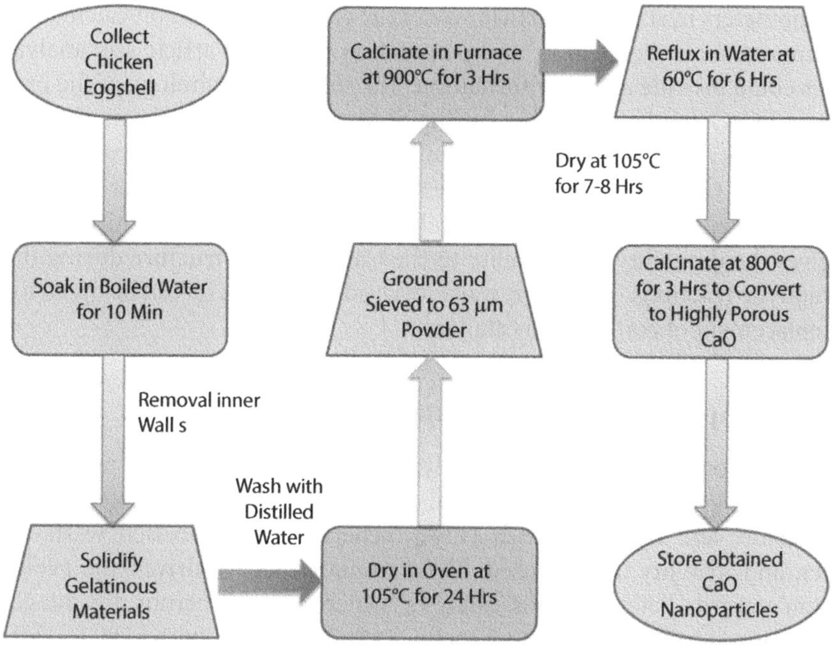

Figure 7.6 Schematic diagram of CaO nanoparticles catalyst preparation from the waste eggshell.

7.4.2 Characterization of CaO Catalysts from Eggshell

Physico-chemical characterization of CaO from eggshell is accountable to ensure efficiency of prepared catalyst. In this section we have discussed various characterization techniques that are aids to understand the structural, morphological and elemental features of catalysts [43, 44].

The pore size and pore diameter of prepared CaO catalyst can be determined using BET analysis. If the pore size of the catalyst ranges between 2 – 50 nm, it is considered as mesoporous, if it is more than 50 nm, then it is considered as macroporous. Microporous catalysts possess tiny pores in the size range of 2 nm in diameter. The CaO derived from the eggshell possesses high porosity, which is responsible for higher diffusion of the reactant molecules on the active sites of the catalyst.

CaO catalyst derived from eggshell exhibits a high surface area due to fine grinding. The large numbers of active sites are present to improve the catalytic performance. These properties can also be found using BET analyzer. Depending upon the preparation method, the size of the particles can be varied. The size of particles can be varied from a few micrometers to nanometers in size. The particle size of the catalyst depends on calcination temperature, process time, and preparation method. Particle size analyzers could provide all this information. The surface morphology of the prepared CaO can be determined using SEM analysis, which shows that the irregularity in shape is due to the synthesis method and various parameters involved. Crystallinity is one of the significant characteristics of the catalyst. The eggshell-derived CaO catalysts possess high crystallinity, which gives the durability and stability to the CaO catalyst structure during the catalytic reactions. Therefore, waste-derived CaO catalysts can be opted for replacement of commercial catalysts.

7.4.3 Applications of CaO Catalysts from Eggshell

Conventional homogeneous catalysts, such as sodium or potassium hydroxide/methoxide, have several drawbacks, including intolerance to moisture and free fatty acids (FFA), generation of processed wastewater, and inability to be reused. These limitations have driven the evolution towards heterogeneous catalysts, which offer numerous advantages in transesterification. These benefits include reusability, which is cost-effective; tolerance to moisture and FFA, allowing the use of lower-quality oils; affordability, mainly when sourced from waste shells; low solubility in methanol; non-corrosiveness; and environmental friendliness [6, 42].

CaO is a heterogeneous base catalyst that is obtained naturally from limestone, crab shells, capiz shells, mussel shells, oyster shells, eggshells etc. Egg shell derived CaO catalyst is widely used in transesterification reactions to transform the oil into biodiesel. In this process, CaO catalyst is mixed with methanol under constant stirring for 45 minutes and then heated below the boiling point of methanol (64.8°C). After this, methanol/CaO mixture was added in oil (3:1 oil to methanol molar ratio) and subjected to agitation for 30 minutes and then irradiated under sunlight. The reaction was performed with different catalyst doses, irradiation time, and reaction temperature to determine the maximum yield. After the reaction, the mixture was allowed to settle for one day for the complete separation of layers [45]. In another study, the efficiency of MM-CaO catalyst was checked in biodiesel production in the presence of sunlight. The electron in the valence band (VB) is excited when the sunlight falls on the MM-CaO surface, and due to excess energy, the electrons from VB move to the conduction band (CB) and generate electron-hole pairs, which are involved in the chemical reaction. The generated hole may attack the methanol absorbed on the surface of the catalyst and form CH_3O^- and H^+, the methanol is thus oxidized. The RCOOH adsorbed on the catalyst surface is reduced to form R-COOH$^-$. The products such as CH_3O^-, H^+, and RCOOH$^-$ react to form intermediates and then follow the trans-esterification process, which involves biodiesel production [43]. The prepared MM-CaO catalyst was used in biodiesel synthesis *via* transesterification reaction and yields about 75–85%. Eggshell-derived MM-CaO catalyst also provides a cost-effective way to prepare green fuel [6].

7.5 Advantages of Green Methods over Traditional Techniques

7.5.1 Sustainable Sourcing

The extraction of fossil fuels such as crude oil and coal has a significant environmental impact mainly on air and water pollution, whereas green methods result in less pollution as compared to traditional methods. The operating temperature and pressure requirements are much lower; this lower energy consumption indicates a smaller carbon footprint. The main resource used in green methods is renewable plant-based biomasses [4]. Hence, it reduces reliance on non-renewable resources and promotes sustainability in the long term. Some green methods even consider using waste materials like wood chips or agricultural leftovers which decreases the dependence on fresh resources and diverts waste from landfills leading to a more circular economy.

7.5.2 Waste Minimization

Traditional catalyst production techniques typically create considerable amounts of leftover materials. Such leftovers might necessitate unique handling or removal, contributing to the environmental load and maybe incurring additional expenses. Green techniques advocate for waste management by converting refuse materials into valuable catalysts. This could involve utilizing woodchips from carpentry sectors or agricultural leftovers like corn stalks or rice husks that might otherwise be thrown away. By integrating refuse materials into manufacturing, green techniques efficiently decrease collective waste production and establish a more sustainable waste management system [4].

7.5.3 Energy Efficiency

The production of catalysts using traditional methods generally involves a high temperature and high-pressure environment, so the energy requirement also rises to a higher level. The greenhouse gas emissions in traditional methods are considerably higher than the green synthesis, whereas the green method environment is milder and thus the emissions are comparatively lower [4]. The carbon footprint is also considerably lower.

7.5.4 Safer Operation

Production of green catalysts employs fewer hazardous chemicals and solvents than traditional methods. This ensures a safer work environment and reduces waste disposal costs generated during production.

7.5.5 Tunable Properties

Some green methods offer finer control over the resulting catalysts size, shape, and porosity. This allows the creation of catalysts with properties optimized for specific applications, leading to enhanced performance and efficiency.

7.6 Conclusion

This comprehensive work emphasized the conversion of agricultural biomass waste into value-added catalysts. Transforming sawdust into carbon-based catalysts tackles waste management issues and contributes to a circular economy by repurposing industrial residues. This catalyst typically possesses high surface areas, requires fewer chemicals and uses less energy than traditional fossil-based carbon materials. This results in a greener and more environmentally friendly production process. Silica is extracted from RHA due to its abundance and renewability. Deriving silica from RHA not only addresses disposal issues but also obtains value from agricultural waste. Silica is a versatile catalyst support material due to its high surface area and thermal stability. Based on these features, it can be employed in various catalytic processes, including synthesizing fine chemicals and environmental catalysis. Eggshells and bones are abundant, renewable waste materials, using them as sources for CaO production promotes waste valorization and reduces environmental pollution. CaO is known for its catalytic properties, especially in processes like biodiesel production, where it acts as an effective base catalyst. Due to high surface area and basicity, it is enabled to carry out an efficient catalytic reaction. Utilizing these waste materials mitigates disposal costs and provides an inexpensive alternative to commercially produced catalysts. Consequently, these sustainable catalysts offer advantages for various catalytic reactions, including adsorption processes and heterogeneous catalysis, making them a cost-effective and performance-enhanced approach to catalysis. These catalysts exemplify the principles of green chemistry by upcycling waste, reducing the environmental footprint, and promoting resource efficiency in industrial processes.

Acknowledgment

The authors wish to acknowledge their sincere thanks to Rajalakshmi Engineering College (Autonomous), Thandalam, Chennai, India for their support in this work.

AI disclosure statement: The author acknowledges the use of AI-based tools for minor language editing or grammatical correction. However, all intellectual contributions, analyses, and conclusions presented in this chapter are entirely the author's own.

References

1. Della, V.P., Kuhn, I., Hotza, D., Rice husk ash as an alternate source for active silica production. *Mater. Lett.*, 57, 4, 818–821, 2002.
2. Wannapeera, J., Worasuwannarak, N., Pipatmanomai, S., Product yields and characteristics of rice husk, rice straw and corncob during fast pyrolysis in a drop-tube/fixed-bed reactor. *Songklanakarin J. Sci. Technol.*, 30, 3, 393–404, 2008.
3. Adegoke, K.A., Adesina, O.O., Okon-Akan, O.A., Adegoke, O.R., Olabintan, A.B., Ajala, O.A., Olagoke, H., Maxakato, N.W., Bello, O.S., Sawdust-biomass based materials for sequestration of organic and inorganic pollutants and potential for engineering applications. *Curr. Res. Green Sustainable Chem.*, 5, 100274, 2022.
4. Rubab, L., Anum, A., Al-Hussain, S.A., Irfan, A., Ahmad, S., Ullah, S., Al-Mutairi, A.A., Zaki, M.E.A., Green Chemistry in Organic Synthesis: Recent Update on Green Catalytic Approaches in Synthesis of 1,2,4-Thiadiazoles. *Catalysts*, 12, 11, 1329, 2022.

5. Morlanes, N., Almaksoud, W., Rai, R.K., Ould-Chikh, S., Ali, M.M., Vidjayacoumar, B., Al-Sabban, B.E., Albahily, K., Basset, J.M., Development of catalysts for ammonia synthesis based on metal phthalocyanine materials. *Catal. Sci. Technol.*, 10, 3, 844–852, 2020.
6. Nadeem, F., Bhatti, I.A., Ashar, A., Yousaf, M., Iqbal, M., Mohsin, M., Nisar, J., Tamam, N., Alwadai, N., Eco-benign biodiesel production from waste cooking oil using eggshell derived MM-CaO catalyst and condition optimization using RSM approach. *Arabian J. Chem.*, 14, 8, 103263, 2021.
7. Sevilla, M., Sanchís, C., Valdés-Soh, T., Morallón, E., Fuertes, A.B., Synthesis of graphitic carbon nanostructures from sawdust and their application as electrocatalyst supports. *J. Phys. Chem. C*, 111, 27, 9749–9756, 2007.
8. Smith, D.J., Pettit, P., Schofield, T., Activated carbon in water treatment. *Water Supply*, 14, 2, 85–98, 1996.
9. Arroyo-Caire, J., Diaz-Perez, M.A., Lara-Angulo, M.A., Serrano-Ruiz, J.C., A Conceptual Approach for the Design of New Catalysts for Ammonia Synthesis: A Metal—Support Interactions Review. *Nanomaterials*, 13, 22, 2914, 2023.
10. Wang, Y., Liu, S., Guo, Q., Zhang, Y., Ni@ C@ CNT catalyst derived from CNT doped Ni-MOF for furfural hydrogenation to tetrahydrofurfuryl alcohol. *Asia-Pac. J. Chem. Eng.*, 17, 2, e2739, 2022.
11. Liu, F., Wang, Q., Zhai, G., Xiang, H., Zhou, J., Jia, C., Zhu, L., Wu, Q., Zhu, M., Continuously processing waste lignin into high-value carbon nanotube fibers. *Nat. Commun.*, 13, 1, 1–10, 2022.
12. Chandrasekhar, S., Pramada, P.N., Praveen, L., Effect of organic acid treatment on the properties of rice husk silica. *J. Mater. Sci.*, 40, 24, 6535–6544, 2005.
13. Anzar, N., Hasan, R., Tyagi, M., Yadav, N., Narang, J., Carbon nanotube - A review on Synthesis, Properties and plethora of applications in the field of biomedical science. *Sens. Int.*, 1, 100003, 2020.
14. Bogeshwaran, K., Kalaivani, R., Ashraf, S., Manikandan, G.N., Prabhu, G.E., Production of Silica from Rice husk. *Int. J. ChemTech Res.*, 6, 4337–4345, 2020.
15. Madduluri, V.R., Mandari, K.K., Velpula, V., Varkolu, M., Kamaraju, S.R.R., King, M., More, S., Rice husk-derived carbon-silica supported Ni catalysts for selective hydrogenation of biomass-derived furfural and levulinic acid. *Fuel*, 261, 116339, 2020.
16. Hassan, S.A., Al-Sabagh, A.M., Shalaby, N.H., Hanafi, S.A., Hassan, H.A., Catalytic performance of organically templated nano nickel incorporated-rice husk silica in hydroconversion of cyclohexene and dehydrogenation of ethanol. *Egypt. J. Pet.*, 22, 1, 179–188, 2013.
17. Cardoso, M., Portugau, P., De Los Santos, C., Faccio, R., Vidal, H., Gatica, J.M., del P. Yesté, M., Castiglioni, J., Torres, M., Catalytic Oxidation of Acetone over MnOx-SiO2 Catalysts: An Effective Approach to Valorize Rice Husk Waste. *Materials*, 17, 24, 1–24, 2024.

18. Lubis, R.A.F., Nasution, H.I., Zubir, M., Production of Activated Carbon from Natural Sources for Water Purification. *Indones. J. Chem. Sci. Technol.*, 3, 2, 67–73, 2020.
19. Erchamo, Y.S., Mamo, T.T., Workneh, G.A., Mekonnen, Y.S., Improved biodiesel production from waste cooking oil with mixed methanol–ethanol using enhanced eggshell-derived CaO nano-catalyst. *Sci. Rep.*, 11, 1, 1–12, 2021.
20. Ghosh, R., A Review Study on Precipitated Silica and Activated Carbon from Rice Husk. *J. Chem. Eng. Process Technol.*, 04, 04, 1–7, 2013.
21. Al Dawery, S.K., Al-Sawai, M.K., Al Muzami, G.M.S., Annamareddy, S.H.K., Al Dawari, M.S., Harharah, R.H., Harharah, H.N., Amari, A., Treatment of Produced Water Using Prepared Activated Carbon-Based Sewage Sludge. *Separations*, 10, 10, 1–17, 2023.
22. Cao, A., Bukas, V.J., Shadravan, V., Wang, Z., Li, H., Kibsgaard, J., Chorkendorff, I., Nørskov, J.K., A spin promotion effect in catalytic ammonia synthesis. *Nat. Commun.*, 13, 1, 1–7, 2022.
23. Sivalingam, S. and Sen, S., Rice husk ash derived nanocrystalline ZSM-5 for highly efficient removal of a toxic textile dye. *J. Mater. Res. Technol.*, 9, 6, 14853–14864, 2020.
24. Park, J.Y., Gu, Y.M., Chun, J., Sang, B.I., Lee, J.H., Pilot-scale continuous biogenic silica extraction from rice husk by one-pot alkali hydrothermal treatment and ball milling. *Chem. Biol. Technol. Agric.*, 10, 1, 1–8, 2023.
25. Islam, M.T., Hossen, M.F., Asraf, M.A., Kudrat-E-Zahan, M., Zakaria, C.M., Production and Characterization of Silica from Rice Husk: An Updated Review. *Asian J. Chem. Sci.*, 14, 2, 83–96, 2024.
26. Kaleli, M.J., Kamweru, P.K., Gichumbi, J.M., Ndiritu, F.G., Characterization of rice husk ash prepared by open air burning and furnace calcination. *J. Chem. Eng. Mater. Sci.*, 11, 2, 24–30, 2020.
27. Peralta, Y.M., Molina, R., Moreno, S., Rice husk silica: A review from conventional uses to new catalysts for advanced oxidation processes. *J. Environ. Manage.*, 370, 122735, 2024.
28. Chandrasekhar, S., Satyanarayana, K.G., Pramada, P.N., Ragjavan, P., Processing, properties and applications of reactive silica from rice husk—an overview. *J. Mater. Sci.*, 38, 3159–3168, 2003.
29. Sidheswaran, P. and Bhat, A.N., Recovery of amorphous silica in pure form from rice husk. *Trans. Indian Ceram. Soc.*, 55, 4, 93–96, 1996.
30. Mehta, A. and Ugwekar, D.R.P., Extraction of Silica and other related products from Rice Husk. *J. Eng. Res. Appl.*, 5, 8, 43–48, 2015.
31. Yalcin, N. and Sevinc, V., Studies on silica obtained from rice husk. *Ceram. Int.*, 27, 2, 219–224, 2001.
32. Krishnarao, R.V., Subrahmanyam, J., Kumar, T.J., Studies on the formation of black particles in rice husk silica ash. *J. Eur. Ceram. Soc.*, 21, 1, 99–104, 2001.
33. Shelke, V.R., Bhagade, S.S., Mandavgane, S.A., Mesoporous silica from Rice Husk ash. *Bull. Chem. React. Eng. Catal.*, 5, 2, 63–67, 2010.

34. Proctor, A., X-ray diffraction and scanning electron microscope studies of processed rice hull silica. *J. Am. Oil Chem. Soc.*, 67, 9, 576–584, 1990.
35. Thuadaij, N. and Nuntiya, A., Preparation of nanosilica powder from rice husk ash by precipitation method. *Chiang Mai J. Sci.*, 35, 1, 206–211, 2008.
36. Yuvakkumar, R., Elango, V., Rajendran, V., Kannan, N., High-purity nano silica powder from rice husk using a simple chemical method. *J. Exp. Nanosci.*, 9, 3, 272–281, 2014.
37. Rafiee, E., Shahebrahimi, S., Feyzi, M., Shaterzadeh, M., Optimization of synthesis and characterization of nanosilica produced from rice husk (a common waste material). *Int. Nano Lett.*, 2, 1, 1–8, 2012.
38. Wang, W., Martin, J.C., Zhang, N., Ma, C., Han, A., Sun, L., Harvesting silica nanoparticles from rice husks. *J. Nanopart. Res.*, 13, 12, 6981–6990, 2011.
39. Azhakesan, A., Yadav, S., Rajesh, V.M., Extraction of silica nanoparticles from Rice Husk Ash and its characterization. *J. Sci. Ind. Res.*, 79, 7, 656–660, 2020.
40. Adam, F., Balakrishnan, S., Wong, P.-L., Rice Husk Ash Silica As a Support Material for Ruthenium Based Heterogenous Catalyst. *J. Phys. Sci.*, 17, 2, 1–13, 2006.
41. Hien, T.T.T. and Long, N.S., Crystallization of Rice Husk Produced Silica and Silicon Carbide. *Key Eng. Mater.*, 1005, 27–32, 2024.
42. Saleem, M., Jamil, F., Qamar, O.A., Akhter, P., Hussain, M., Khurram, M.S., Al-Muhtaseb, A.H., Inayat, A., Shah, N.S., Enhancing the Catalytic Activity of Eggshell-Derived CaO Catalyst and Its Application in Biodiesel Production from Waste Chicken Fat. *Catalysts*, 12, 12, 1627, 2022.
43. Laca, A., Laca, A., Diaz, M., Eggshell waste as catalyst: A review. *J. Environ. Manage.*, 197, 351–359, 2017.
44. Niju, S., Begum, M.M.M.S., Anantharaman, N., Modification of eggshell and its application in biodiesel production. *J. Saudi Chem. Soc.*, 18, 5, 702–706, 2014.
45. Ahmad, S., Chaudhary, S., Pathak, V.V., Kothari, R., Tyagi, V.V., Optimization of direct transesterification of Chlorella pyrenoidosa catalyzed by waste eggshell based heterogenous nano – CaO catalyst. *Renewable Energy*, 160, 86–97, 2020.

8

Metal, Metal Oxide Catalysts and Beyond: Greener Approaches in Organic Synthesis

T. S. Perundevi

PG & Research Department of Chemistry, Ethiraj College for Women, Chennai, Tamilnadu, India

Abstract
Green chemistry utilizes only environmentally friendly chemicals and procedures which decreases the adverse impact of surrounding toxic materials on the environment. Catalysis plays a significant role in the synthesis of new and existing chemicals that are harmless and less hazardous to the environment. The green synthesis engages catalysts that can be recovered and utilized several times to run the reaction. The current study aims to provide a deep knowledge about the catalysts that are employed in various catalytic approaches. The chapter will also explain reactions engaging metal, metal oxide catalysts, phase transfer catalyst, biocatalyst, photocatalyst and many more. Catalytic applications of waste derived catalysts in various reactions and its significance will also be discussed.

Keywords: Green chemistry, biocatalyst, photocatalyst, phase transfer catalyst, metal, metal oxide, waste derived catalysts

8.1 Introduction

Green chemistry or sustainable chemistry is a branch of chemistry that deals with the discovery of non-toxic, environmentally friendly chemicals and procedures which can be applied to all facets of chemistry. Green chemistry, often known as environmentally benign chemistry minimizes toxicity. It decreases harmful effects on the environment, minimizes the generation of non-biodegradable waste, saves energy, and reduces the

Email: perundevi_ts@ethirajcollege.edu.in

Deepti Goyal, Sakshi Kabra Malpani and Renuka Jain (eds.) Catalysis from Waste: Transforming Solid Residues into Value, (193–216) © 2026 Scrivener Publishing LLC

depletion of natural resources. A shift from the traditional 'command-and control' approach to environmental issues, which required waste treatment, control, and clean-up through regulation, and towards preventing pollution at its source is reflected in the current emphasis on green chemistry [1]. It aims to design and implement pollution prevention technologies in addition to waste management [2, 3]. It explains the engineering concept of no waste and prevention of pollution at both industrial and laboratory scales. Treatment and disposal become unnecessary when waste is eliminated at the initial state of the reaction. For this purpose, reactions must be carefully designed in such a way that they do not generate hazardous wastes. However rapid increase in industrialization and urbanization has generated immense solid waste, whose improper disposal will have an adverse effect on the environment. Solid waste is a treasure of renewable resources. Hence, by identifying suitable ways to utilize solid waste, their potential can be utilized and their harmful impact on environment can be minimized.

Large amounts of solid waste can be processed into heterogeneous catalysts and utilized as starting materials, which greatly reduces the amount of solid waste produced as well as the expenses associated with the environmental remediation and creation of sustainable energy. The approach of turning solid waste into heterogeneous catalysts is not only innovative but also more cost effective and environmentally friendly than a pure chemical synthesis of catalysts. At present the principal methods engaged for managing most solid waste employs landfilling or stacking, which utilizes considerable land resources, emits toxic gases, thus leading to severe air pollution and global warming [4]. Therefore, exploring viable approaches for resourcefully utilizing solid waste is imperative. The idea of 'waste-to-wealth' is embodied in solid waste, which is a resource of abundant non-metallic or metallic elements, which can be used as modifiers, or can be used as a precursor in the synthesis of carbonaceous materials [5, 6]. At present, carbonaceous catalysts derived from solid waste (SW-CCs) are increasingly being utilized in environmental clean-up and energy-related applications, like the activation of persulfate (PS) [7]. Green chemistry uses cost-effective and environmentally friendly methods that reduce waste disposal costs after a chemical process, while also increasing yield. In green reactions, catalytic reagents are superior to stoichiometric reagents [8]. Adaptation of green chemistry benefits the society, human health with restoration of non-toxic environment. Catalysis plays a crucial role in sustaining the green synthetic route. Catalysts employed in a chemical process can reduce the waste generation at the end. Reactions can be made more selective and efficient through catalysis, which helps to eradicate a lot of waste chemicals and by-products [9].

The current chapter aims at the introduction of green or sustainable chemistry, catalysis, role of catalyst in prevention of waste. It then extends with types of catalyst. Over fourteen types of catalysts have been elucidated in this chapter. The nature of each catalyst has been briefed, followed by the emphasis of the catalyst employment in suitable reactions. Application of the catalysts has also been discussed, and specific examples have been cited.

8.2 Green Chemistry or Sustainable Chemistry

Green chemistry makes efficient use of ideally renewable raw materials, reduces waste, and avoids the use of toxic and hazardous chemicals and solvents in the production and use of chemical products [10]. The focus of green chemistry is primarily on pollution prevention rather than waste management. The term was formally established with the release of the 12 principles of green chemistry by Anastas and Warner in their 1998 book, Green Chemistry: Theory and Practice, [11] where the central guiding concept was 'benign by design'.

The 12 principles of green chemistry can be summarized as follows [12]:

(1) Waste should be prevented instead of being remedied
(2) Efficiency of atom usage
(3) Employing less hazardous materials
(4) Designing to yield safe end products
(5) Innocuous solvents and auxiliaries
(6) Proposal of reaction to be energy efficient
(7) Usage of sustainable raw materials
(8) Designing the method of synthesis with less steps
(9) Using catalytic reagents instead of stoichiometric
(10) Designing the products for bio degradation
(11) Methodologies for analysis aimed at preventing pollution
(12) Process that is fundamentally safe.

Green chemistry and catalysis are concepts that work hand in hand to decrease the toxic effects of chemical processes on the environment. A crucial component of green chemistry is catalysis, which promotes the efficient and waste-free completion of chemical reactions. Catalysis and green chemistry are vital for fostering environmentally friendly and waste-free chemical processes.

8.3 Role of Catalysts

The use of stoichiometric reagents, especially inorganic metals or compounds for organic synthesis is a potential cause of waste generation in the manufacture of pharmaceuticals and fine chemicals. The use of reducing agents such as metals like Na, Mg, Zn, Fe etc., and metal hydrides such as $LiAlH_4$, $NaBH_4$ and oxidizing agents like permanganate, manganese dioxide, and mineral acids, Lewis acids in organic reactions are also major sources of waste. Chemical reactions involving multi-step are likely to generate more waste. Substitution of stoichiometric methodologies with atom efficiency and minimal steps engaging economical catalytic species might prove a solution to minimize the waste [12]. Use of phase transfer catalyst, biocatalyst, photocatalyst and nanomaterials as catalysts strengthens the sustainable chemistry [13, 14].

Scientists can save energy and resources by gradually switching from stoichiometric chemical procedures to more efficient catalytic alternatives. A catalytic process can be generally referred to as green catalysis when homogeneous and heterogeneous catalytic reactions employing organic, organometallic, inorganic, and biological catalysts are preferred over stoichiometric processes [15, 16].

8.4 Classification of Catalysts

Based on the number of phases in the catalytic reaction that is carried out for a synthesis, homogeneous or heterogeneous catalysis can be cast off. Heterogeneous catalysis is a two or multi-phase catalytic reaction, whereas homogeneous catalysis is a single-phase catalytic reaction that is typically liquid/liquid. Homogeneous catalysts have a number of benefits, such as carrying out reactions at low temperature and preserving energy. At lower temperatures less quantity of undesired and undesirable by-products is produced, which tend to offer more specificity. Drawbacks involved in utilizing these homogeneous catalysts, are the difficulty in removing the catalyst from the reaction mixture, the high cost of recovering the catalyst, to use the catalyst for increased cycles of reaction and the release of significant quantities of heavier transition metals into the biosphere, which have a negative influence on the environment. Heterogeneous catalysis involves systems in which the reactions take place in diverse phases. Typically, the catalysts are in the solid phase, whereas the reactants are in the gaseous or liquid phase [17]. The lifespan of heterogeneous catalysts is significantly shortened by disadvantages such as high operating temperatures which hinder catalysts' capacity to produce high yields, speed up deactivation, and diminish their selectivity. Following types of catalysts are included in heterogeneous catalysts:

8.4.1 Surface Catalysts

In a broad spectrum of chemical processes, surface catalysts are crucial. Atoms are provided for chemical reactions by catalyst surfaces. For example, in a hydrogen fuel cell, hydrogen atoms from a gas are attracted to a metal surface in an electrochemical reaction to produce electricity [18]. Sites interacting with one another might be present on the catalyst surface. Even when both sites are present, the acid or base behavior may prevail on some surfaces [19]. Thin films and two-dimensional single- or polycrystalline bulk surfaces are examples of extended surface catalysts. These surfaces are crucial in many catalytic reactions because of their large surface area [20]. More surface area can be generated by either avoiding or reducing the agglomeration and sintering of tiny catalyst particles.

8.4.2 Acid-Base Catalysts

Solid acid catalysts have been made from waste polymers and clays, although most of these catalysts are made from waste biomass, which is frequently a byproduct of the manufacturing of biodiesel [21, 22]. The cost-effectiveness of the entire biodiesel manufacturing process may be improved by producing catalysts from leftover seed cakes or microalgal residues after bio-oil extraction. Oil seed cakes have been pyrolyzed and then sulphonated to create solid acid catalysts. To create a catalyst for the one-step conversion of the extracted oil into biodiesel, Dawodu *et al.*, first extracted oil from the non-edible seeds of the evergreen tree *Calophyllum inophyllum*, which contained about 15% free fatty acid [23]. Konwar *et al.*, used a similar method to turn the Mesua ferrea tree's seed cake into a solid acid catalyst [24].

Compared to their solid acid counterparts, heterogeneous solid bases usually offer significantly better rates of triacyl glycerides transesterification. Most of these catalysts are made by calcining shells from shellfish, snails, eggs, at high temperatures (usually $\geq$ 800°C) [25]. This method provides an economical way to dispose of agricultural and mariculture waste. This method has been used to create solid bases from shrimp [26] and turtle shells [27] by calcining them at 450 and 500°C, respectively.

By calcining coral fragments at 700°C, Roschat *et al.*, produced CaO, which was then used as a catalyst for transesterification of different oils such as palm, soybean, rice bran, and residual cooking oils—with methanol [28].

8.4.3 Metal Catalysts

Single atoms such as Na, Mg, Zn, Fe etc., nanoclusters, and nanoparticles of various metal species exhibit diverse catalytic behaviors for various processes. Lacanau *et al.*, skipped the typical separation and purification processes by extracting Pd from e-waste and using Pd (II) as a catalyst [29]. Recently, McCarthy *et al.*, showed that complexed metal ions recovered from e-waste can also be utilized as efficient heterogeneous catalysts for a variety of chemical processes, proving that catalyst recovery need not be solely focused on recovering the metal of interest [30]. While recycling used lithium-ion batteries, Paone *et al.*, added a reduction step that resulted in the partial creation of metal particles possessing catalytic activity [31]. There have been several reports of waste-derived catalysts for the generation of hydrogen from different precursors. To prepare catalysts for the synthesis of H_2 through methane dry reforming, Kanamori *et al.*, utilized battery trash as a source of Ni. Ni was present along with La, Ce, Co, Mn, and Nd in the waste of Ni-metal hydride batteries. Acid leaching was used to remove these metals, which were then precipitated with a base and finally calcined [32]. Genova *et al.*, investigated the breakdown of methanol over Co/Fe/C catalysts made from waste biomass [33].

8.4.4 Metal Oxide Catalyst

Metal oxide catalysts play a substantial role in many industrial catalytic processes [34]. Simple oxides like silica, alumina, and perovskite oxide materials act as metal oxide catalysts [34, 35]. Additionally, metal oxides made from industrial metal waste through spark erosion have also been reported to be used as catalysts [36]. These materials, which had low surface areas of 12–70 m^2 g^{-1} and included approximately 50 weight percent Fe and trace amounts of Cr, Cu, and other metals, were investigated as fixed bed catalysts for the oxidation of volatile organic molecules (ethanol and xylene) in continuous flow between 250 and 300°C. An oxide catalyst with 50 weight percent Fe, 1 weight percent Cr, and less than 1 weight percent transition and s-block elements calcined at 600°C produced the maximum conversion, with total ethanol (xylene) oxidation occurring at 300°C. Spark erosion thus offers a potential way to extract catalytically active metals from industrial waste. A waste product obtained from the manufacture of automobiles with high iron content is grinding wheel ash. Such ash was dissolved, precipitated, and aged to produce a catalyst for the degradation of methylene blue [37]. Unique physicochemical characteristics of metal oxide-based catalysts make them efficient for the removal of gaseous pollutants. Metal oxide catalysts are useful and significant for a wide range of chemical processes. Calcite, a mineral that contains CaO, has shown remarkable effectiveness in trans-esterification when used as a catalyst, and it can be easily obtained from various organic or inorganic waste products. Mollusk shells and eggshells are effective organic sources that yield highly active CaO based catalysts. Eggshells, which make up around 10% of an egg, are produced in large quantities by the egg processing industries; as a result, a significant amount of this solid waste is disposed of in landfills as garbage. The issue of organic contamination was brought up by this disposal approach without any pre-treatment [38, 39]. Taufiq-Yap *et al.*, analyzed the effectiveness of an eggshell catalyst in producing hydrogen by wood gasification [40]. Catalysts based on eggshells can also be used to create various bioactive substances such as 2-aminochromenes [41].

Dawodu *et al.*, have used the seed cake to create a catalyst for the one-step conversion of the extracted oil into biodiesel [23]. Sushil *et al.*, utilized red mud waste as precursors of metal/metal oxide catalysts [42]. Teixeira *et al.*, showed that red mud can function as a catalyst for the transformation and storage of petroleum gas, which is a blend of methane and additional hydrocarbons produced and burned off during offshore oil exploration [43]. Waste from bovine bones can also serve as a possible catalytic material in the transesterification process of soybean oil [44]. Grinding wheel ash, a by-product with high iron content generated from automobile manufacturing, is also reported to act as precursor of catalysts. The processes of

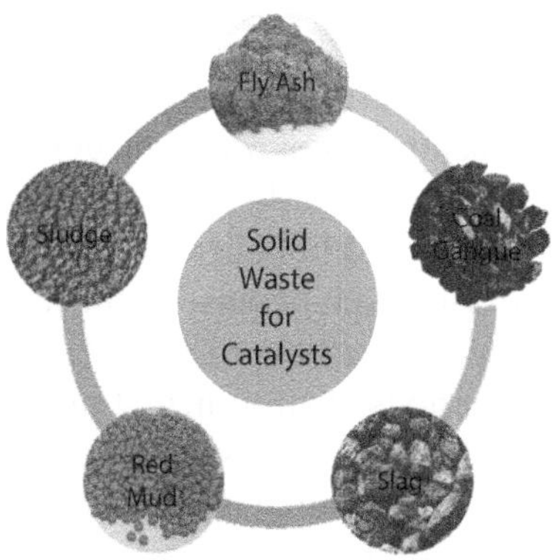

Figure 8.1 Solid waste for metal and metal oxide catalyst.

dissolution, precipitation, and aging of this ash were employed to develop a catalyst for the degradation of methylene blue [37]. Stainless steel slag was utilized as a raw material to produce silica and silica-alumina supports for Pt-based VOC oxidation catalysts [45].

Solid waste such as fly ash, slag, red mud, coal gangue and sludge are some waste which are necessary for replacing natural silica precursors (Figure 8.1). Since then, the abundant and inexpensive silica substitutes that can be freely obtained from such solid waste have been used in the production of heterogeneous catalysts for a wide range of applications. Coal fly ash (CFA), sometimes referred to as bottom ash, is a waste product produced by burning coal, biomass [46]. With aluminosilicates constituting up the majority of CFA's composition, its diverse metal oxide composition exhibits exceptional thermal stability [47]. Therefore, because of its high silica and alumina content, CFA has remarkable potential to be used as commendable porous catalysts by impregnation with other active ingredients and adsorbents. CFA was used as a heterogeneous catalyst by Kastner *et al.* [48], to catalytically oxidize sulfur compounds that have an unpleasant odor. Likewise, Kolar *et al.*, used wood fly ash as a catalyst for aldehyde catalytic ozonation and catalytic air oxidation. In catalytic air oxidation, wood fly ash was used to evaluate two common volatile organic compounds (VOCs), 2-methylbutanal (2-MB) and 3-methylbutanal (3-MB) [49]. The use of CFA trash in South Africa to create biodiesel (methyl esters) by a transesterification process between methanol and sunflower oil was demonstrated by Babajide *et al.* [50]. KNO_3 (5wt.%) loaded CFA-based catalyst with a maximum oil conversion of 86.13% at a reaction temperature of 160°C is reported. Tests were conducted to evaluate the fuel nature of the synthesized methyl esters (biodiesel) with the ASTM biodiesel standard. The simple separation of catalysts and their combined use in a continuous process significantly decreased the overall cost of producing biodiesel.

One of the largest agro-industries in Southeast Asia is the palm oil sector, which produces a vast amount of trash, including empty fruit bunches (EFB), oil palm fibers, and oil palm shells [51]. Meanwhile, when palm oil husk and kernel shell are burned in boilers at palm oil mills, palm oil fuel ash (POFA) is obtained. However, because of the high levels of black soot and ash produced during combustion, this traditional way of disposing of oil palm wastes caused serious environmental problems [52]. The use of oil palm kernel shell (OPKS) as a catalyst support for the fatty acid methyl ester (FAME) generation was explored by Kathirvel *et al.*, at different oil to catalyst loadings and reaction temperatures [53], thus demonstrating the great viability of palm-oil based catalysts. In general, biodiesel is a type of diesel fuel that is mostly made up of FAME, which is created when methanol and vegetable oil undergo a catalyzed trans-esterification reaction. Because FAME is non-toxic, low emission, carbon-neutral, renewable, and biodegradable fuel, it has garnered a lot of interest from researchers and industry [54]. The percentage of FAME generated showed that the NaOH incorporation on OPKS was successful. The molar ratio of oil to methanol (1:6–1:12), reaction temperature (30–60°C), catalyst to oil ratio (1-3 weight percent), and reaction time (3-6h) were all factors analyzed in the batch system experiment. It was demonstrated that an oil palm-based catalyst could be employed as an active catalyst to produce FAME.

The outer sheaths that develop on rice grains throughout their maturation that are removed during processing are known as rice husks (RH). RH is composed of 20% ash, 38% cellulose, 22% lignin, and other organic materials. Burning the RH produced a large amount of rice husk ash (RHA), which was then disposed of as waste [55, 56]. The use of RHA as support to produce a solid base catalyst to facilitate the catalytic conversion of palm oil into biodiesel has been shown by Chen *et al* [57]. Furthermore, Maheswari and colleagues documented the application of bamboo RHA, an environmentally benign, reusable, biodegradable, and effective heterogeneous catalyst, for the synthesis of tetrahydro-4H chromene-3-carbonitriles [58].

Cao *et al.*, produced hydrogen from ammonia using a modified mesoporous red mud [59]. Dehydrogenation catalysts made from recycled tires were tested using methylcyclohexane, another suggested H_2 carrier [60].

8.4.5 Photocatalysts

The term "photocatalysis" was initially used in scientific publications in 1911 [61]. The four basic steps involved in photocatalysis are given in Figure 8.2. Pyrolysis, hydrolysis, gasification, and carbonization are the primary methods used to create biochar (BC), a carbon-rich pyrogenic substance which can be used as an efficient photocatalyst. The large surface area, high porosity, enhanced ion exchange capacity, and abundance of oxygen functional groups are the important properties of biochar-based photocatalysts [62]. Early investigations expanded the uses of photocatalysis, thereby enticing substantial research interest in the 1980's for comparable reactions employing TiO_2 nanoparticles as the photocatalysts in particular. Liu *et al.*, have established the fact that approximately out of 53% of the Ti that is processed into iron-rich concentrations for blast furnace feeding, 25% of the slag produced is TiO_2. With an annual production volume of around three million tons, TiO_2 emerges as a promising and cost-effective candidate for photocatalyst development [63]. Since then, studies have concentrated on comprehending the underlying ideas, improving photocatalytic efficiencies, finding novel photocatalysts, and broadening the range of reactions.

Since ZnO has an identical band gap energy to TiO_2 but better absorption efficiency when exposed to sunlight, it has been suggested as an alternate photocatalyst. Waste zinc dust derived ZnO nanoparticles were employed as a photocatalytic material for degradation of methylene blue (MB) dye [64].

The creation of a ZnO/diatomite hybrid photocatalyst from industrial waste is another noteworthy achievement that deserves attention. Diatomite waste from breweries was utilized as a porous substrate, while zinc dust waste from a hot dip galvanizing process was utilized as a ZnO precursor [65].

The mesoporous materials derived from gibbsite solid waste, γ-Al_2O_3 nanowires modified with 12% Fe_2O_3, were more effective than individual analogue electrodes as an electrocatalyst for MeOH oxidation in an acidic medium [66].

In several processes, sulfides have also been suggested as appropriate photocatalysts. A sulfur source is required to make the equivalent sulfides, in contrast to metallic oxide, where the oxide can be easily produced by calcination in the air. One significant source of sulfur produced from waste is extraction of mining leftover that contains sulfur. An interesting example of sulfur containing waste is waste pig bristles from which S and C can be extracted to form metallic sulfides. The use of pig bristles as precursors of ZnS and Ag/Ag_2S utilized in photocatalysis has been reported in the literature [67, 68].

In recent years, carbon quantum dots (CQDs) have drawn more attention as a desirable substitute for other kinds of multicolor emissive materials.

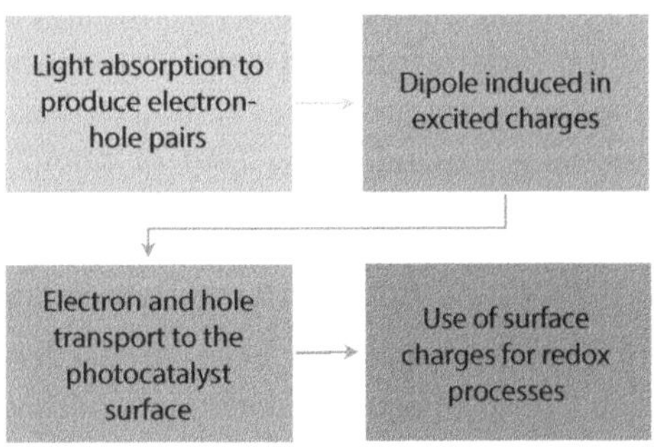

Figure 8.2 Four significant steps in photocatalysis.

Using their high solubility and aqueous affinity, chemical/photo and colloidal stability, notable optical properties, photo bleaching resistance, low cost, low toxicity, and good functionalization possibilities, CQDs have been used for biological purposes as well as photocatalytic materials for pollutant degradation, solar devices, and photo/electrochemical water splitting. Biomass-derived CQDs are exceptionally inexpensive and environmentally beneficial. Although a variety of biomass sources, such as fruit peels, dairy waste whey, spent coffee grounds, pulp and paper industry residues, and cigarette filters, have been used to prepare CQDs, [69–71] citric acid is currently the most widely used starting material to produce highly luminescent CDs, typically in conjunction with nitrogen-containing dopants. For example, a one-step pyrolysis approach was used to prepare CQDs with intriguing optoelectronic properties from citrus limetta waste pulp. These CQDs were reported to be successfully used in photo-electrochemical water splitting, photocatalytic MB degradation, bioimaging, iron ion detection, and antibacterial activity [72].

A family of photoredox N-doped CQDs was recently produced by Prato *et al.*, and their tunable oxidation/reduction potential suggests that they could be used as photocatalysts [73]. A sustainable and environmentally friendly alternative for creating carbon particles in an aqueous media from a readily available biomass residue- orange peel is the hydrothermal carbonization process. Under UV light, a composite material made of the produced CQDs and ZnO was used as a photocatalyst to degrade naphthol blue-black azo dye [74]. Waste from the paper industry has been reported to be converted into CQDs with promising fluorescence qualities using a microwave-assisted treatment employing a solid acid catalyst [75].

Several carbon-based materials generated from biomass, such as hydrochar and carbonized biomass *via* moderate oxidation ($NaOH/H_2O_2$) solution have also been used in the large-scale synthesis of CQDs. CQDs were produced under these circumstances with an exceptional yield of 76.9 weight percent, which is significantly greater than that produced by conventional hydrothermal and strong acid oxidation procedures. Excellent quantum yield (QY) was another feature of the as-synthesized CQDs [76].

Because of superior charge transport capabilities, broad spectrum of solar absorption, and increased photodegradation efficiency under visible light irradiation, plasmonic photocatalysts have drawn interest from researchers [77]. Victor *et al.*, have reported the synthesis of plasmonic responsive gold nanoparticles from the upcycling of e-waste [78]. This type of interesting material architecture can be created by dispersing noble metal nanoparticles onto the semiconductors. By doing this, the advantages achieved are localized surface plasmon resonance (LSPR) and a

Schottky barrier [79]. These features will facilitate the effective separation and transfer of the charge carriers when exposed to visible light. The LSPR, which indicates a notable oscillation on the surfaces of semiconductor photocatalysts and metal nanoparticles, is the most important component of plasmonic photocatalysis.

8.4.6 Nano Catalysts

Amidst the eclectic series of catalysts available, the nano catalysts that are derivatives of nanoparticles behave as exceptional substitutes to conventional bulk catalysts because of their exceptional stability, robustness, and surface area [80, 81]. When compared to bulk catalysts, the nano catalyst has more active surfaces, is cost effective, possesses high surface area and catalytic activity, can easily synthesize *via* conventional methods as well as by using natural plant extracts (Figure 8.3). A larger surface area for reactions at a catalytic site results in a faster rate of reaction; and thus, increased catalytic activity, which improves process efficiency. The rate of catalysis is directly proportional to the effective catalytic sites available for the reaction to occur. A catalytic site with a greater surface area for reactions results in a higher catalytic activity and a faster rate of reaction, which raises the reaction's efficiency. The catalyst's insolubility in the reaction mixture is an additional benefit that facilitates easy separation and reuse of the nano catalysts. Agricultural waste derived nano catalysts have emerged as efficient catalysts due to their low aggregation, stability and high surface area [82].

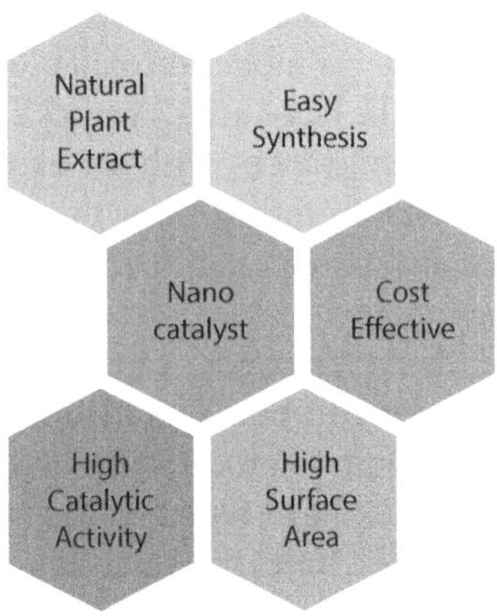

Figure 8.3 Significant features of nano catalyst.

8.4.7 Auto Catalysts

To reduce hazardous emissions, auto catalysts are used in car catalytic converters. In the process of autocatalysis, one of the byproducts also serves as a catalyst for a related or coupled reaction [83, 84]. Auto catalysts are employed in the context of vehicle catalytic converters to transform perilous pollutants in smoke into less dangerous ones. In order to accelerate the chemical reactions that transform pollutants into less dangerous molecules, the auto catalyst is coated with a chemical solution and a combination of platinum, rhodium, and/or palladium metals [85].

8.4.8 Metal Nanoparticles as Green Catalyst

Metal nanoparticles are synthesized in nanoscale dimension (1–100 nm) comprising both organic and inorganic constituents with various metals [86, 87]. The synthesized metal nanoparticles possess a higher surface area-to-volume ratio than the bulk material, thus playing a crucial function as catalyst. These metal or metal oxide nanoparticles can also be synthesized *via* easy, cost-effective, and environmentally friendly ways by using natural plant extracts. Mostly for this purpose, extracts from fruits, flowers, roots, seeds, and leaves have been employed. Protein, amino acids, vitamins, polysaccharides, terpenoids, and organic acids are among the phytochemicals found in such extracts. These compounds function as a natural reducing agent and stabilizer, which is essential for controlling the shape and size of nanoparticles. Noble metals, such as gold and silver nanoparticles, are therefore the most often utilized in the catalytic industry due to their robust catalytic reduction, antimicrobial, and optoelectronic qualities. Pathak and co-workers have synthesized magnetic nanoparticles from the peel of M. acuminata and used it as a solid catalyst for Henry's reaction [88]. Owing to their intriguing characteristics and increased catalytic activity, metal nanoparticles have drawn a lot of attention. The ability to synthesize metal nanoparticles at a low cost and their inherent versatility has made these materials indispensable building blocks for emerging technology.

The synthesis of metal nanoparticles from typical solid waste and its use in the treatment of polluted water was examined by El Hameed *et al.* Numerous benefits of Ni nanoparticle catalysts made from starch-based loose fill were demonstrated. It was inexpensive to produce, simple to separate and reuse, and straightforward to prepare industrially. In terms of the environment, it degraded water contaminants quite effectively and prevented the use of harmful materials. In terms of society, it offered a fresh and cost-effective approach to solid waste management that lessened its negative impacts [89]. Because of these remarkable qualities, metal nanoparticles are being used in a wider range of optical, pharmaceutical, technical, and biological applications.

The primary benefits of the metal nanoparticles are their strong optical and magnetic characteristics, electrical conductivity, high melting and boiling temperatures, and thermal and mechanical resilience. While taking metal nanoparticle toxicity into account; research indicates that an array of aspects, such as size, shape, and surface properties, influence both the toxicity and biological activity of metal nanoparticles [90]. Solubility and chemical composition are two factors that are important in the recycling process of these catalysts. High surface area and more active sites in metal nanoparticles promote faster reactions and upsurge the product yield. These nanoparticles may be widely segregated into two major groups: noble-metal (Au, Pt, Ag, etc.)-based metal nanoparticles and non-noble-metal (Fe, Cu, Ni, Co, etc.)-based nanoparticles. Progress in materials science and nanotechnology in the last two decades have contributed to achieving exceptional control over the production of metallic nanoparticles with different sizes, shapes, and compositions [91].

In general, under mild conditions, metal nanoparticles including Pt, Pd, and Au showed high activity and selectivity towards the selective oxidation of alcohols. Furthermore, new studies show that allylic alcohols can be selectively oxidized by Pd-supported mesoporous alumina and other single-atom catalysts. On the other hand, single-atom Pd supported on mesoporous alumina demonstrated a higher turnover frequency (TOF) than Au nanoparticles while having poor selectivity [92]. However, the high expense and toxicity of catalysts supported by noble metals highlights the need for substitutes in the search for easily accessible, non-toxic, and less costly catalytic materials. Non-noble metal supported nanoparticles (based on Fe, Cu, Ni, etc.) are among the many possibilities that have been extensively explored because of their exceptional catalytic properties and wide availability.

Multiwalled carbon nanotubes were synthesized from low boiling point hydrocarbon tyre pyrolysis oil derived from waste tyre material mixed with ferrocene as the catalyst on a quartz substrate with a flow rate of 20 ml/min at 950 [93].

Despite significant progress in this field, many challenges need to be overcome, including:

- Developing simple methods for producing large amounts of consistent, less hazardous, durable nanoparticles with high surface areas and selectivity characteristics.
- Maintaining the nanoparticle's catalytic characteristics in adverse conditions.
- Gaining an extensive knowledge of the catalytic process's mechanisms in order to design strong catalytic systems.
- Examining the nanoparticle's biocompatibility while considering various factors like particle size, coordination ligands, etc.
- Planning a number of stratagems for mass production. However, some of the issues that come up frequently are surface contamination, capping agent removal, and particle aggregation. For both present-day and foreseeable industrial uses, each of these challenges must be evaluated immediately. While there are still a few obstacles to overcome and a number of procedures that need to be thoroughly assessed. These issues should be resolved soon, which will pave the way for the creation and broader distribution of effective catalysts with numerous real-world uses.

8.5 Conclusion

Green chemistry, a much-needed method for the prevention of contamination of earth's crust with toxic and hazardous chemicals, can achieve its purpose by employing catalysts that promote green synthesis. A broad spectrum of catalyst is available, choosing and utilizing a perfect catalyst for a particular reaction which enhances the yield and fetches the required end product is a skillful job. This chapter caters almost an entire range of catalysts that can be used in sustainable chemistry to reduce the release of harmful and toxic effluents. By recognizing the unique nature of reactions and the products that are aimed for, tailored catalysts such as homogenous or heterogenous can be used. Bronsted-Lowry concept explains the role of acid or base catalyst in green synthesis; transition metal ions serve as suitable catalysts and has augmented the progress of new reactions to access functionalized molecules at cheap and quick production. Metal oxide and metal complex catalysts are being used for various redox reactions. The use of transition-metal catalysis to form stereogenic centers in organic compounds is now a well-established method in organic chemistry. In the past, metals such as rhodium, iridium, ruthenium, and palladium were commonly favored. At present, research is aimed at substituting these metals with more affordable and less toxic alternatives like iron. Unfortunately, simply transferring methods, meaning applying the same chiral ligands or reaction conditions used with well-established catalysts like Rh, Ir, Ru, and Pd to iron, is either not feasible or presents challenges. Therefore, there is a strong demand for innovative catalysts. Organic synthesis makes use of a variety of photocatalytic techniques. This field has recently gained particular interest due to the significant rise in research activity in organic photo redox catalysis. These waste-derived catalysts' potential to promote reactions with less of an impact on the environment—for example, using fewer solvents or reducing greenhouse gas emissions—is closely related to the 12 green chemistry principles. Their catalytic efficiencies often equal, if not surpass, those of conventionally available catalysts in processes ranging from carbon dioxide reduction or building block synthesis to biomass conversion, as several studies have noted. However, challenges continue to persist. Variations in waste composition can lead to inconsistencies in catalyst characteristics, which may decrease their effectiveness;, for instance, due to catalyst poisoning. Moving from laboratory settings to industrial uses will necessitate process enhancement and deeper investigation to completely grasp their enduring stability and recycling capacity. The precise ways in which certain waste-derived catalysts speed up reactions may not be thoroughly understood. Gaining further understanding of the molecular-level interactions can aid in the refinement of the catalytic process. Although some research indicates possible economic advantages, detailed lifecycle assessments are essential to measure both ecological and economic effects throughout the catalyst's lifecycle. This involves accounting for the energy and resources utilized in the stages of production of catalyst, its operation, and disposal. Tackling these challenges demands a multidisciplinary approach involving chemistry, material science, environmental science, economics, and policymaking.

AI disclosure statement: The author confirms that no artificial intelligence (AI)-generated content or AI-assisted writing tools were used in the preparation of this chapter (including summaries and the abstract). All text, analysis, and conclusions are the result of the author's original work.

References

1. Anastas, P.T., Kirchhoff, M.M., Williamson, T.C., Catalysis as a foundational pillar of green chemistry. *Appl. Catal., A*, 221, 1-2, 3–13, 2001.

2. Onuegbu, T.U. and Ekemezie, P.N., Overview of Green Chemistry, an Effective Pollution Prevention. *CSN Mag.*, 3, 3, 23–27, 2006.
3. Onuegbu, T.U. and Ibeh, T.N., *An Overview of Green chemistry for Sustainable Development Proceeding of 31st Annual International Conference of CSN, (Delta Chem 2008)*, pp. 3–8, 2008.
4. Yaashikaa, P.R., Kumar, P.S., Saravanan, A., Varjani, S., Ramamurthy, R., Bioconversion of municipal solid waste into bio-based products: A review on valorisation and sustainable approach for circular bioeconomy. *Sci. Total Environ.*, 748, 141312, 2020.
5. Cheng, N., Wang, B., Chen, M., Feng, Q., Zhang, X., Wang, S., Zhao, R., Jiang, T., Adsorption and photocatalytic degradation of quinolone antibiotics from wastewater using functionalized biochar. *Environ. Pollut.*, 336, 122409, 2023.
6. Wang, N., Wang, B., Zhang, X., TiO_2-loaded phosphor gypsum-modified biochar for the removal of ofloxacin and Cu^{2+}: Performance, mechanisms, and toxicity assessment. *Chem. Eng. J.*, 498, 155441, 2024.
7. Ma, D., Wang, J., Feng, K., Liu, B., Xie, G., Xing, D., A green strategy from waste red mud to FeO-based biochar for sulfadiazine treatment by peroxydisulfate activation. *Chem. Eng. J.*, 446, 136944, 2022.
8. Raksmey, I.M. and Romasco, R., Green Power, the Positive Impacts of a Green Chemistry Service-Learning Project. The magazine for ACS. *Green Chem. Generating React.*, 14, 4, 19, 2005.
9. Simmons, M.S., The Role of Catalysts in Environmentally Benign Synthesis of Chemicals, Green Chemistry, Designing Chemistry for the Environment. *ACS Symposium Series 626*, New York, p. 251, 1996.
10. Sheldon, R.A., Atom utilisation, E factors and the catalytic solution. *C.R. Acad. Sci., Ser. IIc: Chim.*, 3, 541–551, 2000.
11. Anastas, P. and Warner, J.C., *Green Chemistry: Theory and Practice*, Oxford University Press, Oxford, U.K., 1998.
12. Sheldon, R.A. and Woodley, M.J., Role of Biocatalysis in Sustainable Chemistry. *Chem. Rev.*, 118, 2, 801–838, 2018.
13. Wender, P.A., Handy, S.T., Wright, D.L., Toward the ideal synthesis. *Chem. Ind. (London)*, 765–769, 1997.
14. Sheldon, R.A., Catalysis and pollution prevention. *Chem. Ind. (London)*, 12–15, 1997.
15. Sheldon, R.A., Arends, I.W.C.E., Hanefeld, U., *Green Chemistry and Catalysis*, Wiley-VCH, Weinheim, Germany, 2007.
16. Sheldon, R.A., Catalysis: the key to waste minimization. *J. Chem. Technol. Biotechnol.*, 68, 381–388, 1997.
17. Chhangani, M.K., Catalytic green synthesis of chemicals for sustainable future. *Int. Res. J. Manag. Sci. Technol.*, 7, 3, 2250–1959, 2016.
18. Offsets, C. and Calculator, C.F., 3 Research Groups Working on Reducing and Converting CO_2 into Renewable Energy (REC). *Renewable Energy*, 2023.

19. Schneider-Coppolino, M., Gautam, S., Gates, B.D., Nano catalysts for proton exchange fuel cells: design, preparation, and utilization In PEM. *Fuel Cells*, 465–545, 2022.
20. Xie, L.S., Skorupskii, G., Dincă, M., Electrically Conductive Metal–Organic Frameworks. *Chem. Rev.*, 120, 16, 8536–8580, 2020.
21. Alonso-Fagúndez, N., Laserna, V., Alba-Rubio, A.C., Mengibar, M., Heras, A., Mariscal, R., López Granados, M., Poly-(styrene sulphonic acid): An acid catalyst from polystyrene waste for reactions of interest in biomass valorization. *Catal. Today*, 234, 285–294, 2014.
22. do Nascimento, L.A.S., Tito, L.M.Z., Angélica, R.S., da Costa, C.E.F., Zamian, J.R., da Rocha Filho, G.N., Esterification of oleic acid over solid acid catalysts prepared from Amazon flint kaolin. *Appl. Catal., B*, 101, 3–4, 495–503, 2011.
23. Dawodu, F.A., Ayodele, O., Xin, J., Zhang, S., Yan, D., Effective conversion of non-edible oil with high free fatty acid into biodiesel by sulphonated carbon catalyst. *Appl. Energy*, 114, 819–826, 2014.
24. Konwar, L.J., Das, R., Thakur, A.J., Salminen, E., Arvela, P.M., Kumar, N., Mikkola, J.P., Deka, D., Biodiesel production from acid oils using sulfonated carbon catalyst derived from oil-cake waste. *J. Mol. Catal. A: Chem.*, 388–389, 167–176, 2014.
25. Bennett, J.A., Wilson, K., Lee, A.F., Catalytic applications of waste derived materials. *J. Mater. Chem. A*, 4, 3617–3637, 2016.
26. Yang, L., Zhang, A., Zheng, X., Shrimp Shell Catalyst for Biodiesel Production. *Energy Fuels*, 23, 8, 3859–3865, 2009.
27. Jie, X., Zheng, X., Aiqin, D., Zhidong, X., Jinhua, Z., Biont shell catalyst for biodiesel production. *Green Chem.*, 11, 355–364, 2009.
28. Roschat, W., Kacha, M., Yoosuk, B., Sudyoadsuk, T., Promarak, V., Biodiesel production based on heterogeneous process catalyzed by solid waste coral fragment. *Fuel*, 98, 194–202, 2012.
29. Lacanau, V., Bonneté, F., Wagner, P., Schmitt, M., Meyer, D., Bihel, F., Contino-Pépin, C., Bourgeois, D., From Electronic Waste to Suzuki-Miyaura Cross-Coupling Reaction in Water: Direct Valuation of Recycled Palladium in Catalysis. *ChemSusChem*, 13, 5224–5230, 2020.
30. McCarthy, S., Desaunay, O., Jie, A.L.W., Hassatzky, M., White, A.J.P., Deplano, P., Braddock, D.C., Serpe, A., Wilton-Ely, J.D.E.T., Homogeneous Gold Catalysis Using Complexes Recovered from Waste Electronic Equipment. *ACS Sustainable Chem. Eng.*, 10, 15726–15734, 2022.
31. Paone, E., Miceli, M., Malara, A., Ye, G., Mousa, E., Bontempi, E., Frontera, P., Mauriello, F., Direct Reuse of Spent Lithium-Ion Batteries as an Efficient Heterogeneous Catalyst for the Reductive Upgrading of Biomass-Derived Furfural. *ACS Sustainable Chem. Eng.*, 10, 2275–2281, 2022.
32. Kanamori, T., Matsuda, M., Miyake, M., Recovery of rare metal compounds from nickel–metal hydride battery waste and their application to CH_4 dry reforming catalyst. *J. Hazard. Mater.*, 169, 1–3, 240–245, 2009.

33. Genova, I., Tsoncheva, T., Dimitrov, M., Paneva, D., Tsyntsarski, B., Ivanova, R., Zheleva, Z.C., Budinova, T., Kovacheva, D., Mitov, I., Petrov, N., Cobalt ferrite nanoparticles hosted in activated carbon from renewable sources as catalyst for methanol decomposition. *Catal. Commun.*, 55, 43–48, 2014.
34. Lebedeva, O.E. and Sarmurzina, A.G., Industrial wastes as catalyst precursors: VOC oxidation. *Appl. Catal., B*, 26, 1, L1–L3, 2000.
35. Zhao, Q., Huang, C., Li, F., Phosphorescent heavy-metal complexes for bioimaging. *Chem. Soc. Rev.*, 40, 5, 2508–2524, 2011.
36. Shee, J., Loipersberger, M., Hait, D., Lee, J., Head-Gordon, M., Revealing the nature of electron correlation in transition metal complexes with symmetry breaking and chemical intuition. *J. Chem. Phys.*, 154, 19, 194109, 2021, doi: 10.1063/5.0047386. PMID: 34240907.
37. Yuan, B., Xu, J., Li, X., Fu, M.L., Preparation of Si–Al/α-FeOOH catalyst from an iron-containing waste and surface-catalytic oxidation of methylene blue at neutral pH value in the presence of H_2O_2. *Chem. Eng. J.*, 226, 181–188, 2013.
38. Gao, Y. and Xu, C., Synthesis of dimethyl carbonate over waste eggshell catalyst. *Catal. Today*, 190, 107–111, 2012.
39. Oliveira, D.A., Benelli, P., Amante, E.R., A literature review on adding value to solid residues: egg shells. *J. Cleaner Prod.*, 46, 42–47, 2013.
40. Taufiq-Yap, Y.H., Wong, P., Marliza, T.S., Nurul Suziana, N.M., Tang, L.H., Sivasangar, S., Hydrogen production from wood gasification promoted by waste eggshell catalyst. *Int. J. Energy Res.*, 37, 1866–1871, 2013.
41. Mosaddegh, E., Ultrasonic-assisted preparation of nano eggshell powder: A novel catalyst in green and high efficient synthesis of 2-aminochromenes. *Ultrason. Sonochem.*, 20, 1436–1441, 2013.
42. Sushil, S., Scholz, P., Pollok, K., Ondruschka, B., Batra, V.S., Application of industrial waste based catalysts for total oxidation of propane. *Chem. Eng. J.*, 166, 2, 568–578, 2011.
43. Teixeira, I.F., Tayline, P.V., Medeiros, Freitas, P.E., Rosmaninho, M.G., Ardisson, J.D., Lago, R.M., Carbon deposition and oxidation using the waste red mud: A route to store, transport and use offshore gas lost in petroleum exploration. *Fuel*, 124, 7–13, 2014.
44. Smith, S.M., Oopathum, C., Weeramongkhonlert, V., Smith, C.B., Chaveanghong, S., Ketwong, P., Boonyuen, S., Transesterification of soybean oil using bovine bone waste as new catalyst. *Bioresour. Technol.*, 143, 686–690, 2013.
45. Domínguez, M.I., Barrio, I., Sánchez, M., Centeno, M.A., Montes, M., Antonio Odriozola, J., CO and VOCs oxidation over Pt/SiO_2 catalysts prepared using silicas obtained from stainless steel slags. *Catal. Today*, 133-135, 467–474, 2008.
46. Dhokte, A.O., Khillare, S.L., Lande, M.K., Arbad, B.R., Synthesis, characterization of mesoporous silica materials from waste coal fly ash for the classical Mannich reaction. *J. Ind. Eng. Chem.*, 17, 742–746, 2011.

47. Shaheen, S.M., Hooda, P.S., Tsadilas, C.D., Opportunities and challenges in the use of coal fly ash for soil improvements e A review. *J. Environ. Manage.*, 145, 249–267, 2014.
48. Kastner James, R., Das, K.C., Quentin, B., Melear Nathan, D., Low Temperature Catalytic Oxidation of Hydrogen Sulfide and Methanethiol Using Wood and Coal Fly Ash. *Environ. Sci. Technol.*, 37, 2568–2574, 2003.
49. Praveen, K., Kastner, J.R., Miller, J., Low temperature catalytic oxidation of aldehydes using wood fly ash and molecular oxygen. *Appl. Catal., B*, 76, 203–217, 2007.
50. Babajide, O., Petrik, L., Musyoka, N., Amigun, B., Ameer, F., Use of Coal Fly Ash as A Catalyst In the Production of Biodiesel. *Pet. Coal*, 52, 261–272, 2010.
51. Cheng, Y.W., Chang, Y.S., Hoong Ng, K., Wu, T.Y., Cheng, C.K., Photocatalytic restoration of liquid effluent from oil palm agroindustry in Malaysia using tungsten oxides catalyst. *J. Cleaner Prod.*, 162, 205–219, 2017.
52. Acquah, C., Yon, L.S., Tuah, Z., Ngee, N.L., Danquah, M.K., Synthesis and performance analysis of oil palm ash (OPA) based adsorbent as a palm oil bleaching material. *J. Cleaner Prod.*, 139, 1098–1104, 2016.
53. Kathirvel, S., Layek, A., Muthuraman, S., Exploration of waste cooking oil methyl esters (WCOME) as fuel in compression ignition engines: A critical review. *Eng. Sci. Technol. Int. J.*, 19, 1018–1026, 2016.
54. Sharma, Y.C., Singh, B., Korstad, J., Latest developments on application of heterogenous basic catalysts for an efficient and eco-friendly synthesis of biodiesel: A review. *Fuel*, 90, 4, 1309–1324, 2011.
55. Adam, F., Appaturi, J.N., Iqbal, A., The utilization of rice husk silica as a catalyst: Review and recent progress. *Catal. Today*, 190, 1, 2–14, 2012.
56. Fernandes, I.J., Calheiro, D., Sánchez, F.A.L., Camacho, A.L.D., de Campos Rocha, T.L.A., Moraes, C.A.M., de Sousa, V.C., Characterization of Silica Produced from Rice Husk Ash: Comparison of Purification and Processing Methods. *Mater. Res.*, 20, suppl 2, 512–518, 2017.
57. Chen, G.Y., Shan, R., Shi, J.F., Yan, B.B., Transesterification of palm oil to biodiesel using rice husk ash-based catalysts. *Fuel Process. Technol.*, 133, 8–13, 2015.
58. Maheswari, C.S., Ramesh, R., Lalitha, A., Synthesis, Characterization, and Catalytic Behavior of Bamboo Rice Husk Ash. *J. Chin. Chem. Soc.*, 64, 889–895, 2017.
59. Cao, J.L., Yan, Z.L., Deng, Q.F., Wang, Y., Yuan, Z.Y., Sun, G., Jia, T.K., Wang, X.D., Bala, H., Zhang, Z.Y., Mesoporous modified-red-mud supported Ni catalysts for ammonia decomposition to hydrogen. *Int. J. Hydrogen Energy*, 39, 11, 5747–5755, 2014.
60. Zhang, C., Liang, X., Liu, S., Hydrogen production by catalytic dehydrogenation of methylcyclohexane over Pt catalysts supported on pyrolytic waste tire char. *Int. J. Hydrogen Energy*, 36, 15, 8902–8907, 2011.
61. Alexander, E., Action of light on pigments I. *Chem. Ztg*, 35, 753–755, 1911.

62. Ahmaruzzaman, M., Biochar based nanocomposites for photocatalytic degradation of emerging organic pollutants from water and wastewater. *Mater. Res. Bull.*, 140, 111262, 2021.
63. Liu, X.H., Gai, G.S., Yang, Y.F., Sui, Z.T., Li, L., Fu, J.X., Kinetics of the leaching of TiO_2 from Ti-bearing blast furnace slag. *J. China Univ. Min. Technol.*, 18, 2, 275–278, 2008.
64. Sujaridworakun, P. and Natrchalayuth, K., Influence of pH and HPC concentration on the synthesis of zinc oxide photocatalyst particle from zinc-dust waste by hydrothermal treatment. *Adv. Powder Technol.*, 25, 4, 1266–1272, 2014.
65. Tanniratt, P., Wasanapiarnpong, T., Mongkolkachit, C., Sujaridworakun, P., Utilization of industrial wastes for preparation of high performance ZnO/ diatomite hybrid photocatalyst. *Ceram. Int.*, 42, 17605–17609, 2016.
66. Mohamed, M.M. and Amer, A.A., Solid waste-derived γ-Al2O3 nanowires supporting α-Fe_2O_3 as an efficient anodic direct methanol fuel cell electrocatalyst. *J. Alloys Compd.*, 966, 171632, 2023.
67. Cova, C.M., Zuliani, A., Puente Santiago, A.R., Caballero, A., Muñoz-Batista, M.J., Luque, R., Microwave-assisted preparation of Ag/Ag2S carbon hybrid structures from pig bristles as efficient HER catalysts. *J. Mater. Chem. A*, 6, 43, 21516–21523, 2018.
68. Cova, C.M., Zuliani, A., Muñoz-Batista, M.J., Luque, R., A sustainable approach for the synthesis of catalytically active peroxidase-mimic ZnS catalysts. *ACS Sustainable Chem. Eng.*, 7, 1, 1300–1307, 2019.
69. Hsu, P.C., Shih, Z.Y., Lee, C.H., Chang, H.T., Synthesis and analytical applications of photoluminescent carbon nanodots. *Green Chem.*, 14, 4, 917, 2012.
70. Liang, Z., Zeng, L., Cao, X., Wang, Q., Wang, X., Sun, R., Sustainable carbon quantum dots from forestry and agricultural biomass with amplified photoluminescence by simple NH_4OH passivation. *J. Mater. Chem. C*, 2, 45, 9760–9766, 2014.
71. Anmei, S., Qingmei, Z., Yuye, C., Yilin, W., Preparation of carbon quantum dots from cigarette filters and its application for fluorescence detection of Sudan I. *Anal. Chim. Acta*, 1023, 115–120, 2018.
72. Thakur, A., Devi, P., Saini, S., Jain, R., Sinha, R.K., Kumar, P., Citrus limetta organic waste recycled carbon nanolights: photoelectro catalytic, sensing, and biomedical applications. *ACS Sustainable Chem. Eng.*, 7, 1, 502–512, 2019.
73. Rigodanza, F., Đorđević, L., Arcudi, F., Prato, M., Customizing the electrochemical properties of carbon nanodots by using quinones in bottom-up synthesis. *Angew. Chem.*, 130, 5156–5161, 2018.
74. Prasannan, A. and Imae, T., One-pot synthesis of fluorescent carbon dots from orange waste peels. *Ind. Eng. Chem. Res.*, 52, 44, 15673–15678, 2013.
75. Padrón, D.R., Algarra, M., Luis, A.C., Tarelho, Frade, J., Franco, A., de Miguel, G., Jiménez, J., Rodríguez- Castellón, E., Luque, R., Catalyzed

microwave-assisted preparation of carbon quantum dots from lignocellulosic residues. *ACS Sustainable Chem. Eng.*, 6, 6, 7200–7205, 2018.
76. Jing, S., Zhao, Y., Sun, R.C., Zhong, L., Peng, X., Facile and high-yield synthesis of carbon quantum dots from biomass-derived carbons at mild condition. *ACS Sustainable Chem. Eng.*, 7, 8, 7833–7843, 2019.
77. Li, J. and Wu, N., Semiconductor-based photocatalysts and photoelectrochemical cells for solar fuel generation: a review. *Catal. Sci. Technol.*, 5, 1360, 2015.
78. Víctor, O., Carolina, S.G., Pontiggia, R., Melina, B.R., Paula, C., Galo, J.A.A., E-waste upcycling for the synthesis of plasmonic responsive gold nanoparticles. *Waste Manage. (Oxford)*, 117, 9–17, 2020.
79. Zhang, P., Zhang, J., Gong, J., Tantalum-based semiconductors for solar water splitting. *Chem. Soc. Rev.*, 43, 4395, 2014.
80. Hu, H., Xin, J.H., Hong, H., Wang, X., Miao, D., Liu, Y., Synthesis and stabilization of metal nanocatalysts for reduction reactions – a review. *J. Mater. Chem. A*, 3, 21, 11157–11182, 2015.
81. Yang, Q., Xu, Q., Jiang, H.L., Metal–organic frameworks meet metal nanoparticles: synergistic effect for enhanced catalysis. *Chem. Soc. Rev.*, 46, 15, 4774–4808, 2017.
82. Lou, J., Babadi, M.R., Otadi, M., Tarahomi, M., Le, Q.V., Khonakdar, H.A., Li, C., Agricultural waste valorization towards (nano)catalysts for the production of chemicals and materials. *Fuel*, 351, 128935, 2023.
83. Paul, N. and Joyce, G.F., A self- replicating ligase ribozyme. *Proc. Natl. Acad. Sci.*, 99, 20, 12733–12740, 2002.
84. Soai, K., Matsumoto, A., Kawasaki, T., *Advances in Asymmetric Autocatalysis and Related Topics*, pp. 1–30, Academic Press, 2017.
85. Plasson, R., Brandenburg, A., Jullien, L., Bersini, H., Autocatalysis: At the root of self-replication. *Artif. Life*, 17, 3, 219–236, 2011.
86. Crooks, R.M., Zhao, M., Sun, L., Chechik, V., Yeung, L.K., Dendrimer-encapsulated metal nanoparticles: synthesis, characterization, and applications to catalysis. *Acc. Chem. Res.*, 34, 3, 181–190, 2001.
87. Sun, Y. and Xia, Y., Shape-controlled synthesis of gold and silver nanoparticles. *Science*, 298, 5601, 2176–2179, 2002.
88. Gunindra, P., Kalyani, R., Rokhum, S.L., Wealth from waste: *M. acuminata* peel waste-derived magnetic nanoparticles as a solid catalyst for the Henry reaction. *Nanoscale Adv.*, 1, 1013–1020, 2019.
89. Abd El Hameed, M.H. and Wael, H.E., Solid waste-supported nickel nanoparticles: A sustainable, recoverable, and efficient heterogeneous catalyst for organic pollutant degradation. *J. Environ. Chem. Eng.*, 11, 3, 110028, 2023.
90. Sukhanova, A., Bozrova, S., Sokolov, P., Berestovoy, M., Karaulov, A., Nabiev, I., Dependence of Nanoparticle Toxicity on Their Physical and Chemical Properties. *Nanoscale Res. Lett.*, 13, 44, 2018.

91. Saboktakin, M., Ye, X., Oh, S.J., Hong, S.H., Fafarman, A.T., Chettiar, U.K., Engheta, N., Murray, C.B., Kagan, C.R., Metal-enhanced upconversion luminescence tunable through metal nanoparticle-nanophosphor separation. *ACS Nano*, 6, 10, 8758–66, 2012.
92. Narayan, N., Meiyazhagan, A., Vajtai, R., Metal Nanoparticles as Green Catalysts. *Materials (Basel)*, 12, 21, 3602, 2019.
93. Parasuram, B., Sundaram, S., Sathiskumar, C., Karthikeyan, S., Synthesis of multi-walled carbon nanotubes using tire pyrolysis oil as a carbon precursor by spray pyrolysis method. *Inorg. Nano-Metal Chem.*, 48, 2, 103–106, 2018.

Part II

SOLID WASTE-DERIVED CATALYSTS FOR VARIOUS APPLICATIONS

9

From Waste to Catalyst: Advancing Green Chemistry with Sustainable Materials

Munsaf Ali[1*†], Avinash Kumar Srivastava[2†] and Aarif Khan[3]

[1]*Department of Nanoscale Science and Engineering, University at Albany, Albany, New York, USA*
[2]*Department of Chemistry, University of Alabama at Birmingham, Birmingham, Alabama, USA*
[3]*Department of Environmental Science, Central University of Rajasthan, Ajmer, Rajasthan, India*

Abstract
Sustainability has emerged as a central concern in modern society, with growing recognition that anthropogenic waste can serve as a valuable resource. Various types of waste, including agricultural residues, industrial by-products, food waste, and municipal solid waste that can be transformed into energy sources, construction materials, chemical reagents, and high value-added functional products. Waste materials are increasingly utilized as renewable feedstocks for chemical processes and the development of heterogeneous catalysts. Waste-derived catalysts offer enhanced energy efficiency and atom economy for both conventional and innovative chemical transformations. Despite the unappealing origin of these materials, they have shown great promise in driving key reactions, such as biomass conversion and environmental remediation. This chapter explores the potential of waste-derived catalysts, their advantages and limitations, and how they can be compared with traditional catalytic systems.

***Keywords*:** Waste-derived catalysts, green chemistry, sustainable materials, biomass conversion, biodiesel production, photocatalysis, hydrogen evolution

**Corresponding author*: munsafali1988@gmail.com
†*Equal Contribution*

Deepti Goyal, Sakshi Kabra Malpani and Renuka Jain (eds.) Catalysis from Waste: Transforming Solid Residues into Value, (219–252) © 2026 Scrivener Publishing LLC

9.1 Introduction

There is a current resurgence in developing catalytic materials using readily available and sustainable components, particularly those that do not depend on platinum group metals (PGMs). PGMs, such as platinum (Pt), palladium (Pd), and rhodium (Rh), play a crucial role in a wide range of catalytic applications, including automotive exhaust treatment (e.g., three-way catalysts), hydrogenation, fuel cells, and petrochemical refining. However, the reliance on these metals presents several challenges. In addition to their high cost, PGMs are scarce and geographically concentrated, with South Africa accounting for more than 70% of global platinum and 36% of palladium reserves, making them vulnerable to supply chain disruptions and geopolitical instabilities. The economic and strategic implications of PGM use have spurred efforts to minimize metal losses during catalytic operations and improve the recovery and recycling of spent catalysts. Beyond PGMs, there is a growing need to replace conventional catalysts made from rare or hazardous materials in processes such as fine chemical synthesis, biomass valorization, and environmental remediation. In this context, waste-derived materials present a promising alternative, offering both environmental and economic benefits [1, 2]. The utilization of waste biomass has emerged as a prominent strategy for sourcing renewable feedstock in the production of fuels [3–5]. Moreover, it has attracted interest as a crucial source for producing platform compounds.

Although catalysts are usually needed in small amounts for these processes, the expense of obtaining catalysts is a significant concern, especially when waste biomass is used as the major substrate. Researchers are investigating ways to create catalysts from waste materials to improve cost-effectiveness and environmental sustainability. This technique seeks to enhance energy efficiency and atom usage in both traditional and cutting-edge chemical processes. Utilizing diverse catalysts made from waste offers a possible path to achieve a more sustainable direction [6–8]. Integrating waste materials into catalyst production offers several advantages, resulting in a more cost-effective and ecologically friendly solution. Researchers have been studying the creation of heterogeneous catalysts using waste materials more. The unique characteristics of the waste source have a substantial impact on the properties of the catalyst produced. Catalysts developed for certain uses often share similarities in their material makeup. This approach improves resource efficiency and promotes the advancement of sustainable chemical processes. Researchers have made significant advancements in moving towards a more eco-friendly and effective future by using catalysts made from trash [9–12]. This chapter aims to give a detailed summary of modern literature, categorizing waste-derived catalysts based on their catalytic uses. Research efforts have significantly increased in the past twenty years, specifically focusing on creating heterogeneous catalysts using waste materials as substrates. This research commonly uses waste streams that are easily accessible and can add up to significant amounts, typically in the range of millions or even hundreds of millions of tons each year. The magnitude is critical as it demonstrates the potential efficacy of catalysts created from waste materials in addressing essential industrial processes. Plentiful resources consist of waste biomass from agricultural sources, mining, sewage, and metal manufacture, particularly in the steel and iron industry [13–20]. In addition to yielding catalysts that are highly active, selective, and stable, an optimal waste stream should be readily collectible and pose minimal safety risks during handling.

This chapter aims to explore the potential of waste-derived materials as catalysts in various chemical processes. It begins by discussing the types and sources of waste suitable for catalyst development, including agricultural residues, industrial by-products, food waste, and municipal solid waste. The chapter then delves into the synthesis strategies used to convert these waste materials into functional catalysts, highlighting techniques such as thermal treatment, chemical activation, and surface modification. Subsequent sections present case studies where waste-derived catalysts have been successfully employed in key reactions, including biomass conversion, pollutant degradation, and fine chemical synthesis. A comparative analysis of their performance relative to conventional catalysts is also provided. Finally, the chapter evaluates the advantages, challenges, and prospects of using waste derived catalysts, offering insights into how this approach contributes to green chemistry and circular economy initiatives [21–25].

9.2 Applications of Waste-Derived Catalysts

Waste-derived catalysts have gained significant attention for their ability to support diverse chemical transformations while promoting sustainability. These catalysts prepared from materials such as red mud, fly ash, eggshells, biomass, and electronic waste demonstrate promising performance across a wide range of applications. These include hydrogen production, oxidation reactions, biodiesel synthesis, photocatalytic degradation of pollutants, pyrolysis of plastics, and biomass valorization. Their utilization not only addresses the issue of waste disposal but also offers cost-effective and eco-friendly alternatives to conventional catalysts. This section categorizes and discusses the major application areas where waste-based catalysts are making a measurable impact in advancing green chemistry and circular economy principles.

9.2.1 Hydrogen Production

Hydrogen, positioned as a crucial energy carrier, holds significant promise. The compatibility with fuel cell technology results in the emission of zero CO_2 or other hazardous substances, producing only water as a by-product. Hydrogen has a higher energy density than traditional fuels such as natural gas or diesel [26–29]. While traditional production methods rely on fossil fuels or nuclear energy, which conflict with eco-friendly principles, there exists an urgent demand for environmentally sustainable and economically viable hydrogen production processes, particularly from renewable sources such as biomass [30, 31]. Sushil and coworkers have investigated various catalysts made from waste materials to generate hydrogen from different sources, demonstrating continuous endeavors to progress on this essential subject. Red mud, a residue of the Bayer process, has various uses [20]. Challenges like tar formation and ash accumulation must be addressed. Catalytic gas cleaning shows promise for tar removal, offering several advantages. For hydrogen production, researchers have used a Mo-Ni-Co/cordierite monolithic catalyst, achieving notable results: hydrogen yield of 49.3 g/kg, gas yield of 1.28 Nm^3/kg, and a 99.2% tar conversion rate from miscellaneous sawdust [32]. The process of biomass conversion is well-established, employing heat, steam, and oxygen to convert organic materials into hydrogen and other valuable products. Such materials encompass agricultural residues, forest by-products, energy crops, and municipal waste [33, 34]. Zhou *et al.*, researched the hydrogen evolution reaction (HER) efficiency of several metal oxides/carbon catalysts made from used ion-exchange resins. The Fe^{3+}-contaminated resin-derived FeO_x/C catalyst showed the highest HER activity among the several metal ions tested, with an overpotential of 60 mV. The CrO_x/C catalyst, produced from ion-exchange resins with chromium, showed significant catalytic activity ($\eta10$=123 mV), offering a potential way to reuse hazardous wastes [35]. Biomass pyrolysis is a thermal decomposition process carried out at high temperatures (400–900°C) over short periods, offering many benefits. It facilitates the utilization of cost-effective feedstock, contributes to environmental sustainability through CO_2-neutrality, and promotes waste management [36].

Lin *et al.* elucidated a three-step methodology for converting substantial quantities of scrap nickel into three-dimensional Ni_2P nanoparticle catalysts with unique 3D nanostructure and high intrinsic activity. The unique catalytic performance arises from their excellent HER activities in both alkaline and acid electrolytes. Specifically, the overpotentials achieved by these materials are 73 mV and 69 mV for current densities of 10 Ma/cm^2. The highest HER catalytic performance is directly proportional to a photodiode IS with a V A /I ratio of VOL. Application of waste materials greatly promotes HER over single-component transition metal-based catalysts [37]. However, it faces some problems like coke formation over the catalyst, the high cost of the reactor construction, and feedstock availability problems according to the season. Catalysts made of nickel-containing BEA zeolite (NiBEA) have been investigated for the generation of hydrogen-rich gas from lignocellulosic biomass. In this method, cellulose and pinewood are pyrolyzed at 500°C, and the resulting vapor is subsequently filtered through a 700°C hot catalyst bed [38]. Biomass gasification is a thermochemical process that takes place at temperatures over 1000°C and pressures of around 33 bars. The process involves the utilization of oxidizing agents, such as oxygen, air, or steam, to produce syngas, which is a combination of hydrogen and carbon monoxide [39, 40]. Altalhi *et al.*, employed industrial fly ash waste in conjunction with TiO_2 to synthesize an FA-TiO_2 nanocomposite. Following post-cathodic polarization treatment, the activated FA-TiO_2 nanocomposite catalyst exhibited significant performance in HER, with an overpotential (η10) of 125 mV and a Tafel slope of 115 mV dec^{-1} in alkaline electrolytes, comparable to that of the Pt/C catalyst. Despite its exceptional electrocatalytic capabilities, identifying the source of activity within the FA-based composite remains challenging due to the complex crystal structure and intricate chemical composition of fly ash. Additionally, many waste-derived transition metal-based electrocatalysts utilized for HER demonstrate notable oxygen evolution reaction (OER) capabilities, including NiCoMn hydroxides [41]. Nickel catalysts are necessary for breaking carbon-carbon and carbon-oxygen bonds in lignocellulosic biomass, which improves the synthesis of gaseous chemicals and enhances hydrogen production efficiency. Temperature and steam flow are key factors in hydrogen generation and must be fine-tuned to enhance economic and operational efficiency. The efficiency of gasification is affected by the selection of materials [42, 43].

Ceyhan *et al.*, explored the use of metallurgic sludge with oil generated from grinding, sharpening, and milling for hydrogen (H_2) production through hydrolysis of sodium borohydride ($NaBH_4$). The hydrolysis of

$NaBH_4$ was studied as a function of NaOH concentration, catalyst amount, $NaBH_4$ concentration, and temperature by using a metallurgic sludge catalyst for the first time. For characterization, the catalyst of metallurgic sludge was analyzed by X-ray diffraction (XRD), Fourier-transform infrared spectroscopy (FT-IR), and scanning electron microscopy (SEM). The metallurgic sludge catalyst realized the highest hydrogen production rate from $NaBH_4$ hydrolysis was 9366 ml/min.g/cat and activation energy of 48.05 kJ/mol [44]. Rajagopal *et al.*, in Germany looked at harnessing liquid biofuels from household waste over a day or so. Results showed that hydrogen production was significantly enhanced by the combined hydrothermal gasification and hydrothermal liquefaction (HTG-HTL) process. When the bentonite/Nb-TiO_2 catalyst was employed for both the HTG and HTL processes, more than 12 wt.% of hydrogen gas was bio-sourced at 39 wt.% with 4 wt.% catalyst loading in the HTG process and 3 wt.% catalyst loading in the HTL process respectively as opposed to 26 wt.% [45]. Cao *et al.*, utilized bean sprouts as a carbon source to synthesize carbon compounds, exploiting their intrinsic self-doping properties through high-temperature calcination. The resulting nitrogen (N) and sulfur (S) co-decorated carbon catalyst demonstrated significant efficiency in the HER, with an overpotential (η10) of 413 mV and a Tafel slope of 98 mV/dec. Additionally, the catalyst exhibited exceptional stability over 2000 cyclic voltammetry (CV) cycles in acidic conditions. Density functional theory (DFT) calculations were conducted to elucidate the impact of N and S dopants on HER performance. The results revealed that the incorporation of sulfur dopants notably altered the electrical properties and improved the adsorption of hydrogen atom intermediates on the catalysts, thereby enhancing HER activity in comparison to catalysts doped solely with nitrogen [46]. Wyss *et al.*, developed a Joule heating strategy to convert mixed plastic waste to holey and wrinkled flash graphene (HWFG). This particular variant of graphene consists of a large number of three-dimensional (3D) and two-dimensional (2D) cavities and pores, resulting in a relatively high surface area of 874 m^2/g. On the other hand, the HWFG has limited performance toward HER with an overpotential of η10=613mV in acidic solution, since it contains only carbon. This work offers an option to get over the organic restrictions and since inflexible layouts used were easy to customize, consideration can be paid to various other varied types of porous architectures, seeking the very best efficiency from the carbon drivers (Figure 9.1) [47].

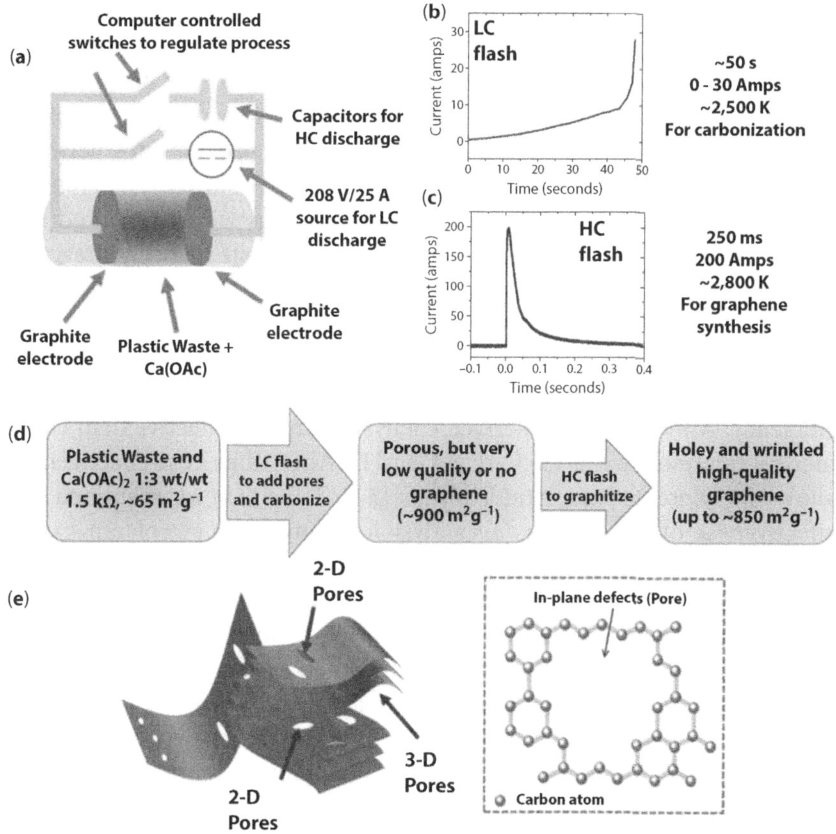

Figure 9.1 Diagram showing (a) an FJH schematic composed of a 30 A, 208 V rectified power supply and a 128 mF capacitor allowing for LC and HC heating to be used in tandem. Current profiles for (b) LC and (c) HC methods and (d) a workflow for the synthesis of HWFG. A conductive additive of 5–8 wt.% such as metallurgical coke or carbon black is necessary to reduce the resistance of the mixture. (e) Diagram showing 2D and 3D pores present in turbostratic HWFG. Adapted from reference [47]. Copyright 2022, with permission from ACS.

9.2.2 Oxidation by Waste-Derived Catalysts

Metal catalysts, [48, 49] metal composites [50, 51] and metal nanoparticles [52] have all been extensively researched for their diverse catalytic applications in producing feedstocks and chemicals. Furthermore, catalysts derived from metal waste from various industries have been explored for their potential in oxidation catalysis [9, 53]. Sushil *et al.*, investigated the repurpose of waste materials from the aluminum, tannery, and electroplating industries as catalyst precursors. Calcined red mud demonstrated efficient catalytic activity for complete propane oxidation at 350°C under continuous airflow. The composition and catalytic performance of red mud varied based on its source, achieving propane conversion rates between 60% and 90% at 500°C. The improved performance at lower temperatures resulted from the combination of tannery shavings with Cr_2O_3 and red mud catalysts. However, pure tannery shavings, despite their high reactivity, were ill-suited to high gas flow rates due to the relatively small size of particles. The red mud catalysts slightly increased the scattering intensity when it was used as an additive, whereas the metal obtained from the acid leaching of electroplating waste sludge led to a significant decrease in its efficiency. Although waste-derived catalysts were less effective than commercial catalysts such as Co_3O_4/CeO_2 and Pt/Al_2O_3, their cost-effectiveness allows for increased catalytic loadings [54]. Teixeira *et al.* examined the effectiveness of red mud as a catalyst for converting petroleum gas, a by-product of offshore oil exploration, primarily comprising methane and other hydrocarbons. They investigated the interaction between methane and metal oxides in red mud, leading to the formation of CO/CO_2 and water, followed by hydrogen generation through interactions with reduced metals. Utilizing a coke-metal catalyst treated with steam, they produced syngas [55]. The researchers conducted experiments by decomposing methane in a continuous flow reactor, maintaining temperatures from 25°C to 900°C in a CH_4/N_2 stream. Significant transformation occurred at temperatures exceeding 650°C. Additionally, they found that metal/carbon composites oxidized to produce hydrogen under steam conditions, while CO was formed by oxidizing carbon deposits at temperatures exceeding 800°C. The study underscores the potential of red mud as a catalyst for converting, transferring, and utilizing petroleum gas [9]. Metal oxides produced from industrial metal waste through spark erosion have been utilized in various applications [53].

Sonkar *et al.* developed a new technique to produce magnetically separable FeO-NC (iron oxide nanoparticles) using leftover printer ink [56]. The FeO-NCs were utilized to catalyze organic processes, transforming styrene into benzaldehyde, an essential intermediate in various fields. The effective utilization of surplus printer ink highlights the considerable capabilities of FeO-NCs as a magnetic catalyst. The catalyst's uncomplicated production technique and recyclability make it an attractive alternative material for industrial applications in diverse catalytic reactions [56]. Li *et al.* reused waste printed circuit boards as a Fenton catalyst for oxidative degradation of organic pollutants in water. Printed circuit boards were composed of

51% silica (SiO_2), plus smaller amounts of other elements and 30% combustible organic material. Catalysts were made in a process involving the breakdown of the boards, grinding and sieving. These catalysts showed good efficiency for pyridine degradation, demonstrating comparable performance to that of pure Cu powder using H_2O_2 under room temperature. To achieve 90% conversion of 0.1 g/L pyridine, it was necessary to apply 4 g/L of the catalyst for 14 hours. A considerable amount of metal ions, such as toxic Cu and Pb, were released during the catalytic process, raising doubts about the appropriateness of these materials for treating wastewater [57]. Yuan *et al.* converted grinding wheel ash, a waste, abundant in iron from vehicle production, into a catalyst for decomposing methylene blue (MB) dye. The catalyst, which includes trace amounts of silicon and aluminum supported on iron oxide, demonstrated comparable efficiency to Fenton-like oxidation in breaking down the dye. Using a catalyst concentration of 1.0 g/L, full conversion of a 20 mg/L MB solution took place within 60 minutes under neutral or basic conditions with the aid of hydrogen peroxide. The presence of silica increased reaction rates, while aluminum reduced iron leaching, with only a slight impact on dye degradation. Stainless steel slags were utilized to produce silica and silica-alumina substrates for Pt-based catalysts, which were used in the oxidation of volatile organic compounds (VOCs). High surface area silica support was obtained through acid digestion and hydrothermal treatment, followed by platinum impregnation to create nanoparticles. The obtained Pt/SiO_2 catalysts showed high activity for both CO and toluene oxidation in a 21% O_2 flow, reaching complete conversion between 200 and 500°C being particularly more active in CO oxidation compared to a commercial Pt/SiO_2, likely due to the presence of extra transition metal promoters in the support derived from wastes. On the other hand, the activity for toluene oxidation was lowered, which might be attributed to the reduced Pt dispersion [58]. Luque *et al.* studied the use of waste eggshells as a template and support for creating Ag/eggshell catalysts to determine their efficacy in accelerating the oxidation of volatile organic compounds (VOCs) [59]. This method is praised for its simplicity, environmental friendliness, and cost-effectiveness.

Biomolecules helped evenly spread Ag nanoparticles of limited size on the eggshell surface. Ag/eggshell catalyst outperformed Ag/com-$CaCO_3$ catalysts in catalyzing benzene oxidation due to its consistent channel topologies and particle size distribution. At the space velocity of 100,000 mL/g h, showed high performance with a T_{90} value of 257°C, and the high stability was maintained over 200 h period of reaction for this catalyst. The activity of catalysts dropped with the inclusion of water vapor and CO_2 in the feed stream but did not alter the catalyst's structure. The work emphasizes the potential of Ag catalysts made from discarded eggshells for VOC oxidation and offers useful insights into creating eggshell-based catalysts/supports for industrial use [59]. After a year, Li *et al.* conducted a noteworthy study focusing on the synthesis of Co_3O_4 nanoparticles utilizing eggshell catalysts *via* the impregnation method. This innovative approach capitalized on repurposing discarded eggshells, known for their robust yet porous nature, as an effective template and support material. Through meticulous adjustment of precursor concentrations, the researchers successfully produced a series of Co_3O_4/Eggshell catalysts denoted as Co_3O_4/Eggshell-1 (33.1%), Co_3O_4/Eggshell-2 (16.7%), and Co_3O_4/Eggshell-3 (9.1%). Notably, among these variants, Co_3O_4/Eggshell-2 exhibited superior catalytic performance, manifesting a remarkable T90% at 256°C, a notably lower threshold compared to Co_3O_4/oyster shell-2 in the context of benzene catalytic oxidation. This observation underscores the significant efficacy of Co_3O_4/Eggshell catalysts in contrast to conventional Co_3O_4 nanoparticles and $CaCO_3$, attributable to the distinctive structure of eggshells and the consequential interaction between $CaCO_3$ and Co_3O_4 [60]. In 2022, a study successfully converted waste eggshells into Hydroxyapatite (HAp) nanomaterials verified by X-ray diffraction. Au-loaded HAp nanomaterials catalyzed formaldehyde oxidation. HAp's specific surface area found to be 85 m^2/g, measured *via* Brunauer–Emmett–Teller (BET) analysis, a notable increase from eggshells. The catalytic efficiency of Au/HAp was tested at varied conditions: temperature (50-260°C), Au loading (0.2 wt.% and 0.5 wt.%), and relative humidity (5% and 65%). Both catalysts exhibited significant activity, achieving a minimum 40% formaldehyde conversion at 150°C. Notably, 0.2 wt.% Au-loaded HAp showed maximum efficiency, reaching 85% conversion at 260°C under 65% relative humidity. Results suggest Au-loaded HAp nanoparticles as effective, sustainable formaldehyde oxidation catalysts, utilizing waste eggshells [61]. Vanaraj *et al.* effectively employed wasted banana material as a catalyst in Fenton's oxidation reaction. Using X-ray diffraction (XRD), researchers identified a metal oxide phase similar to perloffite when dried banana bracts were carbonized through aerial oxidation. Raman spectroscopy and attenuated total reflectance (ATR)-FTIR spectroscopy showed a layered structure with various functional groups in the carbonized catalyst. X-ray fluorescence (XRF) and energy-dispersive X-ray spectroscopy (EDAX) detected main metal elements such as calcium, magnesium, potassium, sodium, and iron. The carbonized banana bract catalyst showed high effectiveness in Fenton's oxidation at a neutral pH, reaching a 95% breakdown rate in four hours and remaining stable for nine cycles. The increased effectiveness in breaking down MB was owing to strong connections between oxygen functional groups and MB molecules. Hydrogen peroxide achieved 99% efficiency because metal oxides helped prevent its self-decomposition [62].

9.2.3 Biodiesel Production

Waste-derived catalysts have been widely applied for the transformations of free fatty acids (FFAs) into esters at low temperatures or the conversions of triacylglycerides (TAGs) *via* transesterification to produce biodiesel. In the creation of those solid acid and base catalysts, various types of waste materials, such as plant residues, algae, eggshells, animal bones, ash, clay, and industrial by-products, were used as precursors. Esterification commonly employs solid acid catalysts, which are frequently derived from activated carbons through processes such as pyrolysis or carbonization followed by sulfonation. In contrast, transesterification processes commonly employ solid base catalysts, which primarily consist of calcium oxides derived from bone, mollusk, or eggshells. The scientific literature extensively covers the utilization of solid-base catalysts obtained from waste in the synthesis of biodiesel [63–67]. Boz *et al.*, synthesized Al_2O_3 nanoparticles which were covered with KF. After that, the nanoparticles were calcined and used as a heterogeneous catalyst for the transformation of vegetable oil into biodiesel. The fuel was converted through a process called trans-esterification with methanol. Various process-related parameters such as the KF to γ-Al_2O_3 feed ratio, calcination temperature, methanol to oil molar ratio, reaction temperature, reaction time and catalyst concentration were investigated throughout the study. The best performance was considered, a methyl ester yield of 97.7 ± 2.14%. These consisted of a 15 wt. % KF loading, calcining at 773°C, 8h reaction at 338°C, 3 wt.%, using 1.1 wt.% of the catalyst but having a molar ratio of methanol/oil of 15:1. This is mainly because γ-Al_2O_3 nanoparticles have a rather high basicity (~1.68 mmol/g), which has been easily transferred to the vegetable oil, leading to the greater vegetable oil transformation into biodiesel [68]. In their study, Dawodu *et al.*, conducted an extraction of oil from calophyllum inophyllum seeds containing 15% FFA. Subsequently, they utilized the seed cake as a catalyst for the one-step biodiesel conversion. The residues were carbonized at a temperature of 400°C using nitrogen, and -SO_3H groups were added through the application of sulfuric acid. The acid site densities of catalysts derived from shorter pyrolysis were found to be greater, resulting in enhanced FFA conversion but also leaching sulfur. The seed-derived catalysts exhibited a high FFA conversion rate of over 75%, but a low Fatty Acid Methyl Esters (FAME) conversion rate of up to 36%. The glucose-derived catalysts produced 99% FAME within a time frame of 4-5 hours, using a loading of 7.6 wt.% and a temperature of 180°C in an autoclave operating at 10 bars [69]. Babadi *et al.*, investigated the possibility of sugar beet pulp as a powerful acid catalyst for transforming palm fatty acid distillate (PFAD) into biodiesel. Based on this approach, we sulfonated partially carbonized activated carbon with concentrated H_2SO_4 to generate a high total acid density-containing catalyst. When ethanol was employed in transesterification of PFAD, 92% of FAME yield was obtained [70]. The study conducted by Rathod *et al.* investigated the utilization of waste cooking oil and dimethyl carbonate (DMC) in conjunction with the Lipase-435 enzyme for biodiesel generation. The resulting by-product of this process was glycerol carbonate. They tackled issues such as extended response times by utilizing microwave technology. The research examined the reaction rate, activation energy, and benefits of microwaves compared to traditional procedures, particularly in terms of reducing the need for DMC.

The results demonstrated a conversion rate of around 94% within a four-hour timeframe using optimum parameters. Lipase-435 maintained an activity level of 88% after undergoing six cycles. This economically efficient and environmentally sustainable strategy provides marketable by-products [71]. In the year 2018, Jindapon *et al.*, undertook the transformation of a dolomite sample into magnesium oxide (MgO) and calcium oxide (CaO), after evaluating their viability for the generation of biodiesel. The influence of the aggregation condition of the oxides (bulk or nanoscale) on the surface basicity of the resultant materials was observed. The material with a high concentration of MgO demonstrated a diminished presence of basic sites, leading to a decrease in the generation of biodiesel from palm oil. In contrast, nanoparticles rich in CaO exhibited a significant abundance of basic sites, resulting in a biodiesel production of 99% under consistent conditions. The research findings indicated that the nano catalyst with a high concentration of CaO enhanced the ability of triglycerides in palm oil to reach its active basic sites and exhibited greater resistance to deactivation in comparison to the catalyst with a high concentration of MgO [72]. Wahyudi *et al.*, explored the deactivation behavior and regeneration potential of red mud catalysts modified with soda lime for biodiesel production using canola oil. Following initial calcination at 700°C, the catalyst exhibited a marked reduction in activity, primarily due to the accumulation of organic residues such as glycerol and diglycerides on its surface. Analytical techniques including XRD, SEM, and FTIR revealed significant losses in surface area and basicity. To recover catalytic efficiency, the authors evaluated three regeneration strategies: calcination alone, hexane washing alone, and sequential hexane washing followed by calcination. Among these, the combined treatment (WUC-cal) proved most effective, successfully restoring both the surface characteristics and catalytic performance to levels comparable with the fresh catalyst. These findings highlight the potential of red mud as a cost-effective, regenerable, and sustainable heterogeneous catalyst for biodiesel synthesis [73]. Yani *et al.*, sought to explore the potential of water hyacinth biomass as a viable raw material for the production of a heterogeneous catalyst. Their research was mainly aimed at solving environmental issues related to the massive use of water hyacinths. The process involved crushing the hyacinth biomass, then calcining it at 600°C to make the catalyst. In addition, impregnation of K_2CO_3 as a co-catalyst was implemented. After the impregnation of K_2CO_3, the same study provided a closer look at surface characteristics, particle sizes and atomic composition of the catalyst to reveal the effects of this specific step. The synthesized catalyst was then investigated for its catalytic activity in biodiesel synthesis by transesterification of palm oil. The best conditions were demonstrated using 15% of hyacinth impregnated with K_2CO_3 10%. This resulted in a yield of 97.57% [74]. The acid catalyst for the manufacture of biodiesel from jupati oil was developed by Bastos *et al.*, using murumuru kernel shell, which is an agricultural waste. The ideal reaction parameters were determined using a response-surface methodology that relied on a 23-central composite design. The catalyst was characterized through the utilization of SEM-EDS, FTIR, Thermogravimetric Analysis (TGA), XRD, and total acid density analysis. After doing optimization, the researchers determined that the optimal molar ratio of methanol to oil, catalyst concentration, and temperature were determined to be 30:1, 6, and 135°C, respectively. The manufacture of FAME achieved a 91.8% yield under the given conditions [75].

Roy and Mohanty successfully converted de-oiled microalgal biomass into a carbon-derived solid acid catalyst to improve the synthesis of biodiesel using microalgae in an integrated biorefinery system. The catalyst has shown exceptional catalytic efficiency, resulting in the production of FAME at yields of 94.23% and 96.25% from AO (Algal Oil) and WCO (Waste Cooking Oil), respectively. The catalytic activity was consistently high, with FAME yields surpassing 90% for a minimum of four cycles. Moreover, the utilization of the de-oiled microalgal biomass in the production of biodiesel complied with global fuel regulations, highlighting its capacity for sustainable biodiesel production *via* effective transesterification methods (Figure 9.2) [76].

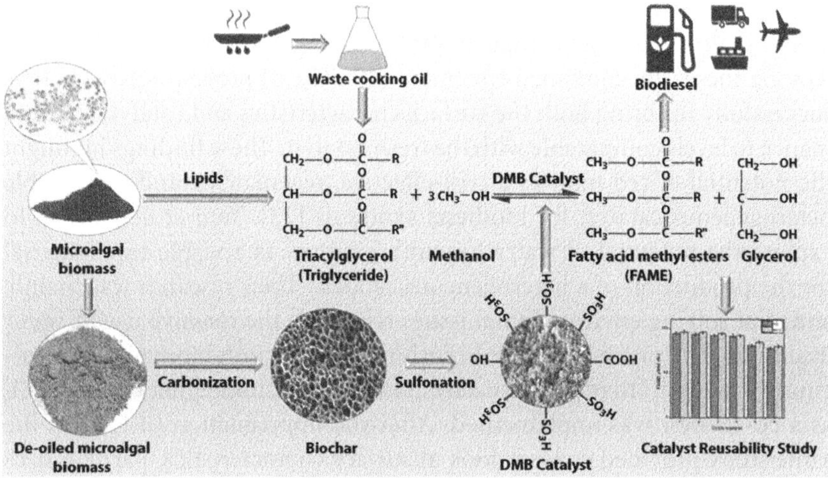

Figure 9.2 Schematic representation of valorization of de-oiled microalgal biomass as a heterogeneous acid catalyst. Adapted from reference [76]. Copyright 2021, with permission from Elsevier.

Leesing *et al.*, employed a comprehensive methodology that incorporated yeast-based and durian peel-derived catalysts to facilitate the synthesis of biodiesel. The catalyst obtained from durian peel showed increased efficiency, resulting in an 81.57% production rate of FAME. This improvement can be attributed to the inclusion of sulfonic sites. The acid density of the catalyst was significantly enhanced by sulfonation, increasing from 0.01 mmol/g to 1.22 mmol/g. This observation suggests the successful incorporation of active sulfonic sites. The energy yield of the resultant FAME was predicted to vary between 32.44 and 34.90 MJ/kg. The biorefinery concept under consideration involves the utilization of 1000 kilograms of waste durian peel, resulting in the production of 16.0 kg of yeast lipid, 101.2 kg of yeast biomass, 717 kg of DP-SO_3H catalyst, 70.4 kg of yeast waste, and 79.7–82.6 kg of FAME [77]. In response to the need for an economically viable catalyst for the synthesis of biodiesel in normal environmental circumstances, Tarigan *et al.*, proposed the utilization of passion fruit peel to create an innovative heterogeneous catalyst. The researchers assessed the catalytic efficacy of calcined waste passion fruit peel (WPFP), which predominantly consists of potassium in the chloride and carbonate states. To evaluate the interactions among the molar ratio of oil to methanol, catalyst weight, and reaction duration under various temperature settings (65°C, 45°C, and room temperature), a factorial design approach was utilized. The leaching investigations revealed a 22% decrease in the weight of the catalyst, emphasizing the necessity of adding precise quantities of catalyst for subsequent use. However, calcined WPFP exhibited the ability to facilitate transesterification at ambient temperature, which could potentially result in cost reductions in the manufacturing of biodiesel [78].

9.2.4 Photocatalytic Degradation

The process of photocatalytic degradation utilizes photocatalysts, commonly semiconductors, to accelerate photochemical reactions. These reactions are employed in various fields, including the elimination of pollutants and pathogens, the conversion of energy, and the production of environmentally friendly hydrogen through the process of water splitting. Such properties are crucial for effective photocatalytic applications in pollutant degradation and hydrogen production [79–82]. The integration of transition metals into titanium dioxide (TiO_2) offers numerous advantages in terms of its photocatalytic efficiency, such as enhanced light absorption and/or prevention of undesirable electron-hole recombination. The compound Cu-doped TiO_2 has garnered significant interest [83, 84]. The utilization of surplus electrical and electronic waste was employed by Liou *et al.*, in the fabrication of a silica support for TiO_2 photocatalysts. The procedure encompassed the utilization of grinding, calcination, and acid-leaching techniques to remove undesirable metals from the waste resin. The resin-derived sodium silicate was subsequently employed in the production of mesoporous silica MCM-48 with CTAB. The provided support demonstrated a substantial surface area, reaching 1153 m^2/g, and successfully stabilized TiO_2 nanoparticles that were considerably lower in size compared to the commercial bulk TiO_2 (2.4 nm for 30 wt.% TiO_2/MCM-48, in contrast to 32.4 nm for pure titania). After undergoing calcination at a temperature of 700°C, the catalyst with a composition of 30 wt.% TiO_2/MCM-48 obtained a conversion rate of 75% for 50 mM MB after being exposed to UV irradiation for 3h. This was achieved by utilizing a catalyst loading of 500 wt.% (0.080 g) (Figure 9.3) [85].

The utilization of waste colemanite ore derived from boron mining procedures has been employed in the production of supports for TiO_2 photocatalysts. The ore residue comprises a range of oxides, such as alkali, alkali earth, and transition metals, with a predominant presence of SiO_2, CaO, and B_2O_3. The mixture of waste oxides was treated with titanium nanoparticles, which had a size range of 5 to 30 nm, using titanium isopropoxide. The resulting supported TiO_2 exhibited a band gap of 3.08 eV, which is similar to the band gaps observed in pure anatase and rutile phases. The titania that received support was subsequently assessed for

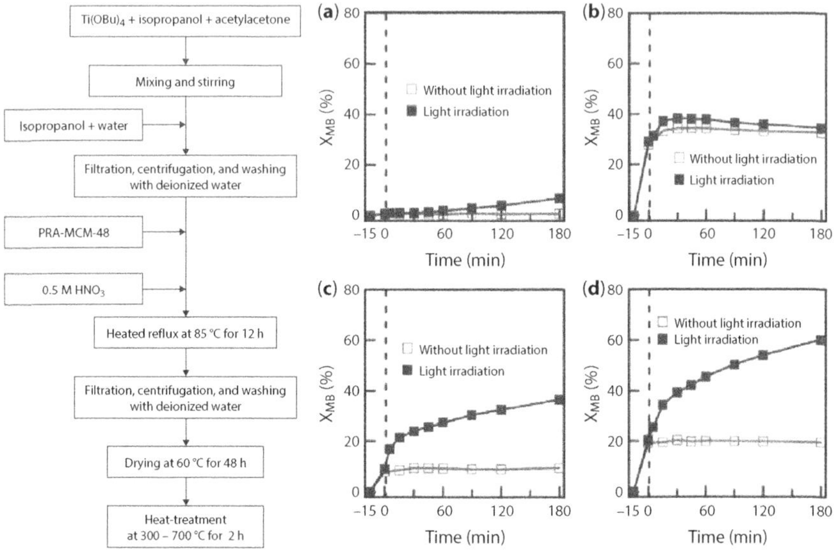

Figure 9.3 Scheme of preparation of TiO_2/PRA-MCM-48 photocatalyst, adsorption and photodegradation of methylene blue on various catalyst types: (a) pure TiO_2, (b) PRA-MCM-48, (c) 50% TiO_2/silica-gel, and (d) 50% TiO_2/PRA-MCM-48. Adapted from reference [85]. Copyright 2012, with permission from Elsevier.

its effectiveness in the photodegradation of pesticides within a simulated wastewater environment. After 70 minutes of UV irradiation, a catalyst suspension with a concentration of 1 g/L effectively eliminated 80% of atrazine from a solution with a concentration of 10 mg/L. In addition, the catalyst demonstrated negligible deactivation throughout four cycles of the reuse process [86]. A catalyst was produced by Sapiña *et al.*, using electric arc furnace dust as the primary constituent. This catalyst consists of 33 wt.% ZnO, 33 wt.% Fe_2O_3, and a mixture of different oxides, including 0.1 wt.% TiO_2. This substance shows reactivity in the process of converting NO into HNO_3. However, its effectiveness was lower compared to pure P25. Additionally, it exhibited photodegradation of Rhodamine B when exposed to both UV and visible light. Increased conversion of Rhodamine B was seen when exposed to UV light, resulting in a 70% conversion of a 1 mg/L aqueous RhB solution utilizing a catalyst with a weight percentage of 750 after 140 minutes. Nevertheless, the process of uncatalyzed photolysis yielded a conversion rate of roughly 20% when subjected to the same conditions. Conversely, the adsorption onto the catalyst in the absence of light led to a reduction in Rhodamine B concentration by nearly 40%, which can be attributed to the high loading of the catalyst. Therefore, the impact of photocatalysis facilitated by electric furnace dust may be similar to or lower than the degradation rate achieved through uncatalyzed UV photolysis [87]. Dionysiou and coworkers proposed a sol-gel method that allows them to produce new W-TiO_2/clay photocatalysts. This methodology facilitates the synthesis of W-TiO_2 nanoparticles (with a size range of 13-16 nm) that are securely bonded to exfoliated layered clay surfaces in a single step. The analysis conducted in this study identified the presence of crystalline anatase, the incorporation of W dopant inside the lattice, and the production of W-O-Ti connections. The results of the photocatalytic evaluations indicated that the W-TiO_2/clay composite, with a 5% water content, demonstrated the highest level of efficiency in the breakdown of atrazine when exposed to solar irradiation. The improvement was ascribed to the integration of W, resulting in a moderate decrease in the band gap of TiO_2. In addition, the incorporation of W resulted in an increased sensitivity of TiO_2 to visible light, albeit this sensitivity was comparatively lower than that observed during solar irradiation [88].

The synthesis of high visible light-responsive anatase NS co-doped, and N-doped nanocrystalline titania was successfully achieved by Sankaran *et al.*, by utilization of a sol-gel precipitation technique. Mechanical stirring and its duration played a crucial role in optimizing catalyst shape and activity parameters. The herbicide aldicarb was degraded in an aqueous solution to undertake photocatalytic evaluations. Significantly, a degradation rate of more than 80% was seen in the insecticide when exposed to visible light for thirty minutes, utilizing a catalyst system co-doped with 1.5 g/L of NS. The increased level of activity is ascribed to the combined impact of the two impurities (N and S) present in their lattice, resulting in a significant reduction in the band gap [89]. Cardoso *et al.*, employed basalt powder, a naturally occurring clay composite, as a potentially effective catalyst for photo-Fenton-like reaction. The objective of this approach was to degrade Atrazine (ATZ) by exposure to both UV and visible light. The optimal conditions for the UV-C process involved a concentration of 5 mg/L ATZ, 1.6 mM H_2O_2, 0.50 g/L basalt, and 0.75 g/L visible light. The pH was maintained at 6.0. The procedure resembling photo-Fenton had a peak removal effectiveness of 96% when exposed to UV-C light and 55.7% when exposed to visible light. The results of reusability studies conducted under UV-C conditions indicated that there was no statistically significant reduction (about 5%) in activity following three treatment cycles. The breakdown of ATZ *via* photocatalytic reactions led to a decrease in toxicity and the generation of dealkylated by-products, namely desethylatrazine (DEA) and desisopropylatrazine (DIA). The results of bioassays, including *E. coli* and *L. sativa* revealed that the application of photo-Fenton-like treatment for 180 minutes, under both UV-C and visible light, did not result in any observed suppression of bacterial growth or inhibitory effects on seeds. In addition, it was shown that the removal effectiveness of ATZ was higher under UV-C irradiation compared to photo-Fenton under visible light irradiation. The formation of $HO^{\bullet}$ radicals played a critical role in determining the ultimate degradation results. The results of this study suggest that basalt within a catalytic system demonstrates chemical stability after undergoing treatment, hence offering the potential for extensive utilization in advanced solar-based oxidation scenarios [90]. Chueca *et al.*, have presented atacamite, a naturally occurring mineral, as a novel photocatalyst that exhibits promising efficacy in the removal of *Enterococcus faecalis* from wastewater. This effectiveness is particularly evident when atacamite is employed in conjunction with various oxidants. This presents a hopeful substitute for traditional catalysts, although additional investigation is required to achieve a thorough comprehension. Atacamite can effectively eradicate *Enterococcus faecalis* within a time frame of 90 minutes when subjected to sun radiation. The efficiency of the system is enhanced when mineral concentrations are lower, as higher concentrations result in the dispersion of radiation, hence reducing the overall effectiveness of the system. The synergistic effect and photocatalytic activity of atacamite and sun radiation surpass their solo applications. Moreover, the atacamite/solar radiation combination exhibits favorable outcomes when employed in conjunction with oxidants such as PMS, PS, and H_2O_2, indicating the occurrence of catalytic reactions between the oxidants and the mineral. The aforementioned reactions yield a range of oxidizing species, with a primary focus on SO_4 and HO radicals [91].

9.2.5 Catalytic Cracking and Pyrolysis

The recycling of used catalysts derived from fluid catalytic cracking (FCC) has been the subject of investigation in the context of polymer catalytic cracking, with a specific focus on waste plastics [92, 93]. Salmiaton *et al.*, conducted a study that investigated the conversion of HDPE into volatile hydrocarbons using fresh, steam-regenerated and deactivated FCC catalysts. The catalysts consisted of rare earth oxides and sodium oxide on Al_2O_3 or SiO_2, and the deactivation catalysts also contained impurities of nickel and vanadium. The catalyst activity was observed to be the highest for fresh catalysts and spent/steam-regenerated catalysts were found to be comparable to poisoned deactivated ones. They maintained a conversion of 70–90% at 450°C. Deactivated catalysts produced a similar gaseous product to the fresh catalyst but with a higher fraction of alkenes regarding alkanes ratio. This shows that the results for the used catalysts that are normally scrapped off are also working catalytically effectively at significantly lower temperatures for the catalytic waste polymer cracking, as in incineration operations [94]. Lin *et al.* [27], have recently reported a comprehensive study on the pyrolysis of some waste plastic feeds of increasing complexity in a fluidized bed reactor using spent FCC catalysts for catalytic pyrolysis. The work using an old FCC catalyst (80%) highlighted the conversion of polypropylene by pyrolysis with high yields on volatile hydrocarbon products (C_1 to C_9) and C_4–C_6 olefins (88%) in only 20 minutes at 390°C. Researchers investigated the effect of the blending ratio of different types of plastic waste on waste plastic conversion and the subsequent distribution of the transformed product. The combination of a spent FCC catalyst plus a 2:1 polyethylene: polypropylene blend or commingled polymer waste (polystyrene, polypropylene, polyethylene, and polyvinylchloride) produced similar recoveries, with yields of 82% and 83%, respectively. The reaction durations were 30 and 20 minutes under the same reaction conditions, respectively [95, 96]. Red mud is considered to be a promising catalyst in the pyrolysis of mixed waste plastics containing various kinds of polymers [8]. However, it was highly effective despite the activity below 440°C, producing 98% of volatiles within the hydrocarbon product at 500°C, outperforming a ZSM-5 zeolite catalyst with considerably improved gas and reduced viscosities of bio-oils. The difference in product selectivity observed in the red mud catalysis relative to ZSM-5, or in the red mud catalysis between Fe_2O_3 and TiO_2, may be at least partially understood as the presence of additional hydrogenation and hydrocracking pathways promoted in the transition metal oxides. Despite red mud's far smaller surface area (28 m^2/g vs. 400 m^2/g for ZSM-5), it proved to be a good solid acid catalyst in reclaiming waste plastic and forming biofuel (Figure 9.4) [97].

Shen *et al.*, studied biomass-derived pyrolysis catalysts based on an impregnation mechanism supported by the surface of rice husk char that was modified with Fe or Ni as the major components [18]. After calcination at a temperature of 700°C, the char was impregnated either with iron or nickel nitrate, which stabilized a char with high silicon content. As investigated efficient tar elimination *in-situ* from biomass pyrolysis using char catalysts with and without pre-calcination at 600°C, the effects showed that metallic states were formed and served as active sites for tar generation. Ni-based catalysts were more active than the Fe-based counterparts, mainly due to the more effective activation of C–H and C–C bonds. The calcination-free preparation of a bimetallic char catalyst resulted in the highest tar reduction (92%) with the greatest portion of usable syngas

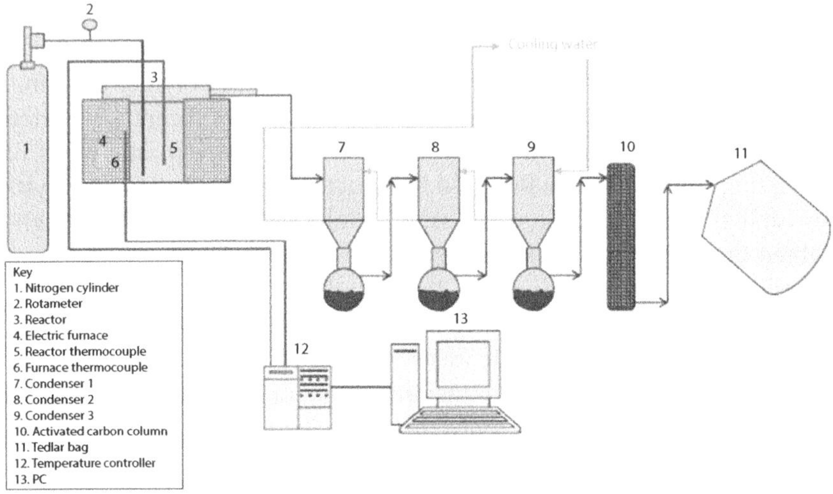

Figure 9.4 The reactor design and liquid fraction composition in the catalytic pyrolysis of waste plastics using red mud and ZSM-5. Adapted from reference [97]. Copyright 2011, with permission from Elsevier.

production and low CO_2 emission. Furthermore, the SiO_2 enhanced the metal oxide reduction but limited the CO_2 absorption, metal ions and tar poisoning during the char-containing supports [98].

The feedstock recycling had been carried out on a grand scale by Akah *et al.*, Pt-containing zeolite Y catalysts showed efficient applications in hydrocracking, the means for the stream upgrading process. An extensive study was carried out to investigate the degradation of high-density polyethylene (HDPE) over Pt/USY (hydrocracking catalyst) and HUSY (cracking catalyst). Utilizing HUSY at a temperature of 310°C and a pressure of 5.5 MPa H_2 led to decreased conversion, and petrol production (with gas to petrol ratio greater than one), as well as significantly increased coke content in comparison to 1.0% Pt/USY. The increased generation of coke was attributed to the lack of a hydrogenation process on a zeolite with large pores, which allowed for the creation of aromatics and polyaromatics (which are precursors to coke). The conversion and yields of USY zeolite were found to be equivalent when the Pt concentration was reduced from 1.5% to 0.5% at a temperature of 270°C, albeit with a little rise in coke content. This indicates that a higher amount of Pt has the potential to improve the stability of the zeolite catalyst by facilitating the hydrogenation of coke precursors [99]. Munir *et al.*, investigated the hydrocracking of a waste plastic mixture using Al-supported mesoporous silica materials and meso-micro-composite catalysts synthesized from zeolite nano-seeds. A series of catalysts based on Al-SBA-16, Al-SBA-15, MZ-15, MZ-16, and MZ-20 were studied. In particular, at 375°C, and corresponding to the reaction conditions of 375–425°C, 2.0 MPa H_2 pressure, 20:1 feed to catalyst, and a residence time of 50–60 minutes, Al-SBA-15 showed superior behavior. With a desilication at 60 min and 400°C, composite catalysts showed comparable activity to that of Al-SBA-15. Among these, MZ-16 containing more mesopores exhibited the best performance. Addition of a higher Si/Al ratio and larger pore size in Al-SBA-16 catalyst showed the highest yield of oil and lowest production of gas compared to other types at a similar conversion rate. Similar trends were observed for micro-mesoporous catalysts at 400°C during 60 min on stream [100].

9.2.6 Catalytic Biomass Valorization

Environmentally friendly fuels and enhanced products can be derived from organic waste materials, such as agricultural wastes, using thermochemical, biological conversion, or catalytic transformation procedures. The hydrogenolysis catalysts were created by Rode *et al.*, using fly ash waste derived from coal-fired power units. Alkali metal hydroxide fusion and acid pre-treatment were used on fly ash, which consists mostly of SiO_2, Al_2O_3, and trace elements, including Fe_2O_3, CaO, TiO_2, K_2O, MgO, and Na_2O. The support was impregnated with a copper compound and then calcined. The surface areas of the catalysts supported on 70 wt.% silica were in the range of 31 to 113 m^2/g, the Cu loadings being adjusted according to the Al: Cu ratio. Glycerol conversion and propanediol selectivity were higher for the 1:1 Al: Cu variation than for the others. The best conversion rate (57% conversion in 5h at 220°C and 52 bar hydrogen) was achieved at 5 wt.% Pt loading. It also showed 95% selectivity for the desired diol. The dispersion and loading of copper were crucial for the optimization of hydrogenolysis. Nevertheless, the use of high pressures of hydrogen endangered the global economic and ecological balance of the process, and it was thus necessary to establish a hydrogen source attracted to the process should be renewable [101]. Carbon catalysts were manufactured by Gonçalves *et al.*, from glycerol and afterward used for glycerol etherification to produce fuel additives. Waste glycerol was treated with sulfuric acid to yield an acidic carbon by refluxing at 150°C for 24 hours. The obtained carbon showed a high acid site density of 3.5 mmol H^+ per gram, and the sulfur content was 3.5 wt.%. The developed carbon catalyst had greater catalytic efficiency and selectivity than commercial catalysts such as Amberlyst-15. The enhanced exposure of acid sites might be the cause of this. The glycerol-derived carbon allowed such ethers to be yielded in a 74% yield within 6 hours at 120°C with a 5 wt.% catalyst loading in 6:1 *t*-butyl alcohol/glycerol molar ratio. The carbon catalyst exhibited stable performance for at least eight cycles without a detectable decrease in product yield. Unfortunately, no additional tests were conducted to confirm its stability through further catalyst recovery examination and hot filtration [102]. Sánchez *et al.*, conducted esterification of glycerol with acetic acid using H_2SO_4-loaded sulfonated activated carbon catalysts generated from the carbonization of sucrose. These catalysts have low acid site densities (0.1–1.4 mmol/g) but large surface areas (up to 800 m^2/g, after carbonization at 850°C), possessing micro- and mesopores. All of them were active to a large extent (>99%), with conversion at 5 wt.% loading within 4 hours at 160°C at a molar ratio of 9 acetic acid with glycerol. The carbonized catalyst at 600°C presented the highest sulfur (4.3 wt. %) and acid site density (1.4 mmol g−1), showing 50% selectivity towards triacetylglycerol. Commercial Amberlyst-15, on the other hand, displayed significantly lower 22% selectivity under the same terms. In addition, the transformation of glycerol waste into propanediols through hydrogenolysis has a vast, versatile use in industries as well (Figure 9.5) [103].

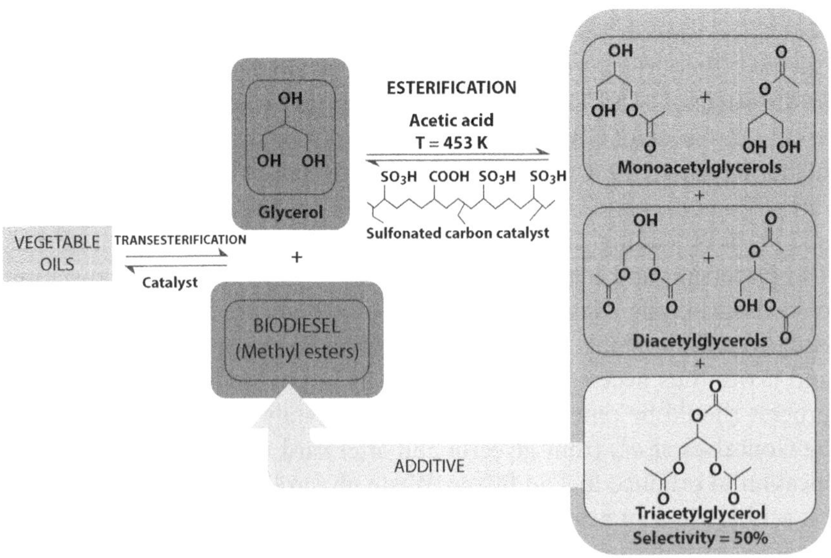

Figure 9.5 Selective esterification of glycerol over carbon-based acid catalysts. Adapted from reference [103]. Copyright 2011, with permission from Elsevier.

Subsequently, Rahim *et al.*, developed a catalytic carbonylation process of glycerol with urea-based on gypsum, which was the first application of this type of catalytic process [43]. The industrial by-product, gypsum, was pretreated physically as well as chemically. Several characterization methods were employed to elucidate the relationships between catalyst structure and activity. Realizing modularity in physicochemical properties had a pronounced effect on catalytic activity, the ß-$CaSO_4$ phase catalyst showed the highest efficiency, with glycerol conversion of 92.8%, selectivity of 90.1%, and glycerol carbonate yield of 83.6%. The recoverability and recyclability of the catalyst were analyzed using a recovered gypsum catalyst. Mechanistic pathways have confirmed glycerol carbonate synthesis during the transformation of waste into value-added products [104]. Calzado *et al.*, reported solid base catalysts for lactose isomerization to lactulose in milk permeate using olive stone waste. The char obtained at 500°C pyrolysis was subjected to chemical or physical activation. The chemical activation was carried out using NaOH and $Ba(OH)_2$ followed by carbonization at 800°C and treatment with HCl. To activate carbon, there are three processes to do carbonization, i.e., physical activation, chemical activation, or wet activation. The physical activation in this study includes carbonization performed in N_2 at 800°C, gasification in (CO_2) and impregnation with Ca. The bases were predominantly microporous with surface areas ranging from 4 to 1960 m^2/g. Assuming gasification at 900°C, the catalyst, that is 5 wt.% Ca/C composition provided the highest number of basic sites. The lactose conversion rate was 28% with 80% selectivity to lactulose at 100°C on 10 wt.% PtIFIEDM for 3 hours. However, leaching studies showed a 1.4% calcium loss, raising concerns about short-term stability [105]. Shin and coworkers conducted a study on the catalytic transformation of glycerol into glycerol carbonate (GC) utilizing Zn/Al oxide catalysts that were supported by activated red mud (ARM), which is a by-product. The ARM-supported catalyst, which had a 50% Zn/Al loading, exhibited a higher GC yield of 58.1% compared to the unsupported catalyst, which had a yield of 49.6%. The catalysts exhibited a volcanic curve in terms of gas chromatography (GC) yield concerning the loading of Zn/Al. The FTIR investigation demonstrated enhanced selectivity in Zn/Al oxide catalysts supported by ARM, leading to an increased yield of GC. The chemical stages in GC synthesis were regulated by the balance of active sites from ARM and Zn/Al oxide, which had an impact on both selectivity and yield. This study highlights the capacity to employ waste materials for effective catalytic processes, contributing to the sustainable manufacturing of chemicals [106].

For biofuel production, the magnetic acid catalyst was developed by de Souza *et al.*, by using red mud and acai seeds collected in the eastern Amazon forest. The catalyst was prepared *via* impregnation, carbonization, and sulfonation and fully characterized by XRD and SEM. After 3h of pyrolysis at 400°C and 3h of sulfonation at 80°C, the results for acai seed–red mud in a 1:1 ratio were the most promising. The results show 88% of the oleic acid was converted to methyl oleate using the 5% loading and the oleic acid to methanol ratio of 1:12 at a temperature of 100°C for 1 hour. The aid of a magnet easily separated the catalyst, and the results show the catalyst can be recycled three times [107]. Maniam *et al.*, developed a red gypsum-derived heterogeneous catalyst for the synthesis of glycerol carbonate from the glycerolysis of urea. The highest activity was observed for the catalyst RG150, and 87% glycerol carbonate reached 92.4% selectivity and 94.1% glycerol conversion. The turnover frequency of RG150 was slightly greater than RG800, most likely due to its more basic nature, possibly caused by the incorporation of anhydrite CaSO4 phases with active hematite (Fe_2O_3) in its structure. The gas employed for ammonia removal did not have any impact on the catalytic process. The synthesis of glycerol carbonate from industrial crude glycerol was effectively achieved by this technique, demonstrating notable resilience to both moisture and contaminants. The efficiency of the catalyst remained consistent over consecutive cycles without the need for pre-treatment [108].

9.3 Summary and Future Perspective

The chapter examines the extensive possibilities of using different waste materials to create diverse catalysts and their use in a broad spectrum of chemical reactions. The process of catalyst preparation frequently entails complex waste treatment procedures to remove certain components. These components can function as either inert supports or active metals/metal oxides inside the catalyst framework. The catalysts can be generated exclusively from waste materials or supplemented with commercially available components to improve their overall performance. Transition metal oxides obtained from industrial waste, as well as carbonaceous and alkaline earth oxides formed from biomass waste, are important objectives for catalyst production. The manufacturing of biodiesel is becoming increasingly popular, often using solid acids/bases, with CaO from discarded shells being recognized as a significant catalyst. Recycling used catalysts from other industries is also highlighted as a highly successful approach.

Nevertheless, it is imperative to enhance the design of waste catalysts to reduce dependence on commercial components and optimize the efficiency of processing procedures. The optimization of catalyst production and application is crucial for the advancement of environmental sustainability and resource efficiency. The proposition of comprehensive catalyst design is put out as a potentially effective strategy for promoting the advancement of environmentally friendly catalytic technologies that make use of waste materials. Researchers want to use this method to reduce environmental impact and facilitate the development of more sustainable and environmentally friendly chemical processes. In the future, catalysts created from waste materials possess significant potential to bring about a transformative impact on chemical processes. Continuous progress in technology is anticipated to enhance the efficiency and diversity of these catalysts, fitting harmoniously with the concepts of a circular economy and fostering resource efficiency and waste reduction. The successful commercialization and scaling up of waste-derived catalysts will heavily rely on the collaboration between academia, industry, and policymakers. Addressing obstacles like scalability and cost-effectiveness necessitates the implementation of inventive strategies and the provision of governmental backing to encourage the adoption of environmentally friendly chemistry methodologies. Waste-derived catalysts offer a sustainable solution for the depletion of resources and pollution of the environment by decreasing dependence on limited resources and limiting the production of trash. The realization of their full potential, the promotion of positive environmental and socio-economic results on a global scale, and the cultivation of a more sustainable future necessitate ongoing study, investment, and policy support.

AI disclosure statement: The author acknowledges the use of AI-based tools for minor language editing or grammatical correction. However, all intellectual contributions, analyses, and conclusions presented in this chapter are entirely the author's own.

References

1. Shukla, M.K., Chauhan, B.V.S., Bhaskar, T., Dhar, A., Vedratnam, A., Recycling of Platinum Group Metals and Alternative Catalysts for Catalytic Converters. *Energy Environ. Sustain.*, 363–398, 2023.
2. Mpinga, C.N., Eksteen, J.J., Aldrich, C., Dyer, L., Direct leach approaches to Platinum Group Metal (PGM) ores and concentrates: A review. *Miner. Eng.*, 78, 93–113, Jul. 2015.

3. Tabatabaei, M., Karimi, K., Kumar, R., Horváth, I.S., Renewable energy and alternative fuel technologies. *Biomed Res. Int.*, 2015, 2015.
4. Yangin-Gomec, C., Sárvári Horváth, I., Martín, C., Energy Production from Biomass Valorization. *Energies*, 16, 11, 16–19, 2023.
5. Guo, M., *et al.*, Advances in biorenewables-resource-waste systems and modelling. *Carbon Capture Sci. Technol.*, 9, Dec. 2023.
6. Sheldon, R.A., Green and sustainable manufacture of chemicals from biomass: state of the art. *Green Chem.*, 16, 3, 950–963, Feb. 2014.
7. Wang, L., Xu, S., Meng, X., Xiao, F.-S., Mesoporous Zeolite for Biomass Conversion, in: *Nanoporous Catal. Biomass Convers.*, pp. 209–230, John Wiley & Sons Ltd., Chichester, UK., Nov. 2017.
8. Brown, T.R., Zhang, Y., Hu, G., Brown, R.C., Techno-economic analysis of biobased chemicals production *via* integrated catalytic processing. *Biofuels, Bioprod. Biorefin.*, 6, 1, 73–87, Jan. 2012.
9. Bennett, J.A., Wilson, K., Lee, A.F., Catalytic applications of waste derived materials. *J. Mater. Chem. A*, 4, 10, 3617–3637, Mar. 2016.
10. Zhang, X., Xu, S., Tang, J., Fu, L., Karimi-Maleh, H., Sustainably Recycling and Upcycling of Single-Use Plastic Wastes through Heterogeneous Catalysis. *Catal.*, 12, 8, 818, Jul. 2022.
11. Ellis, L.D., *et al.*, Chemical and biological catalysis for plastics recycling and upcycling. *Nat. Catal.*, 4, 7, 539–556, Jul. 2021.
12. Chen, H., Wan, K., Zhang, Y., Wang, Y., Waste to Wealth: Chemical Recycling and Chemical Upcycling of Waste Plastics for a Great Future. *ChemSusChem*, 14, 19, 4123–4136, Oct. 2021.
13. Chen, K.T., *et al.*, Rice husk ash as a catalyst precursor for biodiesel production. *J. Taiwan Inst. Chem. Eng.*, 44, 4, 622–629, 2013.
14. Smith, S.M., *et al.*, Transesterification of soybean oil using bovine bone waste as new catalyst. *Bioresour. Technol.*, 143, 686–690, 2013.
15. Abdullah, S.H.Y.S., *et al.*, A review of biomass-derived heterogeneous catalyst for a sustainable biodiesel production. *Renewable Sustainable Energy Rev.*, 70, 1040–1051, Apr. 2017.
16. Liu, Z., *et al.*, Efficient hydrogen production from high-concentration aqueous formic acid over bio-based γ-Mo2N catalysts. *Carbon Resour. Convers.*, 100209, Sep. 2023.
17. Ghorai, S., Nanda, D., Ghosh, A., Dash, P.S., Review on the recent advances in catalytic conversion of carbon dioxide for synthesis of cyclic propylene carbonate. *Mol. Catal.*, 113720, Jan. 2023.
18. Yuan, S.J. and Dai, X.H., Facile synthesis of sewage sludge-derived mesoporous material as an efficient and stable heterogeneous catalyst for photo-Fenton reaction. *Appl. Catal., B*, 154–155, 252–258, Jul. 2014.
19. Zhi, Y., *et al.*, A review of hydrothermal carbonization of municipal sludge: Process conditions, physicochemical properties, methods coupling, energy balances and life cycle analyses. *Fuel Process. Technol.*, 254, Feb. 2024.

20. Balakrishnan, M., *et al.*, Hydrogen production from methane in the presence of red mud –making mud magnetic. *Green Chem.*, 11, 1, 42–47, Jan. 2009.
21. Fei, C., *et al.*, Climate Change and Livestock Production: A Literature Review. *Atmosphere*, 13, 1, 140–159, Jan. 2022.
22. Woznicki, S.A., *et al.*, Climate change and livestock: Impacts, adaptation, and mitigation. *CRM*, 16, 145–163, Feb. 2017.
23. Barange, M., *et al.*, Impacts of climate change on marine ecosystem production in societies dependent on fisheries. *Nat. Clim. Change*, 4, 3, 211–216, Feb. 2014.
24. Thornton, P.K., Livestock production: recent trends, future prospects. *Philos. Trans. R. Soc. B Biol. Sci.*, 365, 1554, 2853–2867, Sep. 2010.
25. Guarino, M., *et al.*, Review: Environmental impact of livestock farming and Precision Livestock Farming as a mitigation strategy. *Sci. Total Environ.*, 650, 10, 2751–2760, Feb. 2019.
26. Sharma, S. and Ghoshal, S.K., Hydrogen the future transportation fuel: From production to applications. *Renewable Sustainable Energy Rev.*, 43, 1151–1158, Mar. 2015.
27. El-Adawy, M., Nemitallah, M.A., Abdelhafez, A., Towards sustainable hydrogen and ammonia internal combustion engines: Challenges and opportunities. *Fuel*, 364, May 2024.
28. Moustakas, K., Loizidou, M., Rehan, M., Nizami, A.S., A review of recent developments in renewable and sustainable energy systems: Key challenges and future perspective. *Renewable Sustainable Energy Rev.*, 119, Mar. 2020.
29. Hosseini, S.E. and Wahid, M.A., Hydrogen production from renewable and sustainable energy resources: Promising green energy carrier for clean development. *Renewable Sustainable Energy Rev.*, 57, 850–866, May 2016.
30. Hosseini, S.E., Abdul Wahid, M., Jamil, M.M., Azli, A.A.M., Misbah, M.F., A review on biomass-based hydrogen production for renewable energy supply. *Int. J. Energy Res.*, 39, 12, 1597–1615, Oct. 2015.
31. Yoo, C.G. and Pan, X., Fuel Ethanol from Lignocellulosic Biomass, in: *Handb. Clean Energy Syst.*, pp. 1–20, John Wiley & Sons, Ltd., Chichester, UK., Jul. 2015.
32. Lu, M., Xiong, Z., Lv, P., Yuan, Z., Guo, H., Chen, Y., Catalytic Purification of Raw Gas from Biomass Gasification on Mo–Ni–Co/Cordierite Monolithic Catalyst. *Energy Fuels*, 27, 4, 2099–2106, Apr. 2013.
33. Pandey, B., Prajapati, Y.K., Sheth, P.N., Recent progress in thermochemical techniques to produce hydrogen gas from biomass: A state of the art review. *Int. J. Hydrogen Energy*, 44, 47, 25384–25415, Oct. 2019.
34. Cao, L., *et al.*, Biorenewable hydrogen production through biomass gasification: A review and future prospects. *Environ. Res.*, 186, 109547, Jul. 2020.
35. Zhou, Y., *et al.*, Metal–Carbon Hybrid Electrocatalysts Derived from Ion-Exchange Resin Containing Heavy Metals for Efficient Hydrogen Evolution Reaction. *Small*, 12, 20, 2768–2774, May 2016.

36. Perego, C., Bosetti, A., Ricci, M., Millini, R., Zeolite Materials for Biomass Conversion to Biofuel. *Energy Fuels*, 31, 8, 7721–7733, Aug. 2017.
37. Lin, Y., *et al.*, Cost-effective and environmentally friendly synthesis of 3D Ni2P from scrap nickel for highly efficient hydrogen evolution in both acidic and alkaline media. *J. Mater. Chem. A*, 6, 9, 4088–4094, Feb. 2018.D.
38. Grams, J., *et al.*, Hydrogen-rich gas production by upgrading of biomass pyrolysis vapors over Nibea catalyst: Impact of dealumination and preparation method. *Energy Fuels*, 34, 12, 16936–16947, Dec. 2020.
39. Atabani, A.E., Pugazhendhi, A., Almomani, F., Rene, E.R., Naqvi, S.R., Recent advances in the thermochemical transformation of biomass to bio-oil, biochar and syngas and its upgrading methods. *Process Saf. Environ. Prot.*, 168, 624–625, Dec. 2022.
40. Taipabu, M.I., Viswanathan, K., Wu, W., Hattu, N., Atabani, A.E., A critical review of the hydrogen production from biomass-based feedstocks: Challenge, solution, and future prospect. *Process Saf. Environ. Prot.*, 164, 384–407, Aug. 2022.
41. Altalhi, T., *et al.*, Cathodic Activation of Titania-Fly Ash Cenospheres for Efficient Electrochemical Hydrogen Production: A Proposed Solution to Treat Fly Ash Waste. *Catalysts*, 12, 5, 466, May 2022.
42. Jara-Cobos, L., Abril-González, M., Pinos-Vélez, V., Production of Hydrogen from Lignocellulosic Biomass: A Review of Technologies. *Catal.*, 13, 4, 766, Apr. 2023.
43. Zhang, Y., Li, L., Xu, P., Liu, B., Shuai, Y., Li, B., Hydrogen production through biomass gasification in supercritical water: A review from exergy aspect. *Int. J. Hydrogen Energy*, 44, 30, 15727–15736, Jun. 2019.
44. Fangaj, E., Ali, A.A., Güngör, F., Bektaş, S., Ceyhan, A.A., The use of metallurgical waste sludge as a catalyst in hydrogen production from sodium borohydride. *Int. J. Hydrogen Energy*, 45, 24, 13322–13329, May 2020.
45. Rajagopal, J., *et al.*, Processing of household waste *via* hydrothermal gasification and hydrothermal liquefaction for bio-oil and bio-hydrogen production: Comparison with RSM studies. *J. Environ. Chem. Eng.*, 10, 2, 107218, Apr. 2022.
46. Cao, X., *et al.*, Synthesis of biomass porous carbon materials from bean sprouts for hydrogen evolution reaction electrocatalysis and supercapacitor electrode. *Int. J. Hydrogen Energy*, 46, 36, 18887–18897, May 2021.
47. Wyss, K.M., Chen, W., Beckham, J.L., Savas, P.E., Tour, J.M., Holey and Wrinkled Flash Graphene from Mixed Plastic Waste. *ACS Nano*, 2022.
48. Sharma, K.N., Ali, M., Srivastava, A.K., Joshi, R.K., (η6-Benzene) Ru(II) half-sandwich complexes of pyrazolated chalcogenoethers for catalytic activation of aldehydes to amides transformation. *J. Organomet. Chem.*, 879, 69–77, Jan. 2019.
49. Ali, M., Srivastava, A.K., Siangwata, S., Smith, G.S., Joshi, R.K., Photo induced alkyne hydration reactions mediated by a water soluble, reusable Rhodium (I) catalyst. *Catal. Commun.*, 115, 78–81, Sep. 2018.

50. Srivastava, A.K., Ali, M., Siangwata, S., Satrawala, N., Smith, G.S., Joshi, R.K., Multitasking FeOCN Composite as an Economic, Heterogeneous Catalyst for 1-Octene Hydroformylation and Hydration Reactions. *Asian J. Org. Chem.*, 9, 3, 377–384, Mar. 2020.
51. Kumar Jaiswal, P., Ali, M., Sharma, S., Kumar Yadav, D., Satyadev Upadhyay, N., Graphene-based Nanocomposite Catalysts: Synthesis, Properties and Applications, in: *Graphene-based Carbocatalysts Synth. Prop. Appl.*, pp. 208–262, Bentham Science Publishers, Sharjah, UAE, May 2023.
52. Sharma, K.N., Satrawala, N., Srivastava, A.K., Ali, M., Joshi, R.K., Palladium(II) ligated with a selenated (Se, CNHC, N−)-type pincer ligand: an efficient catalyst for Mizoroki–Heck and Suzuki–Miyaura coupling in water. *Org. Biomol. Chem.*, 17, 40, 8969–8976, Oct. 2019.
53. Lebedeva, O.E. and Sarmurzina, A.G., Industrial wastes as catalyst precursors: VOC oxidation. *Appl. Catal. B Environ.*, 26, 1, L1–L3, Apr. 2000.
54. Sushil, S., Scholz, P., Pollok, K., Ondruschka, B., Batra, V.S., Application of industrial waste based catalysts for total oxidation of propane. *Chem. Eng. J.*, 166, 2, 568–578, Jan. 2011.
55. Teixeira, I.F., Medeiros, T.P.V., Freitas, P.E., Rosmaninho, M.G., Ardisson, J.D., Lago, R.M., Carbon deposition and oxidation using the waste red mud: A route to store, transport and use offshore gas lost in petroleum exploration. *Fuel*, 124, 7–13, May 2014.
56. Saini, D., Aggarwal, R., Anand, S.R., Satrawala, N., Joshi, R.K., Sonkar, S.K., Sustainable feasibility of waste printer ink to magnetically separable iron oxide–doped nanocarbons for styrene oxidation. *Mater. Today Chem.*, 16, 100256, Jun. 2020.
57. Li, N., Lu, X., Zhang, S., A novel reuse method for waste printed circuit boards as catalyst for wastewater bearing pyridine degradation. *Chem. Eng. J.*, 257, 253–261, Dec. 2014.
58. Domínguez, M.I., Barrio, I., Sánchez, M., Centeno, M.A., Montes, M., Odriozola, J.A., CO and VOCs oxidation over Pt/SiO2 catalysts prepared using silicas obtained from stainless steel slags. *Catal. Today*, 133–135, 1–4, 467–474, Apr. 2008.
59. Guo, Y., *et al.*, Enhanced catalytic benzene oxidation over a novel waste-derived Ag/eggshell catalyst. *J. Mater. Chem. A*, 7, 15, 8832–8844, Apr. 2019.
60. Li, Z., *et al.*, Waste eggshells to valuable Co3O4/CaCO3 materials as efficient catalysts for VOCs oxidation. *Mol. Catal.*, 483, 110766, Mar. 2020.
61. Thien, L.T., Quang Tu, L.N., Trung, B.C., Long, N.Q., Au nanoparticles loaded Hydroxyapatite catalyst prepared from waste eggshell: synthesis, characterization and application in VOC removal. *IOP Conf. Ser. Earth Environ. Sci.*, 964, 1, 012027, Jan. 2022.
62. Alagarsamy, P., Daniel, S., Chinnapparaj, M.I., Kim, S.C., Manivasagam, V.R., Vanaraj, R., Boosting Fenton's Oxidation Reaction by a Food Waste-Derived Catalyst for Oxidizing Organic Dyes: Synergistic Effect of Complex

Iron Oxides and the Layer Carbon Structure. *ACS Appl. Bio Mater.*, 6, 8, 3291–3308, Aug. 2023.

63. Faruque, M.O., Razzak, S.A., Hossain, M.M., Application of Heterogeneous Catalysts for Biodiesel Production from Microalgal Oil—A Review. *Catal.*, 10, 9, 1025, Sep. 2020.
64. Lee, A.F. and Wilson, K., Recent developments in heterogeneous catalysis for the sustainable production of biodiesel. *Catal. Today*, 242, Part A, 3–18, Mar. 2015.
65. Hamza, M., *et al.*, A review on the waste biomass derived catalysts for biodiesel production. *Environ. Technol. Innov.*, 21, Feb. 2021.
66. Chouhan, A.P.S. and Sarma, A.K., Modern heterogeneous catalysts for biodiesel production: A comprehensive review. *Renewable Sustainable Energy Rev.*, 15, 9, 4378–4399, Dec. 2011.
67. Ao, S., Changmai, B., Vanlalveni, C., Chhandama, M.V.L., Wheatley, A.E.H., Rokhum, S.L., Biomass waste-derived catalysts for biodiesel production: Recent advances and key challenges. *Renewable Energy*, 223, Mar. 2024.
68. Boz, N., Degirmenbasi, N., Kalyon, D.M., Conversion of biomass to fuel: Transesterification of vegetable oil to biodiesel using KF loaded nano-γ-Al2O3 as catalyst. *Appl. Catal. B Environ.*, 89, 3–4, 590–596, Jul. 2009.
69. Dawodu, F.A., Ayodele, O., Xin, J., Zhang, S., Yan, D., Effective conversion of non-edible oil with high free fatty acid into biodiesel by sulphonated carbon catalyst. *Appl. Energy*, 114, 819–826, Feb. 2014.
70. Babadi, F.E., Hosseini, S., Masoudi Soltani, S., Aroua, M.K., Shamiri, A., Samadi, M., Sulfonated Beet Pulp as Solid Catalyst in One-Step Esterification of Industrial Palm Fatty Acid Distillate. *J. Am. Oil Chem. Soc.*, 93, 3, 319–327, Mar. 2016.
71. Panadare, D.C. and Rathod, V.K., Microwave assisted enzymatic synthesis of biodiesel with waste cooking oil and dimethyl carbonate. *J. Mol. Catal. B: Enzym.*, 133, S518–S524, Nov. 2016.
72. Jindapon, W. and Ngamcharussrivichai, C., Heterogeneously catalyzed transesterification of palm oil with methanol to produce biodiesel over calcined dolomite: The role of magnesium oxide. *Energy Convers. Manage.*, 171, 1311–1321, Sep. 2018.
73. Wahyudi, A., Kurniawan, W., Hinode, H., Wahyudi, A., Kurniawan, W., Hinode, H., Study on Deactivation and Regeneration of Modified Red Mud Catalyst Used in Biodiesel Production. *Green Sustainable Chem.*, 7, 4, 247–258, Sep. 2017.
74. Lake, V., *et al.*, Utilization of water hyacinth-based biomass as a potential heterogeneous catalyst for biodiesel production. *J. Phys. Conf. Ser.*, 1402, 5, 055005, Dec. 2019.
75. Bastos, R.R.C., da Luz Corrêa, A.P., da Luz, P.T.S., da Rocha Filho, G.N., Zamian, J.R., da Conceição, L.R.V., Optimization of biodiesel production using sulfonated carbon-based catalyst from an amazon agro-industrial waste. *Energy Convers. Manage.*, 205, 112457, Feb. 2020.

76. Roy, M. and Mohanty, K., Valorization of de-oiled microalgal biomass as a carbon-based heterogeneous catalyst for a sustainable biodiesel production. *Bioresour. Technol.*, 337, 125424, Oct. 2021.
77. Leesing, R., Siwina, S., Fiala, K., Yeast-based biodiesel production using sulfonated carbon-based solid acid catalyst by an integrated biorefinery of durian peel waste. *Renewable Energy*, 171, 647–657, Jun. 2021.
78. Tarigan, J.B., *et al.*, Waste Passion Fruit Peel as a Heterogeneous Catalyst for Room-Temperature Biodiesel Production. *ACS Omega*, 7, 9, 7885–7892, Mar. 2022.
79. Park, H., Park, Y., Kim, W., Choi, W., Surface modification of TiO2 photocatalyst for environmental applications. *J. Photochem. Photobiol. C Photochem. Rev.*, 15, 1, 1–20, Jun. 2013.
80. Ryu, J., Kim, W., Kim, J., Ju, J., Kim, J., Is surface fluorination of TiO2 effective for water purification? The degradation vs. mineralization of phenolic pollutants. *Catal. Today*, 282, 24–30, Mar. 2017.
81. Luo, J., Zhang, S., Sun, M., Yang, L., Luo, S., Crittenden, J.C., A Critical Review on Energy Conversion and Environmental Remediation of Photocatalysts with Remodeling Crystal Lattice, Surface, and Interface. *ACS Nano*, 13, 9, 9811–9840, Sep. 2019.
82. Ohtani, B., Photocatalysis A to Z—What we know and what we do not know in a scientific sense. *J. Photochem. Photobiol. C Photochem. Rev.*, 11, 4, 157–178, Dec. 2010.
83. Dahl, M., Liu, Y., Yin, Y., Composite titanium dioxide nanomaterials. *Chem. Rev.*, 114, 19, 9853–9889, Oct. 2014.
84. Liu, L. and Chen, X., Titanium dioxide nanomaterials: Self-structural modifications. *Chem. Rev.*, 114, 19, 9890–9918, Oct. 2014.
85. Liou, T.H. and Lai, B.C., Utilization of e-waste as a silica source for the synthesis of the catalyst support MCM-48 and highly enhanced photocatalytic activity of supported titania nanoparticles. *Appl. Catal. B Environ.*, 115–116, 138–148, Apr. 2012.
86. Yola, M.L., Eren, T., Atar, N., A novel efficient photocatalyst based on TiO2 nanoparticles involved boron enrichment waste for photocatalytic degradation of atrazine. *Chem. Eng. J.*, 250, 288–294, Aug. 2014.
87. Sapiña, M., Jimenez-Relinque, E., Castellote, M., Turning waste into valuable resource: potential of electric arc furnace dust as photocatalytic material. *Environ. Sci. Pollut. Res.*, 21, 20, 12091–12098, Oct. 2014.
88. Belver, C., Han, C., Rodriguez, J.J., Dionysiou, D.D., Innovative W-doped titanium dioxide anchored on clay for photocatalytic removal of atrazine. *Catal. Today*, 280, 21–28, Feb. 2017.
89. Kattiparambil Manoharan, R. and Sankaran, S., Photocatalytic degradation of organic pollutant aldicarb by non-metal-doped nanotitania: synthesis and characterization. *Environ. Sci. Pollut. Res.*, 25, 21, 20510–20517, Jul. 2018.
90. Carmelo da Rocha, A., de Oliveira Sampaio Dantas, A., Angélica Vieira, P., Luiz Cardoso, V., Evaluation of basalt powder as a natural heterogeneous

catalyst in photo-Fenton like treatment of atrazine. *J. Photochem. Photobiol. A Chem.*, 446, 115149, Jan. 2024.
91. Haro, D., García-Muñoz, P., Mola, M., Fresno, F., Rodríguez-Chueca, J., Atacamite (Cu2Cl(OH)3) as catalyst of different AOPs for water disinfection. *Catal. Today*, 429, 114496, Mar. 2024.
92. Vyas, A., *et al.*, Catalytic Cracking of Waste Polypropylene in a Nitrogen Fluidized Bed Reactor. *J. Geosci. Environ. Prot.*, 4, 4, 20–25, Apr. 2016.
93. Liu, S., Kots, P.A., Vance, B.C., Danielson, A., Vlachos, D.G., Plastic waste to fuels by hydrocracking at mild conditions. *Sci. Adv.*, 7, 17, 8283–8304, Apr. 2021.
94. Salmiaton, A. and Garforth, A., Waste catalysts for waste polymer. *Waste Manage.*, 27, 12, 1891–1896, Jan. 2007, doi: 10.1016/J.WASMAN.2006.07.024.
95. Lin, Y.H. and Yang, M.H., Chemical catalysed recycling of waste polymers: Catalytic conversion of polypropylene into fuels and chemicals over spent FCC catalyst in a fluidised-bed reactor. *Polym. Degrad. Stab.*, 92, 5, 813–821, May 2007.
96. Lin, Y.H. and Yang, M.H., Catalytic pyrolysis of polyolefin waste into valuable hydrocarbons over reused catalyst from refinery FCC units. *Appl. Catal. A Gen.*, 328, 2, 132–139, Sep. 2007.
97. López, A., de Marco, I., Caballero, B.M., Laresgoiti, M.F., Adrados, A., Aranzabal, A., Catalytic pyrolysis of plastic wastes with two different types of catalysts: ZSM-5 zeolite and Red Mud. *Appl. Catal. B Environ.*, 104, 3–4, 211–219, May 2011.
98. Shen, Y., Zhao, P., Shao, Q., Ma, D., Takahashi, F., Yoshikawa, K., In-situ catalytic conversion of tar using rice husk char-supported nickel-iron catalysts for biomass pyrolysis/gasification. *Appl. Catal. B Environ.*, 152–153, 1, 140–151, Jun. 2014.
99. Akah, A., Aramco, S., Rallan, C., Garforth, A., Hernandez-Martinez, J., Garforth, A.A., Enhanced Feedstock Recycling of Post-Consumer Plastic Waste. *Chem. Eng. Trans.*, 43, 2015.
100. Munir, D., Abdullah, Piepenbreier, F., Usman, M.R., Hydrocracking of a plastic mixture over various micro-mesoporous composite zeolites. *Powder Technol.*, 316, 542–550, Jul. 2017.
101. Rode, C.V., Mane, R.B., Potdar, A.S., Patil, P.B., Niphadkar, P.S., Joshi, P.N., Copper modified waste fly ash as a promising catalyst for glycerol hydrogenolysis. *Catal. Today*, 190, 1, 31–37, Aug. 2012.
102. Gonçalves, M., Mantovani, M., Carvalho, W.A., Rodrigues, R., Mandelli, D., Silvestre Albero, J., Biodiesel wastes: An abundant and promising source for the preparation of acidic catalysts for utilization in etherification reaction. *Chem. Eng. J.*, 256, 468–474, Nov. 2014.
103. Sánchez, J.A., Hernández, D.L., Moreno, J.A., Mondragón, F., Fernández, J.J., Alternative carbon based acid catalyst for selective esterification of glycerol to acetylglycerols. *Appl. Catal. A Gen.*, 405, 1–2, 55–60, Oct. 2011.

104. Zuhaimi, N.A.S., *et al.*, Reusable gypsum based catalyst for synthesis of glycerol carbonate from glycerol and urea. *Appl. Catal. A Gen.*, 502, 312–319, Aug. 2015.
105. Calzado, M., *et al.*, Lignocellulosic waste-derived basic solids and their catalytic applications for the transformation of biomass waste. *Catal. Today*, 257, Part 2, 229–236, Nov. 2015.
106. Nguyen-Phu, H., Park, C., Eun, W.S., Activated red mud-supported Zn/Al oxide catalysts for catalytic conversion of glycerol to glycerol carbonate: FTIR analysis. *Catal. Commun.*, 85, 52–56, Oct. 2016.
107. Araujo, R.O., *et al.*, Magnetic acid catalyst produced from acai seeds and red mud for biofuel production. *Energy Convers. Manage.*, 228, 113636, Jan. 2021.
108. Mohd, M.H., Zuhaimi, N.A.S., Saud, A.S., Madduluri, V.R., Alshammari, H., Maniam, G.P., Synthesis of glycerol carbonate from industrial by-products by alcoholysis of urea: Crude glycerol and red gypsum. *Fuel*, 357, 129774, Feb. 2024.

10

Waste-Derived Catalysts in Energy Applications: Prospects and Challenges for Green Hydrogen Production

Amina Shehbaz, Abdul Majid* and Alia Jabeen

Department of Physics, University of Gujrat, Gujrat, Pakistan

Abstract

In the quest for sustainable energy to meet future needs, waste treatment has emerged as a robust strategy to provide valuable energy sources through the preparation of materials, processes, and functional products. In this scenario, the production of green hydrogen through water splitting has shown great optimism. The process of water splitting is a highly viable energy-producing mechanism, particularly due to its cost-effectiveness, abundance, cleanliness, and affordability across all aspects. Owing to the future prospectus, the search for efficient materials to split the water and make the entire process practicable has been one of the most attractive research areas over the year. The manufacturing of catalysts frequently utilizes various waste treatment methods, ranging from simple to complex, by recovering valuable components such as active metals, metal oxides, and inert or functional supports. The chapter discusses the synthesis strategies for transforming waste into functional catalysts, including pyrolysis, hydrothermal treatment, wet chemical method, and microwave preparation methods, emphasizing their role in producing materials with large surface area, tunable porosity, and abundant active sites. It examines electrocatalytic performance of these catalysts in both the hydrogen evolution reaction and oxygen evolution reaction by comparing their activity, stability, and scalability. Waste-derived catalysts are being improved through innovative strategies focusing on electroactive sites, mass/charge transfer, structural and electrochemical durability. This work aims to pave the way for sustainable hydrogen production by leveraging waste-based catalysts, offering a dual solution to energy and environmental crises while outlining future research pathways for practical development.

**Corresponding author*: abdulmajid40@yahoo.com

Deepti Goyal, Sakshi Kabra Malpani and Renuka Jain (eds.) Catalysis from Waste: Transforming Solid Residues into Value, (253–290) © 2026 Scrivener Publishing LLC

***Keywords*:** Waste-derived catalysts, water electrolysis, hydrogen evolution reaction, oxygen evolution reaction, performance stability, wastewater treatment, hetero-structure catalyst

10.1 Introduction

The rising energy demand prompts the search for renewable and sustainable energy sources. On the other hand, the increasing climate problems, such as rapid production of CO_2, NO_2, and other air pollutants, along with global warming, consistently demands the development of novel materials with large capabilities for storing alternative energy to address these issues. The United Nations (UN) states that energy production contributes 60% of global greenhouse gas emissions, making it the main cause of climate change [1]. In addition to fossil fuels, there are various eco-friendly and renewable energy sources, such as solar, hydropower, wind, and geothermal. Nevertheless, there are some drawbacks with each making the switch difficult, like windmills producing unsortable electricity, hydropower facing dam construction issues, and geothermal energy has limited lifetime and is costly to operate [2]. Solar energy can generate heat and electricity without using turbines because it is unlimited, renewable, and cost-free. However, owing to its intermittent nature, which depends on geographical location, time of day and low energy density per unit of earth surface, its usage is limited [3]. Hydrogen is increasingly recognized as a promising next-generation energy source [4]. It is a promising alternative energy source, and offers the benefits of being pollution-free, efficient, and cost-effective. However, it has various drawbacks, such as excessive energy and cost needed to store hydrogen as a compressed gas or liquid, limited infrastructure for hydrogen fueling, and dependence on fossil fuels for its generation. To address all these issues, extensive research is being conducted to explore hydrogen production techniques from renewable sources. In this regard, photocatalytic water splitting has emerged as a renowned method for generating green energy *via* dissociation of water. The process of photocatalytic water splitting produces energy without any emission of greenhouse gases, which makes it suitable for household and industrial applications [5]. The process of water splitting refers to breakdown of water molecules to produce hydrogen to be used as fuel, with the process varying based on catalyst's characteristics. The rapid increase in industrialization leads to unprecedented damage to the environment, *via* either direct or indirect dumping of pollutants into the water resources, without any prior treatment [6].

Contaminants like dense metals, phenolic compounds, dyes, pharmaceuticals, and others are being liberated from industries [7]. Such pollutants are carcinogenic along with mutagenic characteristics that have negative impacts on living beings including human beings as well as aquatic life. The various procedures, including photodegradation, advanced oxidation processes, and adsorption, have been utilized and are considered highly effective for pollutant removal [7–9]. There are distinct techniques for the decontamination of wastewater, including electrode ionization, electrodialysis, electrochemical reduction, membrane filtration, adsorption, and precipitation. The above-mentioned techniques generally utilize a great amount of energy and are cumbersome due to the varying pollutants, different wastes and fluids, and by-products produced during wastewater treatment process [10]. From the social and economic development point of view, milder reaction conditions along with the effective catalysts are highly required to remove several pollutants from wastewater [11]. For the past several years, heterogeneous photocatalysis has been investigated and implemented in several related areas like water splitting, nitrogen (N_2) fixation, air/water purification, reduction of CO_2, etc. Photocatalysis is a green technology that operates under mild conditions, follows a simple procedure, and can degrade organic pollutants present in wastewater [12, 13]. The catalyst is exposed to the process of self-etching during photocatalysis due to its instability, consequently, causes deactivation of photocatalysts [14].

Waste biomass grows rapidly and is recognized as a sustainable feedstock for fuel synthesis and production of fundamental chemicals. They are utilized to produce both the catalyst and the desired output, and the cost of a catalyst, even when used in limited quantities, can be substantial, particularly when waste biomass is utilized as a starting material [15, 16]. The type of waste sources often influences the characteristics of an innovative catalyst, leading to the formation of related material groups for specific applications [17]. Major research utilizes annual waste streams of millions of tons, crucial for producing catalysts for significant industrial operations. A wide range of abundant resources, including agricultural waste like rice husks, corn stalks, and sugarcane bagasse, provide readily available sources of carbon and silica, which are used to produce carbon-based electrocatalysts. The industrial byproducts (like fly ash and red mud, which provide metal oxides and other valuable components), sewage, waste materials from mining, metalwork, and especially the iron and steel industry are among the most plentiful resources [18]. Furthermore, an ideal waste product should be highly feasible, stable, selective, convenient, and secure for collection and should ensure it does not cause negative environmental effects.

The conversion of waste resources into useful products can significantly enhance their cost-effectiveness, reduce disposal costs and provide various benefits for practical applications [20]. Waste materials also possess innovative qualities, including excellent adsorption capacity, functional groups, and surface area, which can be enhanced through combining them with nanoparticles. The sustainability of waste materials is crucial for the maintenance and dispersal of nanoparticles [21, 22].

This book chapter critically analyzes the dynamics and promising areas of using waste-derived catalysts for green hydrogen production. The study delivers comprehensive analysis of current advancements in waste-based catalysts, including their characteristics, synthesis techniques, modification methods, and processes for studying solid and acid-base catalysts.

10.2 Methods for Conversion of Waste into Valuable Catalysts

One of the major global issues that affect the environment is the increasing amount of Municipal Solid Waste (MSW) and Plastic Waste (PW). On average, every single human produces 1.42 kg of waste daily, which will become a serious future problem [23]. Conventional methods of waste disposal, like burning and landfills, contribute to climate change and health issues due to greenhouse gas emissions, for example CH_4, CO_2 and N_2O [24]. So, a master plan for waste management is needed that resolves issues associated with MSW and PW [25]. Several sustainable waste management processes are discussed below.

10.2.1 Pyrolysis

Pyrolysis is a method that is carried out under high temperature conditions and decomposes organic materials like biomass and PW in the lack of oxygen [25]. This process is globally recognized for its potential to produce valuable products like biochar (solid), bio-oil [26], and biogas [21], as a sustainable solution to waste management and energy production. Bio-oil serves as a promising bio-based energy source and raw material for chemical production. However, its direct application is hindered by several unfavorable properties, such as high-water content, elevated viscosity, low calorific value, and corrosive nature [27]. In a later section, it is shown that the efficiency of catalysts in thermal cracking is diminished by the combination of N, S, and Cl on its surface. Although the gaseous products primarily contain Cl, S, and Br, the resulting pyrolytic oil reveals a wider elemental composition, including Sb, Zn, and Ca. This indicates that dehalogenation reactions, which release HBr and HCl, contribute to the broader range of elements found in pyrolytic oil. Owing to the latest advancements in biomass energy technologies, the efficient utilization of bio-oil has emerged as a novel challenge within the realm of biomass conversion. The steam reforming of bio-oil for hydrogen production has been recognized as an efficient method and has garnered significant attention among numerous biomass utilization processes [28]. Only 5-8% hydrogen is produced from biomass as compared to dry biomass used, which is very low. To enhance the production of gas, an addition of plastic such as polypropylene (PP), polyethylene (PE), and polystyrene [29] in biomass is a good option [30]. Moreover, plastic can also help in producing valuable chemicals like light olefins, which are used in making other products. This section explores the latest developments in the co-pyrolysis of PW and solid biomass mixtures, beginning with an examination of the physical and chemical characteristics of these feedstocks [31]. The presence of a catalyst in steam reforming facilitates the efficient conversion of bio-oil to gas products at temperatures below those required for non-catalytic bio-oil conversion. A diverse array of catalysts, including calcined natural minerals, iron ores, zeolites, alkali metals, nickel-based catalysts, and noble metals, is predominantly employed in the steam reforming of bio-oil. The use of raw and metal-impregnated chars from biomass and coal as catalysts in the steam reforming of bio-oil and its model compounds has been a significant area of recent scientific inquiry [32–34]. Utilizing char as a catalyst in steam reforming presents several benefits, including its low cost, ability to withstand high temperatures, significant tolerance to sulfur and nitrogen compounds, and the absence of regeneration issues. In addition to its catalytic role, char is utilized across a broad spectrum of applications, including carbon removal, pollutant abatement, gas storage, carbon fuel cells, and cell biology research [35, 36].

10.2.2 Electrochemical Method

Electrochemical synthesis is a sustainable and innovative method for engineering electrocatalysts, utilizing electrolysis to modify catalytic materials derived from waste [37]. This method has the advantage to form metal and metal-oxide catalysts, as it enables precise control over their composition, morphology, and electronic properties [38]. The properties of these electrocatalysts, including their surface chemistry and nanostructure, are influenced by key experimental parameters such as deposition period, potential, and current [39]. For instance, battery industrial wastewater as a precursor produced NiCoMn-layered triple hydroxides (LTHs) with varying nanostructures depending on the deposition time [40]. These structures had evolved from nanoclusters (5 min) and nanospheres (10 min) to nanoflowers (20 min) and nanoplates (30–40 min). Among these, the nanoflowers (20 min deposition) exhibited optimal catalytic performance for both oxygen evolution reaction (OER) and hydrogen evolution reaction (HER) due to their hierarchical nanostructure, small crystallinity, and large metal content (67.33%). Oxidation potential is another electrochemical approach that can functionalize carbon-based waste materials to enhance catalytic performance, particularly in hydrogen production. Additionally, in such electrochemical methods, transformation under potential has been employed to process solid, metal-bearing waste. By the electrochemical process, a positive-bias-driven exfoliation is developed to change spent $LiCoO_2$ electrodes into Co-OOH, a high-performance catalyst for OER [41]. This eco-friendly method disrupted the parent crystal structure while oxidizing electroactive elements to higher valence states, improving their catalytic activity. Electrochemical synthesis offers significant advantages over conventional material recovery methods like hydrometallurgy and pyrometallurgy [42].

10.2.3 Wet-Chemical Methods

Wet-chemical synthesis is a versatile technique that includes various methods based on solutions involving sol-gel process, precipitation, solvothermal, and hydrothermal processes for waste material conversion into functional catalysts [43]. These methods provide significant flexibility in reshaping the properties of catalysts through precise control of reaction parameters, including pH, precursor concentration, and temperature. The sol-gel method allows the foundation of thermally stable metal oxide catalysts from waste extracted metal precursors. This process comprises the hydrolysis and condensation of metal alkoxides, leading to enhanced catalytic performance materials [44]. Solvothermal and hydrothermal methods are especially effective for fabricating nanomaterials under high pressure from waste, which improve the catalytic properties of materials [45]. These methods are widely used to synthesize metal oxides, carbon-based composites, sulfides, phosphides, hydroxides, and heterostructure catalysts. For example, hydrothermal carbonization involves conversion of biowaste into carbon materials in combination with metal salts, followed by post-processing such as pyrolysis to form hybrid catalysts [46].

Alternatively, electroactive metal can be developed on pre-synthesized biowaste-derived carbon *via* hydrothermal process, which further enhances catalytic performance [47]. Another outstanding wet chemical method is boriding, which involves reducing metal species (e.g., Co, Ni, Cu, Fe, Mn, and Sn) in waste materials to produce metal boride nanoparticles [7]. The catalytic performance of these metal borides depends upon boriding conditions like atmosphere, temperature, composition of the waste, and the amount of reductant used. Metal borides with low particle sizes, high dispersion, and an optimal ratio of Ni and Fe exhibit exceptional OER activity. Despite the advantages of wet-chemical synthesis, challenges such as solvent consumption, long processing times, and environmental concerns remain. Addressing these limitations through optimization and innovation could enhance the sustainability and efficiency of wet-chemical methods for electrocatalyst synthesis.

10.2.4 Microwave Synthesis and Beyond

Microwave-assisted synthesis is a rapid and energy-efficient approach for converting waste materials into high-performance catalysts. In this method, the microwave is used to achieve uniform heating that accelerates reaction kinetics, enhances catalyst formation, and ensures improved properties of material homogeneously, controls size of particle, and crystallinity [48]. Compared to conventional heating techniques like hydrothermal processes or calcination, microwave-assisted synthesis offers distinct advantages, including small reaction time, cleanliness, and high utilization energy efficiency. Significantly, the uniform heating provided by microwaves facilitates rapid crystal nucleus growth and crystallization, which is crucial for creating high-quality catalysts. Various waste materials, like PW, agricultural residues, and metal-containing sludge, can be efficiently processed by microwave heating to produce carbon-based or metal oxide catalysts. For example, carbonization of biomass under microwave irradiation results in highly porous carbon catalysts, which are particularly suitable for electrocatalysis. Similarly, oxide catalyst nanostructures can be collected from metal-containing waste by using microwave-assisted hydrothermal synthesis. Several studies highlight the potential of microwave synthesis for advanced catalyst preparation. In 2018, Cova *et al.* developed a microwave-assisted strategy to produce an Ag/Ag_2S-carbon hybrid catalyst from pig bristles [49]. The pig bristles were efficiently decomposed under microwave heating, and the release of sulfur facilitated the formation of Ag_2S. More recently, Miao and co-authors used a microwave hydrothermal process to synthesize a NiFe-borate layered double hydroxide biomass-derived

N-doped carbon (NiFe-BLDH/NC) hybrid catalyst [50]. This catalyst exhibited multistage decentralized architecture, good electrical conductivity, rich active sites, and efficient charge transfer kinetics, which lead to high OER activity. Despite its advantages, microwave-assisted synthesis faces challenges such as high equipment cost, non-uniform heating at larger scale, and the need for optimized reaction conditions. These limitations slow down its widespread adoption for catalyst production. Interestingly, alternative methods like ultrasound treatment have shown comparable or even superior performance for certain applications. In addition to microwave and ultrasound methods, hybrid synthesis strategies have been explored for converting waste into electrocatalysts. Combinations of techniques like hydrothermal-pyrolysis, pyrolysis-hydrothermal, sol-gel-calcination/pyrolysis, and pyrolysis-microwave processes enable construction of diverse high-performance catalysts. These rich combinations give flexibility and innovation in designing catalysts for specific applications, such as water electrolysis and energy conversion.

10.3 Waste-Derived Catalysts for Water Electrolysis

The electrocatalysts have gained significant attention owing to their diverse structural features, such as 2D materials [51], hierarchically structured materials [52], metal-organic frameworks [53], and covalent-organic frameworks [54]. The operating cost of water electrolysis systems would be significantly reduced by using inexpensive electrocatalysts as they are abundant in common waste like industrial waste, biowastes, and electronic wastes, promoting a circular economy and cost-effective waste management. The methodology has the potential to decrease reliance on resource markets and lower waste disposal expenses as compared to linear recycling economy approaches. The circular economy, when implemented across all sectors, could help reduce carbon emissions by 45% by 2030 and achieve carbon neutrality by 2050 [55]. Therefore, utilizing waste resources to develop functional materials is a reasonable method to implement the circular economy and reduce the carbon footprint associated with material fabrication. Because of the significant economic and environmental benefits, scientists have recently become more interested in manufacturing electrocatalysts from waste [56].

For instance, Cao and coworkers used bean sprouts to create porous carbon catalyst for HER that was N and S self-doped by pyrolysis [57]. In acidic environments, the derived carbon catalyst demonstrates sufficient HER activity (η_{10} = 413 mV, Tafel slope = 98 mV dec^{-1}) [58]. Not only can biowastes be transformed into carbon-based catalysts *via* the pyrolysis method, but other waste derived electrocatalysts can also be synthesized using other advanced techniques such as electrochemical synthesis, microwave synthesis, and wet-chemical synthesis. Effective water electrolysis is made possible by circular economy-driven green hydrogen energy systems using strategies such as adsorption, doping, defect/vacancy engineering, alloying, heterostructure formation, and nano structuring, which enhance the performance of waste-derived catalysts [59].

10.3.1 Mechanism of Water Electrolysis

The design of resourceful electrocatalysts requires careful consideration of four general design factors: *abundant electroactive sites, long-term stability, high electrical conductivity, and intrinsic catalytic activity.* Their inherent catalytic activities mainly influence the electrocatalytic performance of these catalysts. The current approaches to improving catalyst intrinsic activity, such as composition optimization, doping, heterostructure creation, defect engineering, and phase engineering, mostly focus on changing the electronic structures of the catalysts. In water electrolysis, HER occurs at cathode, and OER occurs at anode. In different electrolytes, HER and OER operate *via* distinct paths and mechanisms. At this point, experimental and computational research has provided significant information about HER mechanisms. HER naturally follows Volmer/Tafel or Volmer/Heyrovsky routes. There are four fundamental phases in alkaline media, represented in Figure 10.1 (b) [60]. The energy barriers for the H_2O adsorption and dissociation phases in alkaline HER are larger than those for the H_3O^+ adsorption in acidic HER. Therefore, some catalysts (such as Pd and Pt) have substantially higher theoretical activity for acidic HER than for alkaline HER [61]. The predicted enhancement of HER activities in alkaline media for HER catalysts is due to their strong H_2O adsorption and dissociation capabilities [62]. The 4-electron OER process is more complex than the 2-electron HER, with lattice oxygen participation and adsorbate evolution mechanisms (AEM) being the most preferred [63]. The AEM process for alkaline OER typically consists of four phases, with intermediate M-OH generated by oxidation of OH^- on an electrocatalytically active site (M) as represented in Figure 10.1 (c).

After that, a proton-coupling-electron transfer (PCET) procedure converts M–OH into M–O. The M-O undergoes additional transformation into M-OOH intermediate through an OH^- attached *via* a one-electron oxidation process. This ultimately starts a second PCET process that produces O_2 molecules. The adsorption of an H_2O molecule on M is the initial stage

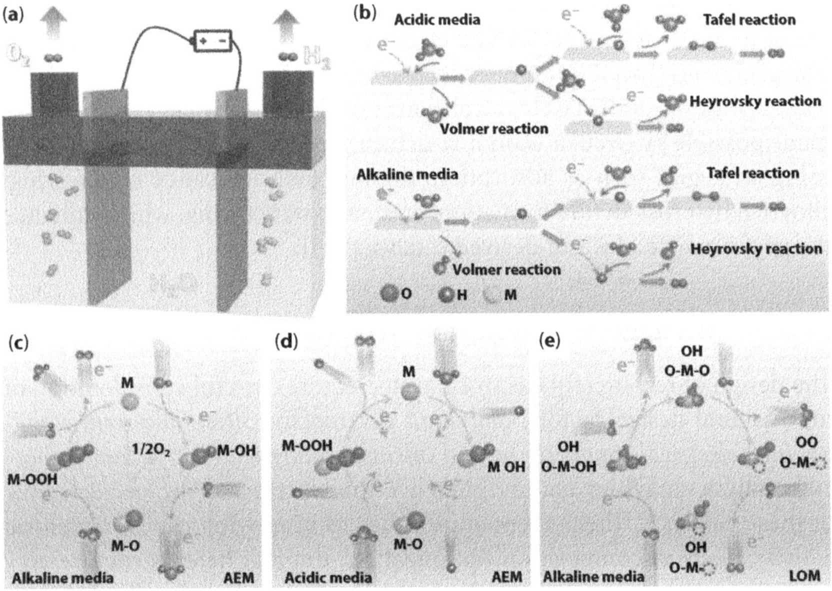

Figure 10.1 (a) Illustrates water electrolyzer; (b) HER processes in alkaline and acidic electrolytes; (c) AEM for alkaline OER; (d) AEM for acidic OER; (e) LOM for alkaline OER; the oxygen vacancy is highlighted by dotted red circles. Reprinted with permission from open access license [59].

of AEM for acidic OER, which is different from alkaline OER Figure 10.1 (d). After that, an H^+ dissociates to produce M–OH, followed by the release of another H^+ to create M–O. Then, H_2O molecule undergoes another nucleophilic attack, producing M–OOH followed by the fourth proton coupling and the desorption of the generated O_2 molecule. Recent research on oxygen origin in O_2 products has revealed that catalyst lattice oxygen plays a crucial role in lattice-oxygen-participation-mechanism (LOM) [64]. In alkaline OER as represented in Figure 10.1 (e), OH^- is adsorbed on an oxygen vacancy-coordinated active site (M–OH/O_v) using LOM, and then another OH^- is absorbed near M, creating the M–OH/–OH species followed by a dehydrogenation process for producing M–OH/–O. However, OH^- dehydrogenation is challenging to initiate, leading to the formation of an unstable transition state (M–OH), which eventually converts into M–OO/O_v. The initial state M–OH/O_v is restored due to the desorption of generated O_2 and filling of OH^-. The OER mechanism is extremely sensitive to the surface properties of catalysts and can be controlled by rebuilding their structural integrity *in situ* under OER conditions [65].

The optimal waste for the conversion of waste to hydrogen (WtH) involves biomass wood waste (for instance, paper or timber waste), MSW, refuse-derived fuel (RDF) obtained from the MSW, sewage sludge, solid recoverable fuel along with the plastics and packages [66, 67]. MSW generally involves the presence of biodegradable waste, electronic waste, recyclable materials, inert material waste, and toxic and harmful waste as scrapped by mankind. The physical and chemical properties, along with relevant parameters of electronic materials are key factors. The tailoring of these parameters can be made to improve efficiency of photocatalysis. The number of protons and neutrons, band gaps and positions of the principal bands of the photocatalysts matter a lot in water splitting. The band gap and positions of valence band [68] and conduction band (CB) of the reported photocatalysts are shown in Figure 10.2. The process of water splitting requires that the band energy of CB should be higher than the redox potential, and VB energy should be lesser than the redox potential. The requirements involve the potential for water splitting as 1.23 V/NHE, band gap dependence, position of CB and VB, and lifetime of electron-hole pair separation.

10.3.2 Designing Efficient Photo-Catalysts for Wastewater Treatment

The development of effective photocatalysts is essential for sustainable wastewater treatment and green hydrogen production. Catalysts from waste sources offer a low-cost and green route by exploiting renewable resources without losing catalytic performance. Improvements in active sites, electronic structures, and surface area through defect engineering, heteroatom doping, and nano structuring contribute significantly to enhancing photocatalytic efficiency. Advanced materials demonstrate superior activity in pollutant degradation and HER. This section reviews the key strategies for optimizing photocatalyst designs to achieve higher efficiency, stability, and long-term applicability in both wastewater treatment and production of green hydrogen.

- **Rich Active Sites**

The effectiveness of photocatalysts produced from waste materials rests on distribution of active sites within the catalyst. Through an increased density of active sites, there is greater adsorption of reactants which enhances the reaction rates and efficiency of catalysis as depicted in Figure 10.2. Certain techniques employed to increase the quantity of active sites include advanced defect engineering, precise heteroatom doping, and nano-structuring of the catalyst surface. Studies demonstrate that oxygen vacancies, which are known for recombination and charge separation loss, can act as electron donors through efficient mitigation [69]. It has also been shown that templating, hierarchical porosity design, and nano-structuring techniques have enhanced surface area and the exposure of active sites. The introduction of heteroatoms can likewise be utilized to alter electronic structure and create supporting active sites. More recent works have emphasized the importance of dual-heteroatom doping, like N-S co-doping, in producing synergistic effects that further increase photocatalytic efficiency. The waste biomass carbon materials are co-doped with N and S and have shown remarkable effectiveness in degrading organic pollutants owing to improved charge transfer and additional active interface. Furthermore, incorporation of transition metals [70], e.g. Fe, Co, Ni, into carbon matrices has been reported to form metal-N_x sites which are very active for pollutant degradation and HER [71]. Active site exposure has been maximized with the use of nano-structuring techniques including the creation of 2D nanosheets and hierarchical porous structures. For example, the fabrication of biochar-based 3D porous networks containing channel-like structures has shown higher mass transfer and better access to the active sites [72]. The photocatalysis has been reported to benefit from the use of sacrificial templates for the fabrication of mesoporous and microporous structures because of improved light penetration and diffusion of reactants.

Waste-derived materials such as biochar-doped metal oxides, metal-organic frameworks (MOFs), layered double hydroxides (LDHs), biomass-derived carbons, and cerium oxide composites exhibit unique structural properties that enhance active sites. The ZnO/biochar composites and Fe-doped waste ceramics show significant improvements in photocatalytic performance due to optimized surface functionalization [73]. There are still some unresolved issues including the proper characterization and quantification of active sites. To capture the active site dynamics during catalytic reactions many researchers are now using *in-situ* X-ray absorption spectroscopy (XAS) and electron paramagnetic resonance (EPR) spectroscopy which actively monitor changes of active sites during operational conditions. These are important for optimizing the performance of catalysts to have specific active site distributions for targeted applications. In addition, active site arrangement prediction and experimental synthesis guidance is done using computer simulation techniques like density functional theory (DFT) [74].

- **Higher Activity for Catalysis**

The ability of photocatalysts derived from waste materials to stir catalysts is dependent on their capability to capture light, producing charge carriers and promoting surface redox reactions. Solid solutions and ternary compounds have been shown to enhance the importance of bandgap engineering, owing to the recently emphasized studies [75].

Figure 10.2 highlights multifaceted influence of catalyst properties on the pyrolysis process, emphasizing how specific attributes can enhance the efficiency and output of pyrolysis reactions. High acidity in catalysts is linked to increased cracking activities, leading to higher gaseous product yields and the isomerization of products, which in turn reduces impurities in the liquid yield. Higher catalytic activity is achieved when materials have suitable band gaps and redox potentials, which enables both strong oxidation and reduction processes. Sludge-activated carbon (SACs) with waste-derived support offers a stimulus to dope single atoms in a nanostructured form and thus improve the atomic and catalytic efficiency. Experiments demonstrated that due to their particular electronic structure and ultrahigh surface area [76], SACs help to maximize photocatalytic

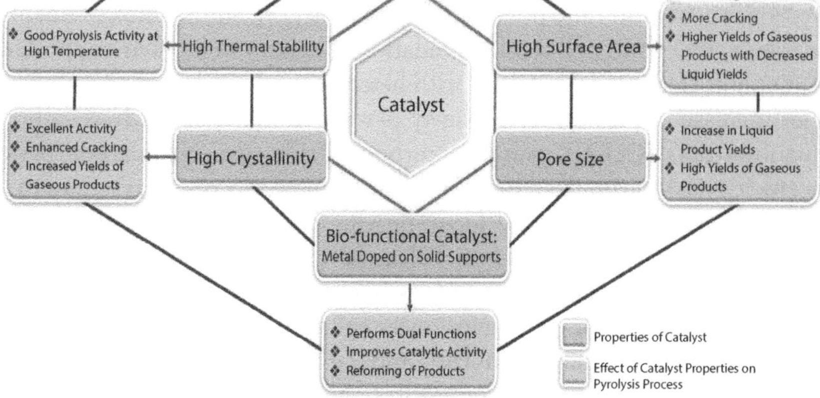

Figure 10.2 The characteristics of catalysts and their influence on the pyrolysis process. Reprinted with permission from [74].

activity in both pollutant degradation and hydrogen production. Plasmonic photocatalysis, which employs the effect of localized surface plasmon resonance (LSPR) of noble metal nanoparticles, has gained popularity due to the ability of metals to enhance light absorption and hot electron generation. Recent reports show that coupling waste-derived carbon support with either Ag or Au nanoparticles increases photocatalytic activity under visible as well as near-infrared illumination. In addition, the creation of bimetallic plasmonic systems (Au-Pd, Ag-Cu) has displayed synergism in the increase of both light absorption and catalytic activity [77]. The optimization of interfacial charge transfer and recombination loss mitigation are still issues to resolve. Understanding the behavior of charge carriers is aided by advanced characterization methods such as transient absorption spectroscopy (TAS) or time-resolved photoluminescence (TRPL), thereby enabling designs for better photocatalysts.

- **Recorded Electrical Conductivity**

The key role of charge separation and transfer in photocatalysts requires high electrical conductivity. Carbon-based materials obtained from waste systems show a powerful ability to increase conductivity. Various carbonaceous materials involving graphene, reduced graphene oxides (rGO), carbon nanotubes (CNTs), and conductive polymers, are widely incorporated into waste-derived photocatalysts to improve electron transport properties and suppress recombination loss. Biochar, activated carbon, and carbon fibers, with inherent graphitic domains and porous structures, play a beneficial role in promoting conductivity. For example, rGO/TiO_2 composites show enhanced electron mobility and photocatalytic performance for hydrogen generation and pollutant degradation [78]. Similar improvements in conductivity and pollutant degradation efficiency of biochar-supported ZnO photocatalysts were found with heteroatom doping, providing a further increase in performance. Newly developed materials such as carbon quantum dots (CQDs), graphitic carbon nitride (g-C_3N_4), carbon nano horns (CNHs), carbonized metal-organic frameworks (C-MOFs), and carbon aerogels have been used for improving conductivity and photocatalytic activity owing to their special electronic properties and large surface areas [79]. Examples of conductive polymers include polyaniline (PANI) and polypyrole (PPy), which effectively enhance their conductance while minimizing recombination in TiO_2 and ZnO matrices [80]. Hybrid systems using PANI-graphene composites show synergistic effects of both materials that enhance photocatalytic activity. Doping with transition metals (TM) like vanadium and rare-earth elements (like lanthanum) provide new energy levels and oxygen vacancies, increasing the mobility of charge carriers. Moreover, the introduction of defect engineering (mainly oxygen vacancies) can significantly improve conductivity and photocatalytic activity. Techniques such as *in-situ* XAS and EPR help provide information about charge transport mechanisms, whereas computational modeling and machine learning are being increasingly employed to optimize photocatalyst designs.

- **Performance Stability for Longer Terms**

Longer stability is important to the practical implementation of photocatalysts. The efficient use of waste photocatalysts is hindered by their long-term stability. Photo corrosion, structural degradation, and leaching of active sites are primary challenges. The most recent advancements target the design of smart materials that passively heal active sites. Protective coatings with defect engineering and self-healing materials are new strategies to enhance stability for longer terms. The pH buffering capacity of layered double hydroxides (LDHs) prevents the leaching of active components and has good stability when they are used as supports for photocatalytic nanoparticles. Encapsulation within porous matrices like MOFs and zeolites increase stability due to prevention of active site leaching. Protective coatings like alumina (Al_2O_3) and silica (SiO_2) that are deposited *via* atomic layer deposition (ALD) also avoid photo corrosion and chemical degradation [80]. Materials like zirconia (ZrO_2) coatings, titanium nitride (TiN) layers, boron nitride [81] encapsulants, tungsten disulfide (WS_2) coatings, and hafnium oxide (HfO_2) layers have shown potential in improving stability under harsh conditions. The polymers containing fluorine atoms as hydrophobic coating promote stability by reducing water adsorption and surface oxidation in water. Defect engineering, particularly oxygen vacancies improves stability by reducing recombination and enhancing charge transport. Strong metal support interactions facilitated by functionalized carbon and LDHs prevent detachment of active components. Regenerative techniques like periodic annealing and *in-situ* self-healing sustain long term activity [82]. Phosphorus-doped carbon, sulfur-doped graphene, nitrogen-doped carbon nanotubes, boron-doped diamond, and silicon carbide (SiC) supports have also been explored for their ability to enhance metal-support interactions and stability. *In-situ* characterization practices like environmental transmission electron microscopy (ETEM) and *in-situ* Raman spectroscopy are vital for understanding the degradation mechanisms with more stable photocatalytic systems. Future research should integrate computational modeling and machine learning to optimize stability under harsh conditions.

10.4 Waste-Derived Catalysts for the Evolution of Hydrogen

The manufacturing of green hydrogen has gained significant interest as a feasible substitute to fossil fuels. The HER plays a critical role in electrolysis, biomass gasification, and catalysis. However, the cost-effectiveness, efficiency and eco-friendliness of these procedures are strongly based on the catalysts. Commonly used metal-based catalysts such as platinum have excellent catalytic efficiency. However, their high cost limits their availability. Consequently, scientists are exploring waste-derived catalysts as an alternative, using industrial, agricultural, and electronic waste to develop cost-effective and feasible catalytic materials. The waste-derived catalytic materials could be categorized into three types: metal-based catalysts, carbon-based catalysts and heterostructure catalysts. Metal-based catalysts such as Ni, Co, Ir, Cu, and Platinum Group Metals (PGMs) have the capability for an effective proton reduction reaction at the cathode [83]. Carbon-based catalysts derived from biomass or waste for HER have been proven an advantageous substitute owing to their reduced cost, enhanced surface area, and tunable electronic properties. The group HHAC700 catalysts, such as rice husk-derived graphene nanosheets exhibit good performance with higher crystallinity [39]. Numerous factors contribute to the effectiveness of waste-derived catalysts, such as active centers for proton adsorption and transfer of electrons in the porous nature of catalysts. Studies demonstrate that waste from palm plants, which forms porous nanosheets with a graphene-like hierarchical structure, serves as an effective HER catalyst. This catalyst exhibits long-term stability under acidic conditions, with over potential of approximately -0.33 V with a 10-mA cm^{-2} current density [84]. The over potential is the additional energy essential for reaction to proceed. It must be minimal so that it lowers the potential at the anode and cathode, improving overall efficiency. The interface and the heteroatom doping over the carbon catalyst give rise to higher stability, i.e., the interface interaction of Au particle covered with the N-doped carbon (Au@NC) supports the charge transfer that results in high activities producing η= 54.1 mV benefiting the HER [85].

10.4.1 Catalysts Based on Carbon

Catalysts based on carbon have appeared as an excellent substitute to noble metal atom catalysts for HER because of their high surface\volume ratio, electrical conductivity and stability. Such catalytic materials are commonly extracted from biomass waste, PW, and industrial by-products that undergo an activation method to enhance catalytic efficiency.

10.4.2 Biomass-Derived Carbon Catalysts

Biomass, activated carbon, and graphene-based materials are frequently synthesized from agricultural and industrial waste, including rice husks and coconut shells. These materials undergo pyrolysis followed by chemical activation by utilizing the several agents such as phosphoric acid (H_3PO_4) or potassium hydroxide (KOH) to increase porosity and surface modification. Nitrogen-doped carbon catalysts extracted from biomass show enhanced catalytic performance due to improved electronic conductivity as well as the active site exposure [86]. The effects of the material like Pd/Ag membrane exhibit the concentration of hydrogen in the nitrogen stream before and after passing through the membrane [87–89]. The percentage of hydrogen was removed by approximately 85-90% by the membrane throughout the work. They observed that ZnO did not extract specific amounts of hydrogen. When the separator was removed from the reactor after 22h, its working efficiency was high, and no indications of the separator being blocked by the other combinations present in the mixture of gas. Hydrogen can also be emitted from other molecules like alcohol and organic acids with an animated catalytic membrane [90]. They found in preliminary observations that a pure Pd separator adsorbs tiny molecules of organic acids and avoids the contamination that must be developed to utilize the latent use of the separators.

10.4.3 Plastic Waste-Derived Carbon Catalysts

In a few years, carbon-based substances have proven highly useful in energy storage, organic synthesis, and environmental remediation owing to their extreme porosity, large surface/volume ratio, large conductivity, along with the generous surface functional groups [91]. Therefore, the synthesis of carbon-based catalysts is highly dependent on the materials that are obtained from fossil fuels with few challenges of high energy consumption, high cost, and damaging effects on the surroundings. Moreover, there is dire need for sustainable management of materials based on carbon. PW such as polyvinyl chloride, polystyrene, polyethylene terephthalate, phenol formaldehyde resin, low-density polyethylene, and high-density polyethylene, are eco-friendly, cost-effective precursors to synthesize the high amounts of catalysts based on carbon. Plastics such as PE, PP, and polyethylene terephthalate (PET) can be pyrolyzed into high-surface-area carbon materials. These materials are further processed into graphene-like structures that act as conductive support for electrocatalysis. The combination of heteroatoms, including N, S, and P further enhances catalytic activity by modifying electronic properties and improving hydrogen adsorption [92]. Utilizing wasted polystyrene foam as a starting point, a new composite electrode material (PCF-MnO_2) was created using the template approach. The hybrid material demonstrated an ultra-high specific capacity of 308 F/g and 247 F/g at 1 mV/s and 1 A/g, respectively. In $LiCl_2$ electrolyte because of the high electrical conductivity and greater specific surface area (1087 m^2/g) of PCF, the specific capacity of MnO_2 is high itself, and the positive synergistic interaction was seen between PCF and MnO_2.

10.4.4 Industrial Waste-Based Carbon Catalysts

Industrial byproducts such as coke waste, asphalt-derived carbon, and spent activated carbon can be repurposed as efficient HER catalysts. These materials often contain inherent defects and functional groups that improve their electrocatalytic behavior. Moreover, combining these waste-derived carbons with 3d metal atoms, like cobalt (Co), iron (Fe) as well as nickel (Ni), can further enhance catalytic performance [93]. In general, the theoretical ion storage capacity of carbonaceous materials is greater than that of graphite. However, there are always disadvantages to nanostructured carbons, including expensive production prices, complex preparation techniques, and environmental concerns. The disadvantages have motivated researchers to create, manufacture, and evaluate more ecologically friendly items. According to their size, shape, and dimensionality, carbon materials are classified into four groups: zero-dimensional (carbon quantum dots), one-dimensional (carbon nanotubes), two-dimensional (graphene), and three-dimensional (graphite) nanomaterials [94]. Molybdenum trioxide (MoO_3) was used as a molybdenum precursor to produce (MoC/Mo_2C) [95, 96]. Magnesium [87] was added to it to aid in its decrease. The two compounds were thoroughly combined in an agate mortar. For the first time, Mo_2C was synthesized using a pipette tip made of leftover plastic from a laboratory.

10.4.5 Catalysts Based on Metals

Catalysts based on metals as derived from waste materials offer an alternative to noble metals, with TM oxides, sulfides, and phosphides playing a significant role in HER. These catalysts can be synthesized from industrial waste, electronic waste (e-waste), and mining residues [97]. One interesting method for improving the catalytic performance of free noble metal atom catalysts in HER is the use of carbon substrate support. Here, we present P-CoNiO2@NC, a waste-derived electrocatalyst that was produced under moderate conditions. In either acidic (η_{10} = 64.6 mV) or alkaline (η_{10} = 96.4 mV) conditions, it demonstrates exceptional HER performance with good stability. In particular, as the current density exceeds 74 mA/cm^2 and 29 mA/cm^2 in acid and alkaline electrolyte respectively, it starts to function better than commercial Pt/C.

10.4.6 Transition Metal Oxide and Hydroxide Catalysts

Industrial waste materials, such as steel slag and red mud (bauxite residue), are rich in TM like iron (Fe), titanium (Ti), and aluminum (Al). These materials can be processed into metal oxides (Fe_2O_3, TiO_2) and hydroxides ($Ni(OH)_2$, $Co(OH)_2$) that exhibit larger electrocatalytic activity for HER. The existence of oxygen vacancies as well as the surface defects in these metal oxides improves electron transfer rates and catalytic efficiency [97]. Moreover, other dual 3d TM oxides, like Co and Fe, Ni and Fe may have their HER performance efficiently enhanced by the waste-derived carbon-based catalysts. According to the techniques of X-ray photoelectron spectroscopy and DFT calculations, transfer of electrons, electronic structural modification, and optimized intermediates adsorption are the sources of the increased HER activity. This study offers a fresh perspective on turning garbage into goods with additional value as well as an environmentally responsible and sustainable method to promote the generation of hydrogen energy.

10.4.7 E-Waste Derived Catalysts

In the current technological period, which is characterized by quick innovation and rising customer expectations, the electronics industry is expanding at a rate that is unusual for other sectors. But this expansion has come with a significant problem: the rise in electronic garbage, or 'e-waste' [98]. A remarkable 70 elements from the periodic table are among the many components found in these abandoned gadgets when their composition is examined. Although a long list of other metals, such as titanium (Ti), Ni, and Al, also make an appearance, precious metals like Cu, Ag, and Au predominate. Electronic waste (e-waste) contains valuable metals, including Ni, Co and Cu, which can be extracted and converted into functional catalysts [99, 100]. Nickel-based catalysts, such as Ni_2P and NiS, have demonstrated excellent HER activity due to their high conductivity and stability. Additionally, recovered palladium (Pd) and ruthenium (Ru) from e-waste can be utilized as catalytic dopants to enhance performance [101].

10.4.8 Heterostructure Catalysts

Heterostructure catalysts, which combine different materials to create synergistic effects, have shown great promise in enhancing HER efficiency. Waste-derived heterostructure catalysts integrate metal oxides, carbon materials, and nanostructured materials to optimize performance [102]. Hybrid catalysts combining TM with carbon-based support (e.g., Ni/C, Co/C, and Fe/C) exhibit improved electron transfer kinetics and stability [103]. Biochar or graphene derived from agricultural waste can be used as conductive support for these metal nanoparticles, enhancing catalytic activity and durability. The enhanced ability to rapidly adsorb and photo degrade organic dyes is demonstrated by hierarchically arranged macro-mesoporous TiO_2-graphene composite sheets. The following outcomes were received from the assembly linked of macrospores in mesoporous films: (1) improved mass transport through the film; (2) increased thin-film accessible surface area; (3) markedly increased photo catalytic activities; and (4) shortened mesoporous channel length. These structures will present an alluring prospect for industrial applications and may be appropriate for air purification and wastewater treatment to eliminate organic contaminants. Interfacial charge transfer increases catalytic efficiency when metal oxides (like TiO_2 and FeO_3) and metal sulfides (like MoS_2 and WS_2) are combined. Mixed metal waste found in industrial waste streams could be utilized to generate these catalysts. These materials perform better in HER because of the stability of oxides and the strong catalytic activity of sulfides. Heterojunctions between semiconductors like TiO_2, CdS, and graphene quantum dots (GQDs) are advantageous for photocatalytic HER processes. High efficiency photocatalysis may be made from semiconductor waste, especially from mining and electronic trash. These material's incorporation with carbon supports improves catalytic activity and charge separation even further.

10.5 Waste-Derived Catalysts for Evolution of Oxygen

The waste-derived catalyst for OER can be categorized into carbon catalysts, TM-based catalysts, and heterostructures consisting of carbon.

One of the key features of these catalysts is that they only work in alkaline media as they get deactivated in acidic media. Mostly, the N doped catalysts prove to be effective for the OER performance because of their favorable defective sites, electrical conductivity, increased structural disorder, etc. The electroactive defect sites can increase by co-doping such as N and P based catalysts. For the transition metal-based catalysts, waste materials like Fe-rich byproducts from industries are great precursors for OER electrocatalysts with 20h of durability in case of RuO_2. The metal (oxy) hydroxides are of great importance such as NiFe and NiCu synthesized from uncycled capacitors have excellent durability of about 24h for alkaline OER and immense electroactive sites. A good substrate and TM-based nanomaterial were fabricated to achieve high OER performance. Metal/alloy, alloy/carbon and sulfides/nitrides or carbides based heterostructures synthesized from waste are potential candidates for the OER catalysts such as $Ni@NiMnCoO_4$, NiFe-borate and Fe_3O_4/NiS heterostructures etc. [40]. OER, which has slow kinetics by nature, is an important obstacle in the water electrolysis system and needs effective electrocatalysts to accelerate the catalytic process. At the moment, inexpensive catalysts made from various wastes (such as biomass and wasted batteries) are essential for improving OER performance [7]. The majority of OER catalysts produced from waste only function in alkaline media because they are at risk for corrosion, dissolution, and deactivation in strong oxidative and acidic environments [37]. OER uses an arrangement of proton-electron linked processes to produce molecular O_2 electrochemically [104]. OER occurs at the anode side (in alkaline media), where the anode releases oxygen gas (O_2). Meanwhile, the cathode undergoes several reactions, including HER, nitrogen reduction reaction (NRR) along with the carbon dioxide reduction reaction (CO_2RR). The issue is the large overpotential, which lowers energy efficiency by requiring more voltage than the theoretical requirement. Performance deterioration over time suggests a stability issue. Therefore, reliable and effective OER catalysts are crucial for improved green hydrogen generation. The reactions HER and OER at the cathode and anode respectively are the two half-cell processes involved in water electrolysis. Water splitting is a non-spontaneous process for which the external energy is prerequisite to continue because of its positive Gibbs free energy (ΔG). By lowering ΔG, this input energy makes it possible for electrical energy to be transformed into chemical energy, which produces oxygen and hydrogen. The following formula is used to get Gibbs free energy, $\Delta G = -n \times E \times F$ where n represents the electrons transferred number throughout the electrode reaction, E represents the theoretical water decomposition voltage, and F represents Faraday›s constant [105].

Although the reaction takes place in an alkaline environment, many studies utilize the acidic pathway mechanism in theoretical calculations to assess the catalytic activity of OER [106]. Following reactions occur in acidic conditions for HER and OER in water electrolysis.

HER $4H^{+}(aq) + 4e^{-} \rightarrow 2H_2(\text{g})$ (Reduction) $E^{o}_{anode} = 0.00\text{V}$

OER $2H_2O(\text{l}) \rightarrow O_2 + 4H^{+}(\text{aq}) + 4e^{-}$ (Oxidation) $E^{o}_{cathode} = 1.23\text{V}$

Overall $2H_2O(\text{l}) \rightarrow O_2(g) + 2H_2(\text{g})$ $E^{o}_{overall} = 1.23\text{V}$

Theoretically, the reaction begins when the anode-cathode voltage hits 1.23 V [107]. The discovery of implicit OER catalysts is necessary to increase the general efficiency of the process because of the nature of the four-electron process that involves OER, which is widely thought to be significantly slower than HER. Since some classes of materials such as spinels, amorphous Co-oxide, perovskites and layered hydroxides have been shown to have excellent efficiency, several investigations have started to concentrate on 3D-TM-based catalysts [105].

10.5.1 Carbon-Based Catalysts

Porous carbon from biomass acts as a conducting support, enhancing electron transfer and showing more active sites, CNTs and graphene derived from PW or organic matter show high activity and corrosion resistance in alkaline OER surroundings. Carbon-based catalysts can be further doped with N, P or S to enhance catalytic performance. Carbon materials derived from waste, such as coconut shells, bagasse from sugarcane, and food waste, are promising given their extensive surface area and conductivity. Numerous research studies have effectively used many ways to transform trash into functional materials based on carbon having customized nano-/micro structures and modifiable surface chemistry. Notably, these materials made from trash have demonstrated encouraging results in a diversity of applications, like energy storage and water purification [108].

Figure 10.3 illustrates the creation and performance analysis of novel carbon-based catalysts designed for electrochemical applications. It is a sustainable method for creating catalysts by synthesizing nitrogen-rich, porous carbon compounds from plant leaves. Figure 10.3 (b) evaluates

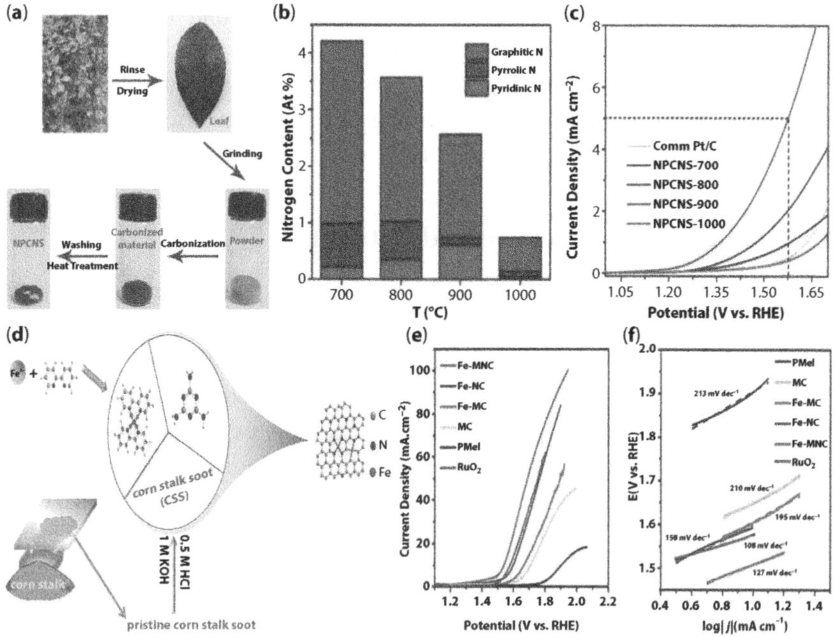

Figure 10.3 (a) Illustrates the production process of nitrogen-enriched porous carbon derived from leaves of plants. (b) Carbon catalysts containing nitrogen contents. (c) LSV graphs of carbon catalysts along with the Pt/C catalyst. Reprinted with permission from ref. [109]. (d) Depiction of the production of N, Fe co-doped microporous along with the mesoporous carbon (Fe-MNC) (e) LSV curves; and (f) comparable Tafel plots of catalysts at the 0.1 M KOH. Reprinted with permission from [110].

the nitrogen content in carbon catalysts and examines the impact of processing variables on the addition of nitrogen functions. The electrochemical performance is then evaluated *via* linear sweep voltammetry (LSV), demonstrating the performance of the developed carbon catalysts compared to a standard platinum/carbon (Pt/C) catalyst, as presented in Figure 10.3 (c). Additionally, Figure 10.3 (d) depicts a strategy to enhance catalytic activity by describing the creation of Fe-MNC, which is a co-doped mesoporous and microporous carbon. In recent years, significant focus has been placed on developing noble metals, metal oxides, hydroxides, and chalcogenides as bifunctional electrocatalysts for OER. Despite the significant advancements, the majority of OER bifunctional metal-based electrocatalysts still have several glaring drawbacks, such as high prices, poor stability, or inadequate selectivity, which have prevented their widespread use in storage devices besides the energy conversion devices. To generate high-performance, low-cost bi-functional carbon electro-catalysts doped with N (N-BCEs), a cost-effective, nitrogen rich polymer waste was chosen as a nitrogen precursor. The benefits of N-doping include improved electrical conductivity, controlled electronic features of surface, improved structural disorder, along with the filling in oxygen vacancies, which can enhance material's performance [111].

10.5.2 Transition Metal-Based Catalysts

Earth-abundant first-row TM is more electrically conductive than their oxide, hydroxide forms, and oxyhydroxide cousins, much like highly conductive carbon-based compounds. Their OER catalytic efficacy is increased by this improved conductivity, which also enhances electron transport inside the catalysts. This enhances OER performance by facilitating quicker electron transport. By combining the short electron route and metal characteristics, cobalt nanoparticles on electrodes increase OER. Catalytic activity is further increased by metal alloys, such as Ni–Fe. However, under alkaline circumstances during OER, metals change into oxides and hydroxides. Thus, the base metal aids in electron transport since the real reaction occurs at the surface of the oxidized metal [112].

Oxides of Metals: TM-based oxides have arisen as promising materials in the field of conversion technologies along with energy storage, owing to their cost-effectiveness, widespread availability in the Earth's crust, and exceptional corrosion resistance. Because of their various oxidation states, TM with +2, +3, and +4 oxidation states are perfect electrocatalysts because they operate as active sites for OER. First-row TM oxides and mixed oxides of Ni, Co, Mn, and Fe have been the focus of recent research due to their exceptional OER activity [113]. Most TM oxides fall into three main categories: spinel oxides, complex perovskite oxides and layered (oxy)hydroxides, although an extensive range of compounds are now under investigation.

Oxides of Perovskite: Perovskite TM oxides are promising OER electrocatalysts because of their tailored electronic structure as well as high activity. Their general formula, ABO_3, consists of large alkaline/rare-earth cations at the A-site, which coordinate with anions of oxygen, and transition metals at B-site, forming octahedral structures with oxygen. This unique arrangement enhances catalytic performance and stability. Substituting, La^{3+} for Sr^{2+} at A site could significantly change the electronic environment of the TM at B site, hence tuning OER catalytic properties of oxides. Notable examples include $Sr_x\ La_{1-x}FeO_3$ and $Sr_x\ La_{1-x}CoO_3$ with $0 \leq x \leq 1$, where the oxidation of the TM at the B site is regulated by A site swapping (La^{3+} for Sr^{2+}) leading to enhanced OER activity.

Oxides of Spinel: Spinel oxide's unique structure has allowed them to demonstrate exceptional OER catalytic activity. These oxides, which have the generic formula AB_2O_4 include TM cations in both octahedral (B site) and tetrahedral (A site) coordination styles. The construction of extremely effective OER catalysts is made possible by the structural flexibility that permits the integration of different TM cations in varied valence states.

TM Layered (Oxy)hydroxides: It has been known that TM oxyhydroxides, which are produced on the surface of TM are extremely active OER catalytic species in alkaline media. Ni oxyhydroxide has garnered significant interest owing to its notably high catalytic activity. It has been shown though that the presence of Fe impurities is the reason for the enhanced performance observed. Mixed Fe-Ni layered double hydroxides (FeNi-LDH) are the main cause of increased OER activity of Ni-based electrodes. A fougérite/pyroaurite-like mineral phase results from the integration of Fe impurities into Ni-based framework. The OER process's stability and catalytic efficiency are greatly increased by this structural change [114].

10.5.3 Heterostructure-Based Catalysts

To achieve excellent OER efficiency, carbon substances are ideal substrates for TM-based electrically active nanoparticles due to their excellent conductivity, broad area of contact, and redox characteristics. By reducing the extreme agglomeration and development of metal/alloy nanoparticles, combining them with carbon can fill electroactive areas and improve the overall strength of catalysis. OER catalysts include carbon compounds combined with the metal borides, sulfides, phosphides, nitrides as well as carbides have been created from waste. Higher conductivity of electricity is the primary advantage of metal sulfides, nitrides, carbides, phosphides, and borides over metallic hydroxides. Because of its favorable structural and electrical characteristics, a utilized methane decomposition catalyst (Ni/CNTs/Al_2O_3) may also function as an efficient OER catalyst. The primary active sites for the oxygen evolution process in this three-part hybrid material are nickel (Ni) nanoparticles. Al_2O_3 serves as a porous support structure, and CNTs facilitate rapid electron transmission by lowering charge transfer resistance. Long-term durability may be impacted by the possibility that Al_2O_3 will dissolve or reach out in strongly alkaline conditions, despite the catalyst's good stability for 20h [7].

Researchers are continuously developing improved catalysts for OER. CuO/CuS heterostructures have emerged as effective and affordable OER catalysts that function better than their constituent parts. They demonstrate outstanding catalytic activity through a small starting potential of 1.49 Volts (vs. RHE), an excess potential of 270 mV, and a slope of Tafel of 67 mV/dec. Furthermore, they are a possible substitute for pricey noble metal catalysts owing to their long-term stability and maintained 10 mA/cm^2 current density within 1.0 M KOH [115]. To maximize electron transport, improve stability, and increase active site exposure, further developments in heterostructure catalysts require exact control of interface topologies. Reaching these objectives can greatly improve catalytic performance in applications involving water splitting. Catalytic performance is improved by optimizing shape with a greater surface area and plenty of active centers. Activity can be further increased by adding chemicals with TM (Fe, Co, Ni, W, Mo) and nonmetals (N, S, P, B), although cautious management is required to preserve structural integrity and prevent undesired phase shifts. Even though it has scaling problems, nanostructure engineering enhances ion/electron transport. Performance is improved by combining doping and hybrid construction, while theoretical studies and *in situ* techniques reveal active sites and charge transfer processes to improve catalyst design [116].

10.5.4 Bifunctional Catalysts

Bifunctional catalysts extracted from waste materials provide a sustainable approach to addressing global energy and environmental challenges. These catalysts are synthesized by waste such as biomass, industrial byproducts, agricultural residues, and electronic waste into high-performance materials [117]. With dual active sites, they can facilitate multiple reactions like OER and HER in water splitting or hydrogenation. Waste-derived catalysts also possess customized properties like high porosity, high conductivity, and greater surface area, making them effective for bifunctional applications. In biomass refining, bifunctional catalysts from waste materials are highly efficient in upgrading oxygen-rich compounds into valuable fuels and chemicals [118]. For example, biomass-derived carbon doped with nitrogen and combination with transition metals like Ni or Fe offers dual active sites. These metal sites facilitate hydrogenation, while the acidic or doped carbon supports deoxygenation or bond cleavage. Zeolite-supported metals derived from industrial waste are useful for hydrodeoxygenation processes, where oxygen removal and molecular rearrangements occur simultaneously [119]. Such applications illustrate the ability of waste-derived materials to improve catalytic performance and reduce dependence upon traditional catalysts. Agricultural residues converted into porous carbon, doped with transition metals like Co or Ni, create eco-friendly and cost-effective electrocatalysts. These catalysts show synergistic effects, with one active site facilitating water reduction while the other supports oxidation. For example, NiFe double hydroxides synthesized from waste sources have demonstrated excellent bifunctional activity owing to their large dispersion of active sites and enhanced charge transfer properties, making them suitable for use in acidic and alkaline electrolysis. Waste materials are also valuable sources for the development of metal-metal oxide catalysts, which combine hydrogenation and deoxygenation functionalities. For instance, electronic and industrial waste can be processed to drive metals like Au, Pt, or Ce, which can then be combined to create catalysts with metal-metal oxide interfaces [120]. These materials are highly effective in the conversion of biomass into gas. A notable example is the Au/CeO_2 catalyst, where Ce dissociates water while gold oxidizes CO during the water-gas shift reaction. This interaction between metal and oxide sites calls attention to the promise of waste-derived catalysts in advanced chemical processes. Despite their promise, bifunctional catalysts from waste materials face challenges in scaling up production and achieving consistent performance due to variations in waste feedstock [121]. The efforts to not only dispose of the animal and plant waste but also to utilize it in catalytic applications for energy harvesting are in progress. In this regard, the conversion of biosolids into a mesoporous material named as SS-Fe-350, which is a high-performance photo Fenton catalyst has been reported through a facile method [117]. The findings of the study offer novel insights to utilize sewage sludge for synthesizing useful materials for catalytical applications.

SS-Fe-350, a robust as well as active heterogeneous photo-Fenton catalyst, was produced from sewage sludge *via* a simplified fabrication technique, resulting in a mesoporous structure as depicted in Figure 10.4 [19]. Mesoporous materials were created using a cost-effective, three-phase synthesis. This technique employed bacterial scaffold templates, inorganic particles, and sewage sludge, resulting in reduced equipment and operational expenses. The mesoporous property of a catalyst is due to the process of calcination, where adsorbed H_2O, organic molecules, and biomacromolecules in sewage sludge evaporate, burn, or carbonize. Further research is needed to improve synthesis methods, such as hydrothermal treatments and microwave-assisted processes, to optimize the structural and functional properties of these catalysts.

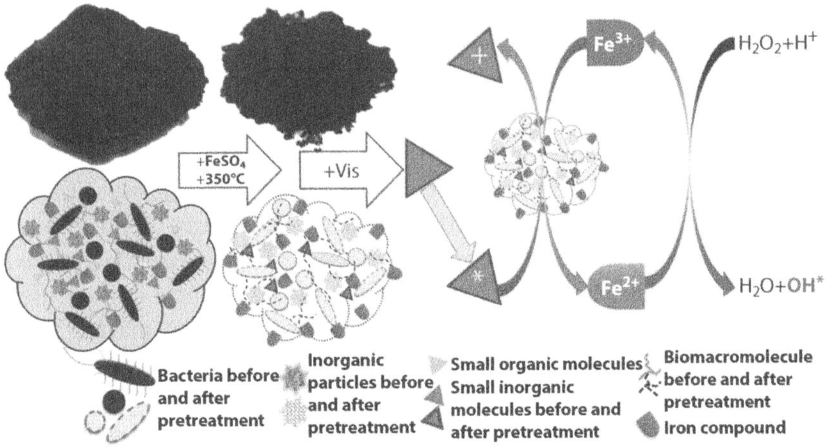

Figure 10.4 The scheme of conversion of biosolids into a mesoporous material *via* photo-Fenton mechanism. Reprinted with permission from Elsevier [19].

10.6 Conclusion

The resourceful use of waste materials as catalyst precursors offers a potential course of action for inscribing dual confronts of increasing waste treatment and development of efficient catalysts. The manufacturing of catalysts frequently utilizes various waste treatment methods, ranging from simple to complex by recovering valuable components such as active metals, metal oxides, and inert or functional supports. Efficient catalysts can be produced by combining waste materials individually or in combination with other industrial products. Biomass waste produces carbonaceous and alkaline earth oxides, while industrial waste supplies TM oxides are commonly used in modern catalysts. Our review on metal and carbon-based catalysts, electrocatalytic, and photocatalytic sourced from waste presents promising opportunities for sustainable and cost-effective hydrogen production. Waste recovery is most effective while using waste biomass and other economic waste to create value-added products, such as red mud, battery waste, and methylcyclohexane. Major advancements in waste catalyst development are needed to reduce demand for economically derived materials and reduce manufacturing steps for final catalyst production. Waste-derived catalysts are being improved through innovative strategies focusing on electroactive sites, mass/charge transfer, structural, and electrochemical durability. The development and optimization of waste-derived catalysts and advanced processing techniques like pyrolysis and wet chemical methods are crucial for a clean and resource-efficient hydrogen economy. The greater redox activity and small cost of waste-derived catalysts make them promising for diverse electrochemical applications in environmental and energy sectors, including pollutant removal and energy conversion reactions. The incorporation of catalysts from waste sources can effectively decrease the environmental impact of fabricating functional materials and promote advanced waste utilization strategies.

AI disclosure statement: The author acknowledges the use of AI-based tools for minor language editing or grammatical correction. However, all intellectual contributions, analyses, and conclusions presented in this chapter are entirely the author's own.

References

1. Hernandez, C., Lopez, F.S., Pacheco, F.A., The sustainable development goals index: An analysis (2000-2022). *Transdiscipl. J. Eng. Sci.*, 13, 111, 2022.
2. Barbier, E., Geothermal energy technology and current status: an overview. *Renewable Sustainable Energy Rev.*, 6, 1-2, 3–65, 2002.
3. Zamfirescu, C., Dincer, I., Naterer, G.F., Banica, R., Quantum efficiency modeling and system scaling-up analysis of water splitting with Cd1– xZnxS solid-solution photocatalyst. *Chem. Eng. Sci.*, *97*, 235–255, 2013.
4. Ishaq, H., Dincer, I., Crawford, C., A review on hydrogen production and utilization: Challenges and opportunities. *Int. J. Hydrogen Energy*, 47, 62, 26238–26264, 2022.
5. Su, T.M., Qin, Z.Z., Ji, H.B., Jiang, Y.X., Y.X., Huang, G., Recent advances in the photocatalytic reduction of carbon dioxide. *Environ. Chem. Lett.*, *14*, 1, 99–112, 2016.
6. Turning Solid Waste into Catalysts: A Path for Environmental Solutions (https://doi.org/10.1002/cplu.202400246).
7. Increasing the efficiency of hydrogen production from solar powered water electrolysis (https://doi.org/10.1016/j.rser.2020.110255)
8. Yadav, G., Yadav, N., Sultana, M., Ahmaruzzaman, M., A comprehensive review on low-cost waste-derived catalysts for environmental remediation. *Mater. Res. Bull.*, *164*, 112261, 2023.
9. Jothi, V.R., Bose, R., Rajan, H., Jung, C., Yi, S.C., Harvesting electronic waste for the development of highly efficient eco-design electrodes for electrocatalytic water splitting. *Adv. Energy Mater.*, *8*, 34, 1802615, 2018.
10. Li, X., Fu, L., Chen, F., Zhao, S., Zhu, J., Yin, C., Application of heterogeneous catalytic ozonation in wastewater treatment: an overview. *Catalysts*, 13, 2, 342, 2023.

11. Dillert, R., Cassano, A.E., Goslich, R., Bahnemann, D., Large scale studies in solar catalytic wastewater treatment. *Catal. Today*, 54, 2-3, 267–82, 1999.
12. Descorme, C., Catalytic wastewater treatment: Oxidation and reduction processes. Recent studies on chlorophenols. *Catal. Today*, 297, 324–334, 2017.
13. Mohapatra, L., Cheon, D., Yoo, S.H., Carbon-based nanomaterials for catalytic wastewater treatment: a review. *Molecules*, 28, 4, 1805, 2023.
14. Zhang, T., Heterogeneous catalytic process for wastewater treatment, in: *Advanced Oxidation Processes*, vol. 2020, p. 17, IntechOpen, London, United Kingdom, 2020.
15. Tabatabaei, M., Karimi, K., Horvath, I.S., Kumar, R., Recent trends in biodiesel production. *Biofuel Res. J.*, 2, 3, 258–267, 2015.
16. Li, W., Dang, Q., Smith, R., Brown, R.C., Wright, M.M., Techno-economic and life cycle analysis of the stabilization of bio-oil fractions for insertion into petroleum refineries. *ACS Sustainable Chem. Eng.*, 5, 2, 1528–37, 2017.
17. Boro, J., Deka, D., Thakur, A.J., A review on solid oxide derived from waste shells as catalyst for biodiesel production. *Renewable Sustainable Energy Rev.*, 16, 1, 904–910, 2012.
18. Chen, K.T., Wang, J.X., Dai, Y.M., Wang, P.H., Liou, C.Y., Nien, C.W., Wu, J.S., Chen, C.C., Rice husk ash as a catalyst precursor for biodiesel production. *J. Taiwan Inst. Chem. Eng.*, 44, 4, 622–9, 1, 2013.
19. Yuan, S.-J. and Dai, X.-H., Facile synthesis of sewage sludge-derived mesoporous material as an efficient and stable heterogeneous catalyst for photo-Fenton reaction. *Appl. Catal., B*, 154, 252–258, 2014.
20. Vermeulen, S.J., Aggarwal, P.K., Ainslie, A., Angelone, C., Campbell, B.M., Challinor, A.J., Hansen, J., Ingram, J.S., Jarvis, A., Kristjanson, P.M., Lau, C., Agriculture, food security and climate change: Outlook for knowledge, tools and action. *CCAFS Report 3. Copenhagen, Denmark: CGIAR-ESSP Program on Climate Change, Agriculture and Food Security*, 2010.
21. Nurfitri, I., Maniam, G.P., Hindryawati, N., Yusoff, M.M., Ganesan, S., Potential of feedstock and catalysts from waste in biodiesel preparation: a review. *Energy Convers. Manage.*, *74*, 395–402, 2013.
22. Oliveira, D.A., Benelli P., P., Amante, E.R., A literature review on adding value to solid residues: egg shells. *J. Cleaner Prod.*, 46, 42–47, 2013.
23. Razzak, S.A., Municipal solid and plastic waste derived high-performance biochar production: A comprehensive review. *J. Anal. Appl. Pyrolysis*, 106622, 2024.
24. Dijkgraaf, E. and Vollebergh, H.R., Burn or bury? A social cost comparison of final waste disposal methods. *Ecol. Econ.*, 50, 3-4, 233–247, 2004.
25. Nanda, S. and Berruti, F., Municipal solid waste management and landfilling technologies: a review. *Environ. Chem. Lett.*, 19, 2, 1433–1456, 2021.
26. Chen, M.F., Chen, Y., Lim, Z.J., Wong, M.W., Adsorption of imidazolium-based ionic liquids on the Fe (1 0 0) surface for corrosion inhibition: physisorption or chemisorption? *J. Mol. Liq.*, *367*, 120489, 2022.

27. Chen, Z., Wei, W., Chen, H., Ni, B.J., Recent advances in waste-derived functional materials for wastewater remediation. *Eco-Environ. Health*, 1, 2, 86, 2022.
28. Duman, G. and Yanik, J., Two-step steam pyrolysis of biomass for hydrogen production. *Int. J. Hydrogen Energy*, 42, 27, 17000–17008, 2017.
29. Franchini, M., Philipsen, P.H.T., Visscher, L., The Becke fuzzy cells integration scheme in the Amsterdam Density Functional program suite. *J. Comput. Chem.*, 34, 21, 1819–1827, 2013.
30. Alvarez, J., Kumagai, S., Wu, C., Yoshioka, T., Bilbao, J., Olazar, M., Williams, P.T., Hydrogen production from biomass and plastic mixtures by pyrolysis-gasification. *Int. J. Hydrogen Energy*, *39*, 21, 10883–10891, 2014.
31. Uzoejinwa, B.B., He, X., Wang, S., Abomohra, A.E.F., Hu, Y., Wang, Q., Co-pyrolysis of biomass and waste plastics as a thermochemical conversion technology for high-grade biofuel production: Recent progress and future directions elsewhere worldwide. *Energy Convers. Manage.*, *163*, 468–492, 2018.
32. Ashok, J., Dewangan, N., Das, S., Hongmanorom, P., Wai, M.H., Tomishige, K., Kawi, S., Recent progress in the development of catalysts for steam reforming of biomass tar model reaction. *Fuel Process. Technol.*, 199, 106252, 2020.
33. Reyes-Cabrera, P., Palomo, J., García-Mateos, F.J., Ruiz-Rosas, R., Rosas, J.M., Rodríguez-Mirasol, J., Cordero, T., Sustainable carbon-based nickel catalysts for the steam reforming of model compounds of pyrolysis liquids. *Fuel Process. Technol.*, *253*, 108028, 2024.
34. Santamaria, L., Lopez, G., Fernandez, E., Cortazar, M., Arregi, A., Olazar, M., Bilbao, J., Progress on catalyst development for the steam reforming of biomass and waste plastics pyrolysis volatiles: a review. *Energy Fuels*, *35*, 21, 17051–17084, 2021.
35. Liu, W.-J., Jiang, H., Yu, H.-Q., Development of biochar-based functional materials: toward a sustainable platform carbon material. *Chem. Rev.*, 115, 22, 12251–85, 2015.
36. Qian, K., Kumar, A., Zhang, H., Bellmer, D., Huhnke, R., Recent advances in utilization of biochar. *Renewable Sustainable Energy Rev.*, 42, 1055–1064, 2015.
37. Jin, W. and Zhang, Y., Sustainable electrochemical extraction of metal resources from waste streams: from removal to recovery. *ACS Sustainable Chem. Eng.*, 8, 12, 4693–4707, 2020.
38. Cuenya, B.R., Synthesis and catalytic properties of metal nanoparticles: Size, shape, support, composition, and oxidation state effects. *Thin Solid Films*, 518, 12, 3127–3150, 2010.
39. Bentley, C.L., Kang, M., Unwin, P.R., Nanoscale surface structure–activity in electrochemistry and electrocatalysis. *J. Am. Chem. Soc.*, 141, 6, 2179–2193, 2018.

40. Chen, Z., Yun, S., Wu, L., Zhang, J., Shi, X., Wei, W., Liu, Y., Zheng, R., Han, N., Ni, B.J., Waste-derived catalysts for water electrolysis: circular economy-driven sustainable green hydrogen energy. *Nano Micro Lett.*, *15*, 1, 4, 2023.
41. Zeng, H., Liu, Z., Qi, J., Chen, J., Zeng, Y., Yang C, C., Li, Z., Wang, C., Gu, L., Zhang, Y., Shu, M., Dynamic Cation Intercalation Facilitating Chemical Oxidation of Water and Surface Stabilization During the Oxygen Evolution Reaction. *Energy Environ. Mater.*, 8, 2, e12813, 2025.
42. Petersen, H.A., Myren, T.H., O'Sullivan, S.J., Luca, O.R., Electrochemical methods for materials recycling. *Mater. Adv.*, *2*, 4, 1113–1138, 2021.
43. Jadhav, A., Wet Chemical Methods for Nanop article Synthesis, in: *Chemical Methods for Processing Nanomaterials*, pp. 49–58, CRC Press, Boca Raton, 2021.
44. Lu, Z.-I., Lindner, E., Mayer, H.A., Applications of sol– gel-processed interphase catalysts. *Chem. Rev.*, 102, 10, 3543–3578, 2002.
45. Shandilya, M., Rai, R., Singh, J., Hydrothermal technology for smart materials. *Adv. Appl. Ceram.*, 115, 6, 354–376, 2016.
46. Kömür, AI., Kızıl, C., Karaman, C., Nature's blueprint for energy: biomass-derived heteroatom-doped graphene materials for advanced energy applications. *Carbon Lett.*, 1–43, 2025.
47. Palma, V., Barba, D., Cortese, M., Martino, M., Renda, S., Meloni, E., Microwaves and heterogeneous catalysis: A review on selected catalytic processes. *Catalysts*, *10*, 2, 246, 2020.
48. Cova, C.M., Zuliani, A., Santiago, A.R.P., Caballero, A., Muñoz-Batista, M.J., Luque, R., Microwave-assisted preparation of Ag/Ag 2 S carbon hybrid structures from pig bristles as efficient HER catalysts. *J. Mater. Chem. A*, *6*, 43, 21516–21523, 2018.
49. Miao, J., Zhao, X., Hu, H.Y., Liu, Z.H., Hierarchical ultrathin NiFe-borate layered double hydroxide nanosheets encapsulated into biomass-derived nitrogen-doped carbon for efficient electrocatalytic oxygen evolution. *Colloids Surf., A*, *635*, 128092, 2022.
50. Zhou, Y., Li, J., Gao, X., Chu, W., Gao, G., Wang, L.W., Recent advances in single-atom electrocatalysts supported on two-dimensional materials for the oxygen evolution reaction. *J. Mater. Chem. A*, *9*, 16, 9979–9999, 2021.
51. Majid, A., Jabeen, A., Khan, S.U.D., Almutairi, Z., On the prospects of layeredness in tantalum pentoxide. *Mater. Sci. Eng., B*, 272, 115349, 2021.
52. Pan, Y., Abazari, R., Wu, Y., Gao, J., Zhang, Q., Advances in metal–organic frameworks and their derivatives for diverse electrocatalytic applications. *Electrochem. Commun.*, 126, 107024, 2021.
53. Ruidas, S., Mohanty, B., Bhanja, P., Erakulan, E.S., Thapa, R., Das, P., Chowdhury, A., Mandal, S.K., Jena, B.K., Bhaumik, A., Metal-Free Triazine-Based 2D Covalent Organic Framework for Efficient H_2 Evolution by Electrochemical Water Splitting. *ChemSusChem*, 14, 22, 5057–5064, 2021.

54. Yang, M., Chen, L., Wang, J., Msigwa, G., Osman, A.I., Fawzy, S., Rooney, D.W., Yap, P.S., Circular economy strategies for combating climate change and other environmental issues. *Environ. Chem. Lett.*, 21, 1, 55–80, 2023.
55. H, Ala'a, Jamil, F., Osman, A.I., Myint, M.T., Kyaw, H.H., Al-Hajri, R., Hussain, M., Ahmad, M.N., Naushad, M., State-of-the-art novel catalyst synthesized from waste glassware and eggshells for cleaner fuel production. *Fuel*, 330, 125526, 2022.
56. Chen, Z., Wei, W., Zou, W., Li, J., Zheng, R., Wei, W., Ni, B.J., Chen, H., Integrating electrodeposition with electrolysis for closed-loop resource utilization of battery industrial wastewater. *Green Chem.*, 24, 8, 3208–3217, 2022.
57. Cao, X., Li, Z., Chen, H., Zhang, C., Zhang, Y., Gu, C., Xu, X., Li, Q., Synthesis of biomass porous carbon materials from bean sprouts for hydrogen evolution reaction electrocatalysis and supercapacitor electrode. *Int. J. Hydrogen Energy*, 46, 36, 18887–97, 25; 2021.
58. Chen, Z., Zheng, R., Zou, W., Wei, W., Li, J., Wei, W., Ni, B.J., Chen, H., Integrating high-efficiency oxygen evolution catalysts featuring accelerated surface reconstruction from waste printed circuit boards *via* a boriding recycling strategy. *Appl. Catal., B*, 298, 120583, 2021.
59. Chen, Z., Yang, H., Kang, Z., Driess, M., Menezes, P.W., The Pivotal Role of s-, p-, and f-Block Metals in Water Electrolysis: Status Quo and Perspectives. *Adv. Mater.*, 34, 18, 2108432, 2022.
60. Chen, Z., Duan, X., Wei, W., Wang, S., Ni, B.-J., Recent advances in transition metal-based electrocatalysts for alkaline hydrogen evolution. *J. Mater. Chem. A*, 7, 25, 14971–15005, 2019.
61. Zhang, L., Lei, Y., Zhou, D., Xiong, C., Jiang, Z., Li, X., Shang, H., Zhao, Y., Chen, W., Zhang, B., Interfacial engineering of 3D hollow CoSe2@ ultrathin MoSe2 core@ shell heterostructure for efficient pH-universal hydrogen evolution reaction. *Nano Res.*, 15, 4, 2895–2904, 2022.
62. Song, J., Wei, C., Huang, Z.-F., Liu, C., Zheng, L., Wang, X., Xu, Z.J., A review on fundamentals for designing oxygen evolution electrocatalysts. *Chem. Soc. Rev.*, 49, 7, 2196–2214, 2020.
63. Yoo, J.S., Rong, X., Liu, Y., Kolpak, A.M., Role of lattice oxygen participation in understanding trends in the oxygen evolution reaction on perovskites. *ACS Catal.*, 8, 5, 4628–4636, 2018.
64. Chen, Z., Duan, X., Wei, W., Wang, S., Ni, B.-J., Electrocatalysts for acidic oxygen evolution reaction: Achievements and perspectives. *Nano Energy*, 78, 105392, 2020.
65. Bala, R., Gautam, V., Mondal, M.K., Improved biogas yield from organic fraction of municipal solid waste as preliminary step for fuel cell technology and hydrogen generation. *Int. J. Hydrogen Energy*, 44, 1, 164–173, 2019.
66. Lui, J., Chen, W.-H., Tsang, D.C.W., Yu, S., A critical review on the principles, applications, and challenges of waste-to-hydrogen technologies. *Renewable Sustainable Energy Rev.*, 134, 110365, 2020.

67. Maya, D.M.Y., *et al.*, Gasification of municipal solid waste for power generation in Brazil, a review of available technologies and their environmental benefits. *J. Chem. Chem. Eng.*, 10, 6, 249–255, 2016.
68. Tang, Y.Q., *et al.*, Synthesis and electrochemical properties of Li2FeSiO4/C/Ag composite as a cathode material for Li-ion battery. *J. Cent. South Univ.*, 26, 6, 1443–1448, 2019.
69. Qi, L., H., B., Kaiser, U.J.S., Perspective towards atomic-resolution imaging of two-dimensional polymers. *SmartMat*, 2, 2, 131–138, 2021.
70. Li, Y., *et al.*, Ordered macroporous structured TiO_2-based photocatalysts for CO_2 reduction: a review. *Chin. Chem. Lett.*, 34, 12, 108417, 2023.
71. Chen, X., Zhai, H., Wang, W., Huang, Z., Fabrication of Cr2AlC-Fe based composites by *in-situ* reaction technique. *J. Chin. Ceram. Soc.*, 41, 3, 309–313, 2013.
72. Cameron, J., Kanibolotsky, A.L., Skabara, P.J., Lest we forget–the importance of heteroatom interactions in heterocyclic conjugated systems, from synthetic metals to organic semiconductors. *Adv. Mater.*, 36, 20, 2302259, 2024.
73. Bakht, M.A., Ali, I., Singh, G., Green synthesis, density functional theory (DFT), molecular docking, molecular dynamics simulation, and biological activity of 1,2,3,4-tetrahydropyrimidine-5-carbonitrile derivatives as PLA2 and proteinase K inhibitors. *Arabian J. Chem.*, 17, 6, 105798, 2024.
74. Mishra, R., Kumar, A., Singh, E., Kumar, S., Recent research advancements in catalytic pyrolysis of plastic waste. *ACS Sustainable Chem. Eng.*, 11, 6, 2033–2049, 2023.
75. Yang, Z., Li, Y., Zhang, X., Cui, X., He, S., Liang, H., Ding, A., Sludge activated carbon-based $CoFe_2O_4$-SAC nanocomposites used as heterogeneous catalysts for degrading antibiotic norfloxacin through activating peroxymonosulfate. *Chem. Eng. J.*, 384, 123319, 2020.
76. Arshad, N.S., Anwar, S., Wahab, R., Hussain, A., Alam, M., Ali, W., Awan, T.I., Nabi, G., Highly Sensitive Plasmonic Au Bowtie Sensors with Extraordinary Optical Absorbance for SERS Applications. *Plasmonics*, 20, 1427–1442, 2025.
77. Yan, Y., Zhang, C., Jia, T., Li, H., Shen, K., Efficient Hydrogen Production by an rGO/TiO_2 Composite Material from Ammonia Borane Hydrolysis in a Photocatalytic Reactor. *Energy Fuels*, 35, 19, 16065–16074, 2021.
78. Phang, S.J., Study, design and development of carbon quantum dots-based graphic carbon nitride (g-C_3N_4) nanocomposites in photocatalysis for dye wastewater treatment. Doctoral Thesis, 2022, http://hdl.handle.net/10399/5023.
79. El-Bery, H.M., Salah, M.R., Ahmed, S.M., Soliman, S.A., Efficient non-metal based conducting polymers for photocatalytic hydrogen production: comparative study between polyaniline, polypyrrole and PEDOT. *RSC Adv.*, 11, 22, 13229–13244, 2021.
80. George, S.M., Sneh, O., Dillon, A.C., Wise, M.L., Ott, A.W., Okada, L.A., Way, J.D., Atomic layer controlled deposition of SiO_2 and Al_2O_3 using ABAB... binary reaction sequence chemistry. *Appl. Surf. Sci.*, 82, 460–467, 1994.

81. Sharuddin, S.D.A., Abnisa, F., Daud, W.M.A.W., Aroua, M.K., A review on pyrolysis of plastic wastes. *Energy Convers. Manage.*, 115, 308–326, 2016.
82. Xu, M. and Wei, M., Layered double hydroxide-based catalysts: recent advances in preparation, structure, and applications. *Adv. Funct. Mater.*, 28, 47, 1802943, 2018.
83. Bennett, J.A., Wilson, K., Lee, A.F., Catalytic applications of waste derived materials. *J. Mater. Chem. A*, 4, 10, 3617–3637, 2016.
84. Nabgan, W., Abdullah, T.A.T., Nabgan, B., Jalil, A.A., Nordin, A.H., Ul-Hamid, A., Hassan, N.S., Catalytic biohydrogen production from organic waste materials: a literature review and bibliometric analysis. *Int. J. Hydrogen Energy*, 46, 60, 30903–30925, 2021.
85. Vuppaladadiyam, A.K., Vuppaladadiyam, S.S.V., Awasthi, A., Sahoo, A., Rehman, S., Pant, K.K., Biomass pyrolysis: A review on recent advancements and green hydrogen production. *Bioresour. Technol.*, 364, 128087, 2022.
86. Nielsen, A.T., Amandusson, H., Bjorklund, R., Dannetun, H., Ejlertsson, J., Ekedahl, L.G., Hydrogen production from organic waste. *Int. J. Hydrogen Energy*, 26, 6, 547–550, 2001.
87. Delgado, H.E., Cappello, V., Zang, G., Sun, P., Marcinkoski, J., Techno-economic analysis and life cycle analysis of e-fuel production using nuclear energy. *J. CO2 Util.*, 72, 102481, 2023.
88. Bianchi, F.R., Risso, R., Cardona, L., Bove, D., Cannizzaro, F., Bonardi, L., Palmisani, E., Bosio, B., Feasibility analysis of e-Hydrogen, e-Ammonia and e-Methanol synthesis compared with methane to fuel production. *Fuel*, 384, 133938, 2025.
89. Singh, H., Li, C., Cheng, P., Wang, X., Liu, Q., A critical review of technologies, costs, and projects for production of carbon-neutral liquid e-fuels from hydrogen and captured CO_2. *Energy Adv.*, 1, 9, 580–605, 2022.
90. Rahimpour, M.R., Samimi, F., Babapoor, A., Mohebi, S., Palladium membranes applications in reaction systems for hydrogen separation and purification: A review. *Chem. Eng. Process. Process Intensif.*, 121, 24–49, 2017.
91. Ren, S., Xu, X., Hu, K., Tian, W., Duan, X., Yi, J., Wang, S., Structure-oriented conversions of plastics to carbon nanomaterials. *Carbon Res.*, 1, 15, 2022.
92. Ma, Z., Wang, L., Wang, Z., Cui, H., Tang, E., Hou, H., Xu, G., Song, T., Wang, Q., Recent Advances of Plastic Waste-Derived Carbon Materials for Energy Storage, Environmental Remediation and Organic Synthesis Applications. *ChemCatChem*, 16, 23, e202401072, 2024.
93. Soffian, M.S., Halim, F.Z.A., Aziz, F., Rahman, M.A., Amin, M.A.M., Chee, D.N.A., Carbon-based material derived from biomass waste for wastewater treatment. *Environ. Adv.*, 9, 100259, 2022.
94. Mir, R.A. and Pandey, O.P., Waste plastic derived carbon supported Mo2C composite catalysts for hydrogen production and energy storage applications. *J. Cleaner Prod.*, 218, 644–655, 2019.

95. Song, C.M., Cao, W.C., Bu, C.Y., He, K., Zhang, G.H., Preparation of ultrafine molybdenum carbide (Mo_2C) powder by carbothermic reduction of molybdenum trioxide (MoO_3). *J. Aust. Ceram. Soc.*, 56, 1333–1340, 2020.
96. Zhang, X., Strzelecki, A.C., Cockreham, C.B., Goncharov, V.G., Li, H., Sun, J., Sun, H., Guo, X., Wu, D., Thermodynamics of molybdenum trioxide encapsulated in zeolite Y. *AlChE J.*, 67, 12, e17464, 2021.
97. Zhao, P., Zhao, Y., Liang, H., Song, X., Yu, B., Liu, F., Ragauskas, A.J., Wang, C., Novel waste-derived non-noble metal catalysts to accelerate acidic and alkaline hydrogen evolution reaction. *Chem. Eng. J.*, 466, 143140, 2023.
98. Robinson, B.H., E-waste: an assessment of global production and environmental impacts. *Sci. Total Environ.*, 408, 2, 183–191, 2009.
99. Seif, R., Salem, F.Z., Allam, N.K., E-waste recycled materials as efficient catalysts for renewable energy technologies and better environmental sustainability. *Environ. Dev. Sustainability*, 26, 3, 5473–5508, 2024.
100. Meloni, E., Martino, M., Palma, V., A short review on Ni based catalysts and related engineering issues for methane steam reforming. *Catalysts*, 10, 3, 352, 2020.
101. Chan, F.L. and Tanksale, A., Review of recent developments in Ni-based catalysts for biomass gasification. *Renewable Sustainable Energy Rev.*, 38, 428–438, 2014.
102. Ansari, A.S., Azzahra, G., Nugroho, F.G., Mujtaba, M.M., Ahmed, A.T.A., Oxides and metal oxide/carbon hybrid materials for efficient photocatalytic organic pollutant removal. *Catalysts*, 15, 2, 134, 2025.
103. Liu, X., Liu, J., Yang, Y., Li, Y.W., Wen, X., Theoretical perspectives on the modulation of carbon on transition-metal catalysts for conversion of carbon-containing resources. *ACS Catal.*, 11, 4, 2156–2181, 2021.
104. Zhang, C., Wang, B., Shen, X., Liu, J., Kong, X., Chuang, S.S.C., Yang, D., Dong, A., Peng, Z., A nitrogen-doped ordered mesoporous carbon/graphene framework as bifunctional electrocatalyst for oxygen reduction and evolution reactions. *Nano Energy*, 30, 503–510, 2016.
105. Chen, M., Kitiphatpiboon, N., Feng, C., Abudula, A., Guan, G., Recent progress in transition-metal-oxide-based electrocatalysts for the oxygen evolution reaction in natural seawater splitting: A critical review. *EScience*, 3, 2, 100111, 2023.
106. Liang, Q., Brocks, G., Bieberle-Hütter, A., Oxygen evolution reaction (OER) mechanism under alkaline and acidic conditions. *J. Phys.: Energy*, 3, 2, 026001, 2021.
107. Kim, J.S., Kim, B., Kim, H., Kang, K., Recent progress on multimetal oxide catalysts for the oxygen evolution reaction. *Adv. Energy Mater.*, 8, 11, 1702774, 2018.
108. Ma, Q., Yu, Y., Sindoro, M., Fane, A.G., Wang, R., Zhang, H., Carbon-based functional materials derived from waste for water remediation and energy storage. *Adv. Mater.*, 29, 13, 1605361, 2017.

109. Huang, Y., Wu, D., Cao, D., Cheng, D., Facile preparation of biomass-derived bifunctional electrocatalysts for oxygen reduction and evolution reactions. *Int. J. Hydrogen Energy*, 43, 18, 8611–8622, 2018.
110. Luo, X., *et al.*, Biomass derived Fe, N-doped carbon material as bifunctional electrocatalysts for rechargeable Zn-air batteries. *J. Alloys Compd.*, 888, 161464, 2021.
111. Xuan, J. and Liu, Z., High-performance N-doped bifunctional carbon electrocatalysts derived from polymer waste for oxygen reduction and evolution reaction. *Int. J. Electrochem. Sci.*, 12, 11, 10471–10483, 2017.
112. Lu, F., *et al.*, First-row transition metal based catalysts for the oxygen evolution reaction under alkaline conditions: basic principles and recent advances. *Small*, 13, 45, 1701931, 2017.
113. Vazhayil, A., *et al.*, A comprehensive review on the recent developments in transition metal-based electrocatalysts for oxygen evolution reaction. *Appl. Surf. Sci. Adv.*, 6, 100184, 2021.
114. Wang, H., *et al.*, The electronic structure of transition metal oxides for oxygen evolution reaction. *J. Mater. Chem. A*, 9, 35, 19465–19488, 2021.
115. Khan, N.A., *et al.*, Effective CuO/CuS heterostructures catalyst for OER performances. *Int. J. Hydrogen Energy*, 48, 80, 31142–31151, 2023.
116. Zheng, D., *et al.*, Structural advantages and enhancement strategies of heterostructure water-splitting electrocatalysts. *Cell Rep. Phys. Sci.*, 2, 6, 2021.
117. Munyentwali, A., Li, H., Yang, Q., Review of advances in bifunctional solid acid/base catalysts for sustainable biodiesel production. *Appl. Catal., A*, 633, 118525, 2022.
118. Iliopoulou, E.F., Triantafyllidis, K.S., Lappas, A.A., Overview of catalytic upgrading of biomass pyrolysis vapors toward the production of fuels and high-value chemicals. *Wiley Interdiscip. Rev.: Energy Environ.*, 8, 1, e322, 2019.
119. Zhang, W., *et al.*, Recent progress on the synthesis and applications of zeolites from industrial solid wastes. *Catalysts*, 14, 10, 734, 2024.
120. Lang, R., *et al.*, Single-atom catalysts based on the metal–oxide interaction. *Chem. Rev.*, 120, 21, 11986–12043, 2020.
121. Zou, X. and Zhang, Y., Noble metal-free hydrogen evolution catalysts for water splitting. *Chem. Soc. Rev.*, 44, 15, 5148–5180, 2015.

11

Sustainable Synthesis of Nano-Zeolite Catalysts from Industrial Waste with Insights from Machine Learning

Sivamani Sivalingam*, Niranjan R. and Srinidhi V.

Department of Chemical Engineering, Rajalakshmi Engineering College, Thandalam, Chennai, Tamilnadu, India

Abstract

The ever-growing demand of zeolite catalysts with diverse applications in separation, catalysis, and adsorption is facing synthesis challenges due to the cost and environmental impact of conventional methods. Large scale solid waste is offering a cheaper and renewable feedstock to produce zeolite material. Waste-derived zeolites could also be synthesized at the nanoscale from top-down and bottom-up methods. Nowadays, the application of machine learning in screening the materials for nano-zeolite synthesis significantly reduces the amount of trial-and-error experimentation. This chapter aims to provide an overview of nano-zeolite catalyst and machine learning approach for examination of the feasibility and potential benefits of utilizing industrial solid waste as a raw material for synthesis of the same. In addition, the challenges associated with conventional synthesis methods, such as high cost, environmental impact, and limited feedstock source are also discussed. Finally, this chapter provides future research directions, emphasizing the importance of sustainable waste management and resource utilization in advancement towards a circular economy.

Keywords: Catalyst, industrial waste, nano-zeolite, sustainable synthesis, screening, machine learning, circular economy

**Corresponding author*: sivamchem@gmail.com; sivamani.s@rajalakshmi.edu.in

Deepti Goyal, Sakshi Kabra Malpani and Renuka Jain (eds.) Catalysis from Waste: Transforming Solid Residues into Value, (291–320) © 2026 Scrivener Publishing LLC

11.1 Introduction

Zeolites are three-dimensional aluminosilicates containing alkali and alkali earth metals where the framework is composed of primary building blocks of TO_4 units, such as SiO_4 and AlO_4 connected in periodic patterns [1]. The tetrahedral units of Si^{4+} combine with an ordinary oxygen atom to form a neutral structure [2, 3]. However, substituting a silicon atom (Si^{4+}) with aluminum (Al^{3+}) results in emergence of a negative charge in the lattice that needs to be compensated by a cation. These charge-balancing cations occupy the micropore space by coulombic forces. Therefore, they can be exchanged with other cations, expanding their functionality and making them very effective as ion exchange minerals.

Zeolites exist in both natural and synthetic forms. Till date, over 40 naturally occurring zeolites have been identified; however, their commercial use is limited due to the presence of impurities. As of now, around 150 frameworks of synthetic zeolites have been developed, which offer greater purity and tailored properties but are more expensive to produce [4]. Zeolites are claimed to be non-toxic and safe to consume by both the International Agency for Research on Cancer (IARC) and Food and Drug Administration (FDA). Therefore, they are widely used in fields of biotechnology and medicine. They can eliminate toxins from the bloodstream effectively in patients suffering from the final stages of uremia. A study by Wernert *et al.*, confirmed that nearly 75% creatinine could be removed by adsorption onto acidic mordenite, and similarly, 60% cresol could be removed by silicate-1 zeolites [5]. Zeolites are also being considered as potential candidates for drug delivery and biosensing. For instance, the ZSM-5 catalyst was developed and patented by Mobil Oil Company (present ExxonMobil) [6]. In ZSM-5, the proton (H^+) acts as the cation to neutralize the negative charge created due to the substitution of Si^{4+} by Al^{3+}, which increases the acidity of the material. Therefore, the acidity of the material is regulated by the Al^{3+} or the Si/Al ratio and affects the catalytic properties of the zeolite. The first application of ZSM-5 was the isomerization of xylene isomers into para-xylene. Para-xylene has a higher diffusion coefficient within the zeolite channels and, therefore, can diffuse out faster than meta or ortho-xylene. Therefore, both ortho-xylene and meta-xylene interact with the Bronsted acid sites to convert into p-xylene before they can diffuse out of the pore network of ZSM-5. Yet another application of ZSM-5 is to convert Methanol to Gasoline (MTG) studied by Kuo [7]. Similar to the ZSM-5 used for the MTG process, notable attempts were being done by Marchi and Froment [8], to convert Methanol to Olefins (MTO). This process used another zeolite catalyst called SAPO-34, which has 3D channels, moderate acidic strength and could achieve selectivity close to 90% or more. SAPO-34 exhibited high-temperature steam stability and good regenerability, making them useful for such applications. Finally, they were being studied for controlling SOx and NOx emissions. Natural zeolites have been studied for application as catalytic converters to deal with emissions from diesel engines and for various other pollution control cases.

In recent years nanoscale zeolites are gaining interest due to their structural flexibility, stability, and high surface area. Nano-zeolites are widely utilized across various industries, like in construction by adding them as additives in concrete mixtures to improve their compressive strength by around 10-20% [3]. Nano-zeolite cation exchange capacity makes them useful for wastewater treatment and radiation containment, as seen in the Chornobyl and Three Mile Island nuclear incidents [1].

A machine learning (ML) approach in nano-zeolite synthesis leverages data-based technologies to fine-tune intricate synthesis parameters, thereby improving the productivity and outcome of zeolite and nano-zeolite manufacturing. The ML correspondence aids in the integration of experimental results with chosen synthesis descriptors and their defining relationships to zeolite structures towards optimizing the synthesis process.

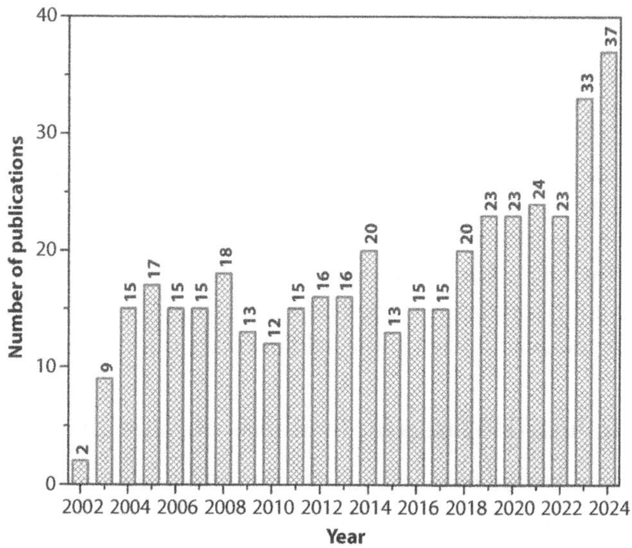

Figure 11.1 Publication trend of zeolite catalyst (Scopus Search: zeolites from industrial waste) until 25th March 2025.

The subsequent sections discuss the framework of this strategy [9, 10]. The objective of this chapter is to present literature review on preparation of nano-zeolite from industrial waste and to provide an overview of ML to screening the optimizing parameters. Figure 11.1 depicts the significance of zeolites from industrial waste till 2024. This clearly shows that the demand of zeolite research has been increasing every year and their applications have also been evolving as well.

11.2 Nano-Zeolite Catalysts

Nano-zeolites are a fascinating class of materials combining the unique characteristics of zeolites with the advantage of nanoscale. Essentially, nano-zeolites are similar to their bulk counterparts and retain the properties of zeolites [11, 12]. However, their dimensions in the nanometer scale give them unique properties, opening more sectors of applications. In bulk zeolites with micropores and micrometer-scale crystallites, the internal surface area is much larger than the external surface area. Therefore, in the case of catalysis, the reactants should be able to access these sites by diffusion. However, in bulk zeolites, the pore size limits the mass transfer of bulky molecules, i.e., only molecules with kinetic diameters less than 1.5 nm can often access the internal surface. This results in lower activity; also, since the active sites are dispersed in both the internal and external surfaces, the pore mouths get plugged easily by coke formation [3], and thus their lifetime is reduced. To avoid this and promote faster mass transfer, there are two strategies: i) increase the number of mesopores within the zeolite or ii) reduce the crystal size to the nanoscale. Nano-zeolites possess a high surface-to-volume ratio; therefore, much of their surface is exposed to the environment and diffusion length and accessible active sites gets shorter, resulting in higher activities, better performance, and higher lifetimes. They have been reported to show postponed coking and catalyst deactivation compared to their bulk counterparts. All these properties make them good candidates in the field of catalysis [1, 13, 14].

11.2.1 Industrial Wastes for Synthesis of Nano-Zeolites

Nano-zeolites derived from commercially available silica precursors (such as fumed silica, colloidal silica, silicic acid, and silicon alkoxides) are often costly and environmentally hazardous. These commercial silica precursors are available in different forms, and some are cheaper, but they suffer from unpredictable reactivity and selectivity. These are some of the most significant limitations in nano-zeolite production using commercial raw materials [15, 16]. Recent studies have been reported that using commercial raw materials for zeolite synthesis could cost around 40% of the total production cost. Therefore, it is desirable to identify alternative feedstocks, at least from an economic standpoint. Several natural clay materials, as well as industrial and agricultural wastes, have been considered as possible alternatives to synthesizing nano-zeolites.

Synthesizing zeolites from industrial wastes has significant advantages for all stakeholders involved. The industries generating these wastes would not have to pay for disposing the waste but are also incentivized to promote zeolite production and follow environmental regulations. However, zeolite synthesis from industrial wastes still faces some challenges that need to be addressed. Removing non-zeolitic phases and residual waste is necessary to improve yield. More importantly, impurities in the final zeolite structure are unsuitable for catalytic applications since they could act as poisons and complicate the process [17, 18].

Some industrial waste used for synthesis of nano-zeolites are summarized below:

- **Coal Fly Ash**

Coal Fly Ash (CFA) is a by-product generated during pulverized coal combustion at thermal power plants. It is a glassy powder primarily composed of silicon dioxide, aluminum, and ferric oxide along with some other metal oxides in trace amounts. It is removed by particle removal technologies and is mixed with water to deposit in dumps. CFA contains meso and microporous particles and can cause respiratory hazards upon exposure. It contains heavy metals, which is a significant threat when released into the atmosphere unchecked. Also, the cost of disposal is significant and adds to the cost of electricity produced. Therefore, it is desirable to identify potential uses of this waste. Of course, various potential solutions to utilize CFA have been researched, such as partial replacement of cement in construction, soil amendment, and production of lightweight concrete [1, 12]. Countries such as the US, UK, and India have already reported reusing a significant proportion of the generated CFA rather than disposing it off. Also, significant improvement has been made in utilizing CFA to synthesize zeolites. One important point to be noted is that CFA has a low Si/Al ratio, which makes it a good candidate for synthesizing high silica zeolite synthesis; and therefore, it is more prominently used for the synthesis of low to medium silica zeolites.

- **Municipal Solid Waste Fly Ash (MSW-FA)**

The by-product of waste incineration comprises fine particulates primarily containing silica, calcium oxide, aluminum oxide, and iron oxide, along with trace amounts of heavy metals like lead, cadmium, and mercury. MSW-FA is used in the construction industry as an additive in cement and concrete, enhancing durability and reducing raw material needs. Additionally, it aids in stabilizing hazardous wastes and is being researched for use in ceramics and glass production, promoting sustainable waste management. One of the main challenges with MSW-FA is that it needs more silica and alumina compared to CFA.

The progress made with zeolite synthesis from MSW-FA is summarized in Table 11.1. Fan *et al.*, attempted to synthesize zeolite from ash generated by the co-combustion of coal and MSW [19]. During this attempt, alkali fusion was used, where the ash was directly mixed with NaOH and heated at 550°C for 1h. The mixture was left to age for 24h at room temperature. Hydrothermal synthesis was done at different temperatures and time periods ranging from 40–180°C and 0.5–26h. After vacuum filtration and multiple washings with distilled water, the solid was dried at 105°C. According to the XRD data, Zeolite X, indicated at NaOH/ash ratio of above 1.2, gets transformed to Zeolite HS (hydroxysodalite) and the ratio was increased to 1.8 due to the difference in solubility between Si and Al components and the quantity of NaOH. Microwave-assisted hydrothermal synthesis method was also adopted for the synthesis of zeolite from MSW-FA. In this study [20], the MSW-FA was mixed with NaOH (0-2 M) and Na_2SiO_3 (0-50 wt.%) under constant stirring to improve the dissolution of the Al and Si. Then, it was placed in a microwave assisted heating apparatus at 180°C for different durations (0.5-2h). Upon increased magnetic stirring, the XRD analysis shows an increase in the intensity of the sodalite and kaolinite phases due to the conversion of the quartz phase in the zeolites. Also, after adding Na_2SiO_3 in different proportions, the hydrated sodalite and kaolinite phases were transformed into the Na-P1 and Na-P2 zeolite phases. Initially, only Na-P1 phase existed and when Na_2SiO_3 was increased to 30 wt.%, and then both Na-P1 and Na-P2 coexist. However, only Na-P2 phase was observed when the Na_2SiO_3 was further increased to 50 wt.%.

- **Bagasse Fly Ash (BFA)**

BFA is a waste from sugar and ethanol plants which is burnt to generate power. After burning, the ash left behind in large quantities needs to be handled. It is mainly composed of unburnt carbon, SiO_2, and Al_2O_3 and, therefore, is considered as a potential low-cost alternative for zeolite

Table 11.1 Si-Al source for zeolite synthesis from MSW-FA using hydrothermal treatment with different types of product. (modified from [21] Creative Commons Attribution (CC BY) license).

Al, Si sources				**Pre-treatment**	**Hydrothermal method**					**Product**
Type	**Si**	**Al**	**Initial Si/Al**	**Type**	**Type**	**Additives**	**L/S**	**Temp**	**Time**	
							mL/g	**°C**	**h**	
MSW-FA	9.5% SiO_2	5.5% Al_2O_3	1.5	DW, Cal, Acid, DI	Autoclave w/ST	1-2M NaOH	25	60	20-30	Zeolite A
						0.5-2M NaOH	25	100-120	20-48	Zeolite P
75μm, WMS	22.0% SiO_2	10.2% Al_2O_3	1.9	No	-	3.5M NaOH	5-15	130-190	24	Gismodine
						6M NaOH				Gmelinite
MSW-FA, WGP, Al_2O_3	0.8% Si	2.2% Al	1.0	No	Heater w/ ST	2.5M NaOH	10	60	24	Zeolite Y, A and L

(*Continued*)

Table 11.1 Si-Al source for zeolite synthesis from MSW-FA using hydrothermal treatment with different types of product. (modified from [21] Creative Commons Attribution (CC BY) license).

Al, Si sources				Pre-treatment	Hydrothermal method					Product
			Initial Si/Al	Type	Type	Additives	L/S	Temp	Time	
Type	Si	Al					mL/g	°C	h	
MSW-FA	12.2% Si	0.8% Al	40	AF	HYD	DI	100	105	24	-
					Sealed reactor	1.2% CTAB[2]		105	24	-
					Cal	-	-	550	3	ZSM-23
MSW-FA and CFA	45.5% SiO_2	22.6% Al_2O_3	1.8	Gr, AF	Aging w/ ST	DI	9	Room	24	-

Calcination at 800°C for 2h and acid treatment with 3 M HCl for 30 min; pH adjusted to 10 using 1M H_2SO_4
WGP = waste glass powder; WMS = wet magnet-separated; Cal = calcination; Gr = grinding; AF = alkali fusion; DI = deionized water washing; DW= distilled water washing, Acid = acid treatment; CTAB = cetyltrimethylammonium bromide; HYD = hydrolysis; w/ST = with stirring or shaking; MWA = microwave-assisted; CRY = crystallization; N.S.: Not specified.

production [21]. Sugarcane is a predominant crop in India and Brazil; therefore, more BFA is possible in these regions.

In a study, the BFA was washed thoroughly and dried before being mixed with NaOH for alkali fusion. The fusion was done at 550°C for 1.5h. After cooling the fused solid, it was ground to reduce size and then allowed to age for 12h by mechanical stirring in double distilled water. The resulting solid was washed and dried. This work concluded that zeolitic materials obtained from the hydrothermal treatment of as-received BFA mainly contain Zeolite P and analcime as the dominant phases (as shown by the PXRD data) [22]. The same procedure was utilized in the other studies involving the use of zeolites for the sorption of phenol, dye, p-nitrophenol and aniline. More importantly, the last work, i.e., zeolite employed for the sorption of aniline, utilized the microwave-assisted hydrothermal treatment. In this method, researchers first sieved BFA and mixed it with 1.5 M NaOH and 1.5 M $BaCl_2.2H_2O$ solutions. The resulting mixture was irradiated with 900 W energy, at a frequency of 2.45 GHz for 17 min. The solid residue was thoroughly washed several times before being dried and then reduced into powder form. This was followed up with another study [23], using microwave-assisted hydrothermal synthesis for the sorption of o-chlorophenol. One of the additional requirements to process BFA for zeolite synthesis is to extract Si and Al or, in other words, remove the unburnt carbon [24]. Moises *et al.*, used BFA for the synthesis of zeolite A by alkaline fusion method. These studies used the alkaline fusion method to dissolve the Si and Al in the solid matrix selectively. However, it is also possible to remove the unburnt carbon by calcination [25]. Then, direct hydrothermal treatment was done to synthesize zeolite Na-A which was used for copper uptake from aqueous solutions.

11.2.2 Methods for Synthesis of Nano-Zeolite from Silica Enriched Waste

The synthesis of zeolites from non-commercial sources involves some general steps with slight deviations depending on the raw materials, SiO_2/Al_2O_3 ratio, and the target zeolite. The pre-treatment involves cleaning and drying, mechanical pre-treatment, and removal of impurities. Mechanical pre-treatment involves size reduction by milling or grinding and is often the first step in separating various impurities from the Si and Al fractions by mere size difference [3, 12]. However, several impurities, such as metals and volatile organic compounds, could be present even after this. These impurities can be dealt with by acid-leaching. Initial calcination steps remove volatile organic compounds while acid leaching helps in removal of metal and other impurities, before starting the actual synthesis. Several methods have been reported to synthesize waste-derived nano-zeolite catalysts such as hydrothermal, alkali fusion and molten salt methods etc. The hydrothermal synthesis is one of the most widely reported methods in literature, but it involves one more step to improve reactivity. The alkaline fusion method is also exclusively used to synthesize zeolites involving impure precursors.

- **Alkali Fusion Method**

In this method, the impure sample containing the Si or Al source is mixed with NaOH at high temperatures for a few hours, followed by stirring to allow the sample to age. It is often used as a pre-treatment method before hydrothermal synthesis to accelerate the dissolution of silicates and aluminates by transforming the stable quartz phase into an amorphous state. Most of the phases are converted to sodium silicate or sodium aluminate (upon treatment with NaOH), i.e., the raw precursor may be composed of several different phases of Si and Al, which dissolve and become available for synthesis at different rates. Thus, it improves the reactivity of the feedstock sources and enhances the overall reaction rate and final quality of products [26, 27].

The steps involved in the alkaline fusion method are detailed in Figure 11.2. Initially, the precursors are mixed with NaOH, often in solid form, to obtain a homogenous mixture. This mixture is then heated at high temperatures to undergo a fusion reaction. As a result, a fused solid containing sodium silicate and sodium aluminate is obtained. This fused solid is then

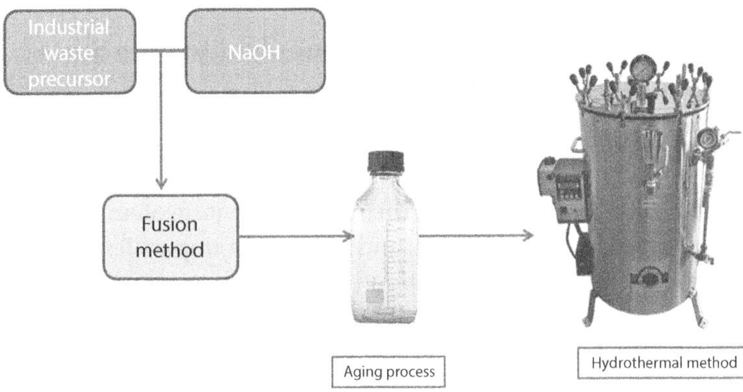

Figure 11.2 Flow chart showing synthesis of zeolites from industrial wastes using alkaline fusion and hydrothermal synthesis methods.

dissolved and aged for several hours before typically being synthesized into the desired zeolite phase by hydrothermal method. The elaboration of this method is available in the sections below since it is the most widespread activation method for zeolite synthesis from waste streams.

- **Molten salt method**

The molten salt method is similar to the alkaline fusion method, except using solid materials (rather than aqueous medium) to start the synthesis process. However, unlike the alkaline fusion method, it does not require an aqueous medium while synthesizing and utilizes molten salt as a growth medium for crystal formation. It also does not generate any waste supernatant stream, reducing the downstream operations required. The general procedure for zeolite synthesis *via* this method includes mixing the ground precursor with a base and salt before heating the mixture to a molten state. In its molten state, salt provides optimal growth conditions for the zeolite crystals. After this, the solid was cooled and washed several times to remove residual bases or salts.

One interesting study with the molten salt method involves zeolite synthesis from sewage sludge. In this study, the sludge was initially calcined at 550°C for 3h and powdered using a ball mill. The sewage sludge, sodium hydroxide, and sodium sulfate were mixed and heated to nearly 950°C to achieve the molten state. It is reported in the literature [28], that adding a zeolite seed reduced the time required for crystallization. Different bases, such as NaOH, KOH, or NH_4F, and salts, such as KNO_3, $NaNO_3$, or NH_4NO_3, were employed for the synthesis. The salt is occluded in the pores in zeolites which are synthesized by molten salt method. This could be advantageous specifically in the field of agriculture, as noted by another study [29]. The molten salt method for zeolite synthesis is efficient, offering shorter synthesis times and eliminating alkaline liquid waste. However, it has limitations, such as insufficient contact between reactants, which leads to lower conversion rates and irregular structures, along with limited control over crystal size, which can be problematic for specific applications.

11.2.3 Utilization of Different Raw Materials

Natural clays are readily available, cheap, and non-toxic sources for silicon and aluminum and, therefore, serve as a great alternative to commercial sources in zeolite synthesis. Their origin, abundance, and other relevant information have been presented in Table 11.2. The choice of the natural clay depends primarily on its Si/Al ratio and the framework to be synthesized. As mentioned in Section 1, this is one of the reasons for the

Table 11.2 Characteristics of important natural clay (source from [32].).

Clay mineral	Si/Al ratio	General formula	Abundance/ Origin	Current applications
Kaolinite	1.0-1.1	$Al_2Si_2O_5(OH)_4$	One of the six most abundant minerals in Earth's crust; abundant in soils	Mainly used as a filler and coating agent in paper production. Other uses include ceramics, adhesives etc.
Bentonite	4.0-5.0	$Al_2O_3.4(SiO_2).H_2O$	Worldwide production in 2018 stood at 20,400,000 metric tons	Used in drilling mud to lubricate and cool the cutting tools; Used as a filler in adhesives, cosmetics, rubber etc.
Feldspar	3.0	(Ca,Na,K)(Al,Si) $AlSi_2O_8$	The most abundant mineral in Earth's crust making up to nearly 60%	Used as a fluxing agent in glass production.
Obsidian	6.42	-	Abundant areas are near recent volcanic activity.	Used as weapons and tools in ancient times; currently, its applications are limited.

(Continued)

Table 11.2 Characteristics of important natural clay (source from [32].). (*Continued*)

Clay mineral	Si/Al ratio	General formula	Abundance/ Origin	Current applications
Pyrophyllite	3.5	$Al_2Si_4O_{10}(OH)_2$	Found in abundance in various countries like the USA, New Zealand, China, South Africa, Brazil, India.	Used as filler in rubber, paper, soap, cosmetics etc.

Table 11.3 Summary of various zeolites synthesis from different natural clays (modified from [32]).

	Synthesis conditions					
		Ageing		Heating		
Clay mineral	Initial treatment	Temp (°C)	Time (h)	Temp (°C)	Time (h)	Product
Kaolin Mordenite	Mordenite + kaolin calcined at 900°C for 30 min	-	-	100	120	Zeolite X Zeolite A
Kaolin	50 g kaolin + 60 g of NaOH mixture heated at 550°C for 1.5h	25	0.5	90	8	Zeolite X
Bentonite	The bentonite mixed with 8 M of NaOH; heated at 200°C for 3h	25	0.5	90	120	Zeolite X

(*Continued*)

Table 11.3 Summary of various zeolites synthesis from different natural clays (modified from [32]). (*Continued*)

Clay mineral	Synthesis conditions					
	Initial treatment	Ageing		Heating		Product
		Temp (°C)	Time (h)	Temp (°C)	Time (h)	
Bentonite	Bentonite reacted with HCl for 8h at 95°C	25	20	97	24	Zeolite Y
Diatomite	Diatomite was crushed; treated with 6 M H_2SO_4 at 100°C for 24h	25	24	100	48	Zeolite Y
Bentonite	Bentonite treated with HCl with 1 g/4 ml ratio, in an oil bath, for 8h at 95°C	25	24	97	24	Zeolites Y
Kaolin	Calcination of kaolin at 800°C for 6h	25	24	90	48	Zeolite Y
Kaolin and Diatomite	Quasi-solid-phase activation; 50 g of kaolin + 75 g of NaOH + 10.2 ml of H_2O at 100°C for 30 min	60	12	100	24	Zeolite Na-Y
Kaolin	Calcination of kaolin at 600°C for 6h	25	96	95	72	Zeolite Na-Y
Kaolin	Kaolin treated with NaOH and H_2O; calcined at 250°C for 2h	70	4	170	48	Microporous ZSM - 5

importance of the Si/Al ratio-based classification. Table 11.3 summarizes the synthesis of various zeolites derived from distinct natural clays, including ZSM-5, Zeolite X, and Zeolite Y. Since diatomite is primarily composed of silicon, Zeolite X powder was synthesized with $Al(OH)_3$ added to compensate for the low aluminum content. In this study by Yao *et al.*, the two compounds were mixed with NaOH solution, and the slurry was left to age for 0 to 120 min at 30 to 60°C. Then, it was allowed to crystallize in an autoclave for 3 to 9h at 90 to 120°C. The samples were then washed with deionized water until the pH comes close to neutral. The wet filtered solid was dried at 105°C overnight. Similarly, in another study with diatomite [30], to synthesize ZSM-5, a high silica zeolite, was discovered that diatomite could function as a templating agent on its own, negating the need for an additional structure directing agent (SDA). Template-free Zeolite Y was synthesized from diatomite [31], to use as support for dimethylglyoxime (DMG) in the removal of Ni(II) from aqueous solutions.

11.2.4 Post-Synthesis Modification of Zeolite

The post-synthetic modification step refers to modifying synthesized zeolites to enhance their performance for specific applications. Post-synthetic modification of zeolites involves different processes such as calcination, ion exchange, loading with metals, reinsertion of tetrahedral framework atoms, dealumination, desilication, etc. Therefore, post-synthetic modification is another critical strategy in synthesizing zeolites for various applications, especially catalysis. Here, the need for dealumination and desilication is considered. Several studies have applied dealumination-desilication to control the Si/Al ratio for these various applications. The different desilication and dealumination strategies in literature have been summarized in Table 11.4.

11.2.5 Applications of Nano-Zeolites

The major applications of nano-zeolites are mainly focused on catalysis. As mentioned above, nano-zeolites offer lower diffusion resistance due to a larger surface area [13]. This makes them a suitable candidate for various processes, especially catalytic conversion. Since zeolites are commonly used as the go-to material in the petroleum and petrochemical industry, nano-zeolites with enhanced performance and lifetime hold great potential to replace them.

A nano-zeolite and macro-zeolite of crystal size approximately 150-200 nm and 1.5 μm respectively have been used for n-hexane cracking and

Table 11.4 Post-synthetic modification of zeolites.

Zeolite type	Si/Al ratio	Modification type	Si/Al ratio	Modification procedure	References
Na-Y		Dealumination		Dehydration; addition of $SiCl_4$ vapor at 450 – 550°C.	[33]
ZSM-5	17	Desilication + Ion Exchange	14.5	1g zeolite + 30 mL of 0.2 M NaOH for 30 min at 65°C, washed, dried and calcined.	[34]
ZSM-5	17	Dealumination	14.2	1g zeolite + 90 mL of 0.1 M oxalic acid for 3h at 80°C; washed, and calcined at 450°C.	[34]
Zeolite Y	2.5	Dealumination	2.5	Adsorption of NH_3 at 150°C; cooled.	[35]
Zeolite Y	2.5	Dealumination	5.0	Washed with 0.1 M Na_2H_2-EDTA at 62°C for 24h; ion-exchanged with 1 M NH_4NO_3 at 80°C for 24h.	[35]
Zeolite ß	11.0	Dealumination	1300	13 M HNO_3 solution for 4h at 180°C. Incorporation of V atoms by mixing with NH_4VO_3 at 25°C	[36]
CZ-25	26.0	Desilication	18.0	Alkaline treatment in 0.2 M NaOH for 30 min.	[37]
CZ-35	37.0	Desilication	24.0		
CZ-200	176	Desilication	133		
CZ-1000	1038	Desilication	560		

acetone-to-olefin reactions. A study was experimented with zinc modified ZSM-5 for the aromatization of 1-hexene [38]. The study extensively tested the conversion and yield % in unmodified nano ZSM-5 and at various zinc loadings. All Zn-modified samples maintained stable conversion for 10h, with the top-performing sample yield close to 65%. Interestingly, although light olefin synthesis and aromatics production seem to compete against one another, both benefit from faster diffusion rates. Another exciting application of nano-zeolites in catalysis is the degradation of plastic waste. Plastic waste has been accumulating in the environment and significantly threatens the ecosystem. Moreover, rather than reusing the plastic, methods to break them down into individual monomers to use them back is a lucrative option. However, the proposed methods are costly and energy-consuming. Here, catalytic degradation offers a unique alternative. Various solid catalysts have been explored for plastic degradation, and the material is desired to possess high porosity and acidity. One such application of nano-zeolite is in the catalytic degradation of polyethylene. Nano ZSM-2 was studied for this application [39]. Sticking with nano ZSM-5 and plastic degradation, another study by H. Zhang *et al.*, utilized the former in cracking low density polyethylene (LDPE). The nano ZSM-5 prepared by seed-induced method could achieve near complete conversion within 30 minutes, while a commercial nano ZSM-5 and traditional ZSM-5 achieved only 40% and 1%, respectively. The nano ZSM-5 was far better at cracking LDPE than the bulk counterpart. Also, the study examined the importance of the synthesis method in improving the acidic sites accessibility and the strength of the acidic sites.

11.3 Machine Learning-Based Material Screening

Computational methods have been applied heavily in material research for various applications. This involves Density Functional Theory (DFT), Grand Canonical Monte Carlo (GCMC) simulations, and ML. High Throughput Computational Screening (HTCS) is an approach that utilizes the above-mentioned computational tools and involves screening materials for preferred performances from a vast material space that is impossible with brute-force experimentation. The evolution of these computational tools in material screening is quite an exciting topic, overcoming multiple challenges over the years outlined well [40]. Of the different computational tools used for HTCS, ML is more straightforward to develop and screen faster the vast chemical space, unlike the DFT and GCMC methods. This makes ML vastly more accessible to researchers with limited funding or support. ML has been successfully applied across various fields to solve different problems. In the context of material chemistry, ML models are primarily utilized to reduce the number of hit and trial experimentation required, thereby reducing the cost and time to develop a successful material. For instance, ML models have successfully identified Metal-Organic Frameworks (MOFs) with optimal performance for carbon capture while searching through a material space of millions of hypothetical materials. Application of ML in catalysis and nano-zeolite synthesis can further pave the best ways for utilizing it in waste-derived zeolite synthesis. To the best of our knowledge, no ML or other computational tool has been developed to aid in utilizing waste for zeolite synthesis and its application in catalysis.

11.3.1 Mechanism of ML

ML is a data-driven approach to learn patterns and develop predictions for various problems. Mathematically it represents the relationship between any two data as $y = f(x)$, where y represents the output variable, x represents the input variable(s), and f is the function that maps x to y. In ML, we could call the function 'f' a pattern since the mapping learned by ML models is not the same as the function derived in empirical models [41, 42].

ML models are initially trained on a significant input and output dataset. The model thus attempts to learn the pattern and map the two. Various metrics are used to check if the model has successfully learned the system before deployment. They are Data Collection, Dataset Pre-processing, Model Selection and Training, Model Evaluation, and Deployment. The workflow does not need to be linear; ML model development is dynamic and may require constant changes in various steps to improve performance.

11.3.2 Machine Learning Applications in Material Screening

In the context of material screening, ML is applied at various stages of the research workflow. The general workflow for material screening is listed below, regardless of the approach taken:

1. Problem Identification and Setting Objective
2. Literature Review
3. Material Selection
4. Experimental Design and Synthesis
5. Material Characterization
6. Applications

While moving straightforwardly, the brute force experimentation method presents several challenges throughout the material research workflow. For instance, identification of the suitable material is only possible upon its successful synthesis and testing. Synthesis itself may often require several iterations to achieve the desired properties, and for each iteration, characterization is an essential step for confirmation. Moreover, even after successful synthesis, the material may exhibit inadequate performance. Therefore, significant resources and time are consumed right from identifying the suitable material to its successful synthesis and testing. Even after successful synthesis, satisfactory performance is not guaranteed, requiring additional resources for further optimization [40, 43].

In the case of catalysts, which are pivotal in various industrial processes, practical applicability necessitates extensive testing beyond simulated conditions. Catalyst regeneration, long-term stability assessment, and evaluation under real-life conditions are fundamentals to determining practicality. Therefore, identifying the suitable material, synthesizing it successfully, and ensuring its practical viability demands significant time and resources that may not be accessible to everyone.

Computational tools present an exciting and more straightforward solution to this practical difficulty. They help to analyze an ample chemical space and identify potential application candidates. With a higher chance of success, researchers can focus on synthesis, reducing the required number of iterations. However, it is not uncommon for a material identified by computational methods to be challenging or impossible to synthesize. Several methods exist to analyze if a structure can be synthesized in the lab, and in recent times, ML models have been deployed to support researchers in the synthesis part as well. In combination with a screening of materials and their synthesis, ML models have contributed to reducing the actual number of experimentations required while also providing tremendous success [44]. In this way, resources can be devoted to testing the material for its practicality in real-life conditions. This section provides an overview of the diverse applications of ML and other computational tools throughout the material research workflow.

An example of generative model in action was studied [42], in which the researchers developed a generative model which can generate MOFs using topology, metal nodes, and organic linkers to represent its structure. But this may differ in case of zeolites. Several initiatives have been taken to construct a sizable hypothetical material database, which is more evident in MOFs development. At present, databases are available for hypothetical MOF materials that researchers can utilize for various applications.

Theoretically generated zeolite structures are energetically accessible or stable. Therefore, several criteria have been proposed to reduce the number of hypothetically generated zeolites and include exclusively only those zeolites that are feasible to synthesize under real conditions. A study [45], suggested the separate upper energy limit value for zeolites, which was around ~ 0.17 eV per SiO_2 formula units (f.u.) above the quartz phase (the most stable phase for SiO_2). Several databases containing thousands of hypothetical zeolite structures already exist, namely, Michael Deem's Predicted Crystallography Open Database (PCOD). This database contains around 2.9 million hypothetical zeolite structures, out of which 10% are thermodynamically accessible. Several of the new additions to zeolite frameworks by the International Zeolite Association (IZA) were initially predicted in this database, thus proving the value that they offer [46]. Monte Carlo Tree Search seems the most popular while constructing hypothetical databases from existing structure's data. To understand generative ML models, first we should understand how Michael Deem's PCOD was made. Around 230 different crystallographic space groups exist which are utilized along with other features to construct a crystal structure. For these 230 space groups, several constraints are present on the possible unit cell parameter values. Unit cell sizes a, b, and c were varied in steps of 3 A° with edges from 3 – 30 A°. Less than 3 A° is impossible since the distance between Si atoms is 3.1 A°, and the step size of 3A° was chosen to insert new Si atoms.

Similarly, the unit angle and the atomic density were also varied, considering the characteristics of real zeolites as given in the IZA Structure Commission Atlas. Now all the possible combinations of unit cell size (a, b, c), unit angles, atomic density, and other structural features should be run to predict different structures. For this, 100 zefsaII biased Monte Carlo runs were performed to generate the unit cells. The zefsaII biased Monte Carlo scheme was devised long back to use PXRD data to randomly arrange the T-atoms (where T = Si, Al, etc.) in such a way that the cost function is minimized, indicating that it is closer to the experimental structure which is noted in this study [47]. Even now, it is one of the simplest and most popular methods to identify the structure of a zeolite with readily available experimental data. However, zefsaII can also be used to generate hypothetical structures without PXRD data, which was done to create Michael Deem's PCOD. These hypothetical databases can now serve as an excellent platform for identifying the best material using predictive ML modeling.

11.3.3 Inverse Material Design

The Inverse Material Design approach is the most extensive application of generative models in the field of material research. Inverse design on is related to generative ML and most applications of generative models involve separately to provide a contrast for their individual potential application.

The property-to-structure models use the performance required as the input features and the representation of the material structure as the output. Therefore, the model learns to map specific structure characteristics with respect to the performance. After training, a need arises to provide the desired performance as an input, and the model would generate a hypothetical material that should be able to provide its performance in agreement to theory. The three main categories of generative models are Variational Autoencoders (VAE), Recurrent Neural Networks (RNN), and Generative Adversarial Networks (GANs) used for inverse design applications. The inverse design of zeolites for specific applications does not seem to be widespread in literature as compared to MOFs. This could be because MOFs are more diverse than zeolites.

A GAN was implemented after getting trained on a dataset comprising 31,273 methane-accessible zeolites to produce 121 crystalline porous materials, with the methane heat of adsorption as the user-defined variable [48], (property to generate the structure). The model takes inputs in the form of energy and material inputs. The GAN algorithm consists of two different entities, the generator and the discriminator, where the former aims to generate realistic data after training, and the latter tries to differentiate between real and fake data. Both work against each other to solve the problem together. Initially, the fake data from the generator may be easily differentiable by the discriminator and sent back. The generator then tries harder to deceive the discriminator, which also tries harder. As the iterations go on, both the generator and discriminator get better at their jobs until the generated data comes closer to accurate data; at this point, differentiating them is no longer possible, i.e., the generated data is statistically like the actual data it was trained with. In more technical terms, GAN uses two different neural networks that contest each other in a zero-sum game where one agent's gain is another agent's loss. This type of arrangement enables GAN models to generate reliably realistic output.

11.3.4 Material Design

The synthesis of a material could be more straightforward and challenging. It involves complex chemical transformations involving the silicate/aluminate condensation in solution, nucleation, and crystal growth. It is very much desired to optimize the synthesis, reduce the number of iterations required to arrive at a structure, and subsequently reduce the quantity of characterization and testing required. Zeolites are primarily synthesized with the help of SDAs, which direct the growth of their skeleton. Understanding the interaction between different SDAs and the zeolite framework is essential for screening the right combination. The most widespread application of ML for zeolites is identifying the most effective SDA(s) for a target zeolite material. DFT calculations have provided insights into the initial stages of zeolite formation, but it should be computationally more efficient to analyze the polymerization. Several algorithms have been developed with the idea that the most effective SDA must have a high binding energy with the target zeolite skeleton. These algorithms include Zeo++, ZEOMICS, and Zebeddee. They serve as valuable tools in all aspects of zeolite and other porous materials research. Schwalbe-Koda and group suggested that the SDA must possess a high binding affinity for the target zeolite and exhibit a low binding affinity for other zeolite structures [49].

Daeyaert, Ye, and Deem [50], developed an evolution-based algorithm to study the stabilization energies of several organic structure directing agents (OSDAs) to synthesize zeolite beta. The model was trained with roughly 4800 OSDA samples and was then deployed to sample all feasible OSDAs. It identified nearly 469 OSDAs with stabilization energy lower than -17 kJ/(mol Si), better than already known OSDAs for zeolite beta. Essentially, the model predicted 469 better OSDA candidates that could perform better than now. Moreover, this was developed with little experimentation, by using already existing data. Another slightly different approach [51], utilized generative ML models to analyze the zeolite-OSDA interaction and identify the "right" OSDA for a given zeolite as input. The weighted holistic invariant molecular (WHIM) descriptor was used to represent the structure of the OSDAs to capture the structural and non-structural properties that affect the zeolites. Then, the generative neural network was utilized to "generate" alternative OSDA in the Simplified Molecular Input Line Entry System (SMILES) format, using given zeolite phase and gel chemistry as inputs. This model generated potential OSDA candidates for the Chabazite (CHA) zeolite system. Nearly 408 unique OSDAs were obtained for CHA zeolites. To screen the 408 samples, an existing OSDA for CHA zeolites, i.e., N, N, N-trimethyladamantammonium (TMAda), was converted to a similar format (i.e., the WHIM descriptor) for efficient comparison. Nearly 57 OSDAs cut this screening approach. They are like TMAda in their structure and thus, could be suitable in using as alternatives. The last two studies prove that ML is suitable for screening existing OSDA material space, generating new potential candidates, and comparing them with existing OSDAs to filter out candidates with more significant potential, improving the diversity in choice.

Another critical study [52], revealed some long-standing questions regarding the zeolite formation and the influence of feed Si:Al:P ratios, pH, SDAs, crystallization temperature, and time. Using an ML-based global optimization technique, they evaluated the global potential energy surface of zeolitic systems. The model analyzed more than 10^6 minima data to gain insights into stable zeolite's energetics and local bonding patterns. The results from the model are exciting and significant as they revealed the roles of SDAs, T-atom ratios, and pH values on zeolite stability. A plot for the total energy of minima vs the framework density (FD) provides some insights.

The FD plot is used to show three different regions: at the low densely packed region (to the left of the plot), having a 2D layered phase, and at the extreme right, the stable quartz phase. Sandwiched between the two regions, with medium FD, are cage structures (i.e., zeolites). The plot clearly defined the energy criterion for zeolite formation, i.e., the energy of the zeolite must be lower than the 2D-Layered phase but higher than the quartz phase to exist, which is at around 0 – 0.17 eV per f.u. This criterion was identified for zeolites of AlPO type but is valid universally, as indicated by the presence of the 2D-Layered phase. This provides an apparent reason for the disparity between the number of synthesized zeolites and the number of hypothetical zeolite structures available. Another insight provided is that the radius has to be optimal to obtain zeolites rather than the other two phases. Too small radius would push it towards the quartz phase, while too big radius would push it towards the 2D-Layered phase. It must be between 3.5 and 5.5A°. For example, the optimal radius for ATO-type zeolite was suggested at 4A°, which matches the experimental value.

Moreover, more importantly, this explains the role of SDAs and the reasoning for identifying the right one among the large spaces. For ATO-type AlPO zeolite, dipropylamine is utilized as the SDA, with a diameter of around 8A°. Therefore, choosing SDAs with suitable sizes is critical to condensing the TO_4 units to form zeolites. ML can also guide the synthesis of zeolites by optimizing the synthesis conditions (as shown in Figure 11.3). Pan *et al.*, extracted data regarding the synthesis of zeolites by the

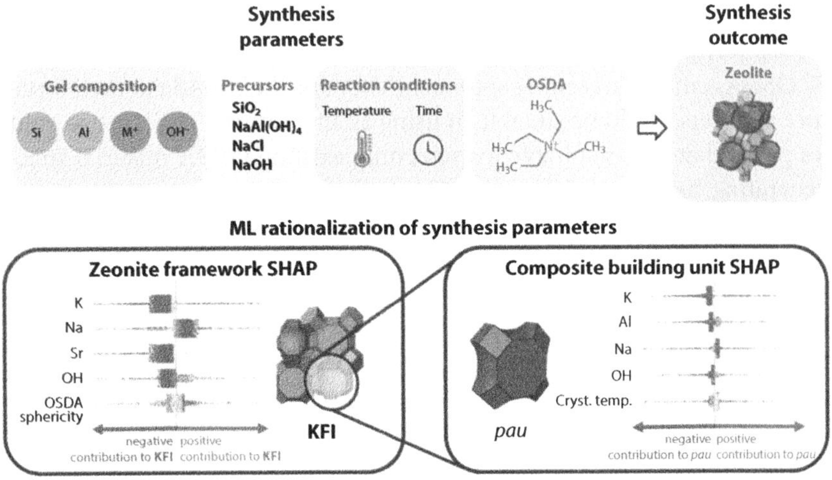

Figure 11.3 Workflow of ML model for guided zeolite synthesis (reproduced from [10] under creative commons license).

hydrothermal method from nearly 50 years of literature to create an extensive database, the ZeoSyn [10]. Each synthesis route in the database consists of the gel composition, reaction conditions, inorganic precursors, OSDAs, and the resulting zeolite structure. Then, a classifier model was trained to predict the zeolite framework product (output as probabilities) given a synthesis route with about 70% accuracy. Another recent attempt was done [53], to optimize the synthesis of Faujasite zeolite by OSDA-free synthesis. They identified the highest possible Si/Al ratio for Faujasite among all direct (not seeded) OSDA-free methods.

These are some notable examples covering various aspects of zeolite synthesis by ML. Another important instance is a study [54], which trained a ML model to predict the success of a reaction from failed experiment data. Although this is not related to zeolites, it is important to consider and incorporate it in future studies since all models are trained by utilizing datasets with only successful synthesis data. The failed experiments are discarded. They could quickly reduce the number of experimentations required to curate a dataset if appropriately utilized.

11.4 Conclusions

This chapter explored the idea of zeolite synthesis from wastes, which is relevant in the context of a circular economy and the context of the economics of the synthesis by replacing costly commercial precursors with cheaper raw materials. This would also incentivize proper waste management in several industries and requires careful coordination with utilization of waste. However, zeolites from waste also need to stay in line with the current research in the field of zeolites as well. Nano-zeolites have been gaining widespread attention for several years in various fields, replacing conventional zeolites. Thus, understanding how different factors affect the crystal size is essential to synthesize nano zeolites from waste precursors. The primary rationale behind the need for waste-derived zeolites is reducing waste disposal and re-utilizing for various purposes. However, one should also remember that reducing waste must reduce the overall waste and not put further stress on the environment. Therefore, focusing on more feasible synthesis methods is as important as the feed materials nature. Moreover, using computational tools, particularly ML, would be beneficial along the research chain for a successful, less costly, and faster synthesis of new zeolites. This chapter have covered the prospects of waste-derived zeolites and all the necessary information to keep them in line with other research areas in zeolites, along with the potential of modern tools to help speed up the process and reduce the economic burden.

Acknowledgements

The authors wish to acknowledge their sincere thanks to Rajalakshmi Engineering College (Autonomous), Thandalam, Chennai, India for the support to this work.

AI disclosure statement: The author acknowledges the use of AI-based tools for minor language editing or grammatical correction. However, all intellectual contributions, analyses, and conclusions presented in this chapter are entirely the author's own.

References

1. Sivalingam, S. and Sen, S., Optimization of synthesis parameters and characterization of coal fly ash derived nano-crystalline mesoporous zeolite X. *Appl. Surf. Sci.*, 455, 903–910, 2018.
2. Boer, D.G., Langerak, J., Pescarmona, P.P., Zeolites as Selective Adsorbents for CO2 Separation. *ACS Appl. Energy Mater.*, 6, 5, 2634–2656, 2023.

3. Sivalingam, S. and Sen, S., Valorization of coal fly ash into nanozeolite by sonication-assisted hydrothermal method. *J. Environ. Manage.*, 235, 145–151, 2019.
4. Ponnaiah, G.P., Lakshmi, S., Sivalingam, A.B.L.S., Sivalingam, S., A review of current practices on lead ions removal from different aqueous streams. *Braz. J. Dev.*, 10, 3, e68334, 2024.
5. Wernert, V., Schäf, O., Ghobarkar, H., Denoyel, R., Adsorption properties of zeolites for artificial kidney applications. *Microporous Mesoporous Mater.*, 83, 1–3, 101–113, 2005.
6. R.A.J. and G.L.R., Crystalline zeolite zsm-5 and method of preparing the same. US Patent, 1972.
7. Kuo, Conversion of Methanol to Gasoline Components. US Patent, 1976.
8. Marchi, A.J. and Froment, G.F., Catalytic conversion of methanol to light alkenes on SAPO molecular sieves. *Appl. Catal.*, 71, 1, 139–152, 1991.
9. Li, X., Han, H., Evangelou, N., Wichrowski, N.J., Lu, P., Xu, W., Hwang, S.J., Zhao, W., Song, C., Guo, X., Bhan, A., Kevrekidis, I.G., Tsapatsis, M., Machine learning-assisted crystal engineering of a zeolite. *Nat. Commun.*, 14, 1, 1–12, 2023.
10. Pan, E., Kwon, S., Jensen, Z., Xie, M., Gómez-Bombarelli, R., Moliner, M., Román-Leshkov, Y., Olivetti, E., ZeoSyn: A Comprehensive Zeolite Synthesis Dataset Enabling Machine-Learning Rationalization of Hydrothermal Parameters. *ACS Cent. Sci.*, 10, 3, 729–743, 2024.
11. Sivalingam, S. and Sen, S., Rice husk ash derived nanocrystalline ZSM-5 for highly efficient removal of a toxic textile dye. *J. Mater. Res. Technol.*, 9, 6, 14853–14864, 2020.
12. Sivalingam, S. and Sen, S., Efficient removal of textile dye using nanosized fly ash derived zeolite-x: Kinetics and process optimization study. *J. Taiwan Inst. Chem. Eng.*, 96, 305–314, 2019.
13. Mintova, S., Gilson, J.P., Valtchev, V., Advances in nanosized zeolites. *Nanoscale*, 5, 6693–6703, 2013.
14. Tao, H., Wu, T., Aldeghi, M., Wu, T.C., Aspuru-Guzik, A., Kumacheva, E., Nanoparticle synthesis assisted by machine learning. *Nat. Rev. Mater.*, 6, 8, 701–716, 2021.
15. Zhang, H., Goeppert, A., Olah, G.A., Prakash, G.K.S., Remarkable effect of moisture on the CO2 adsorption of nano-silica supported linear and branched polyethylenimine. *J. CO2 Util.*, 19, 91–99, 2017.
16. Wang, G., Wu, W., Zan, W., Bai, X., Wang, W., Qi, X., Kikhtyanin, O.V., Preparation of Zn-modified nano-ZSM-5 zeolite and its catalytic performance in aromatization of 1-hexene. *Trans. Nonferrous Met. Soc. China*, 25, 5, 1580–1586, 2015.
17. Xu, B., Bordiga, S., Prins, R., van Bokhoven, J.A., Effect of framework Si/Al ratio and extra-framework aluminum on the catalytic activity of Y zeolite. *Appl. Catal., A*, 333, 2, 245–253, 2007.

18. Corma, A., Rey, F., Valencia, S., Jordá, J.L., Rius, J., A zeolite with interconnected 8-, 10- and 12-ring pores and its unique catalytic selectivity. *Nat. Mater.*, 2, 7, 493–497, 2003.
19. Fan, Y., Zhang, F.-S., Zhu, J., Liu, Z., Effective utilization of waste ash from MSW and coal co-combustion power plant—Zeolite synthesis. *J. Hazard. Mater.*, 153, 1–2, 382–388, 2008.
20. Zhou, Q., Jiang, X., Qiu, Q., Zhao, Y., Long, L., Synthesis of high-quality Na P1 zeolite from municipal solid waste incineration fly ash by microwave-assisted hydrothermal method and its adsorption capacity. *Sci. Total Environ.*, 855, 158741, 2023.
21. Vogelsang, C. and Umar, M., Municipal Solid Waste Fly Ash-Derived Zeolites as Adsorbents for the Recovery of Nutrients and Heavy Metals—A Review. *Water*, 15, 21, 3817, 2023.
22. Shah, B., Tailor, R., Shah, A., Sorptive sequestration of 2-chlorophenol by zeolitic materials derived from bagasse fly ash. *J. Chem. Technol. Biotechnol.*, 86, 10, 1265–1275, 2011.
23. Shah, B.A., Pandya, D.D., Shah, H.A., Impounding of ortho-Chlorophenol by Zeolitic Materials Adapted from Bagasse Fly Ash: Four Factor Three Level Box-Behnken Design Modelling and Optimization. *Arabian J. Sci. Eng.*, 42, 1, 241–260, 2017.
24. Moises, M.P., da Silva, C.T.P., Meneguin, J.G., Girotto, E.M., Radovanovic, E., Synthesis of zeolite NaA from sugarcane bagasse ash. *Mater. Lett.*, 108, 243–246, 2013.
25. Oliveira, J.A., Cunha, F.A., Ruotolo, L.A.M., Synthesis of zeolite from sugarcane bagasse fly ash and its application as a low-cost adsorbent to remove heavy metals. *J. Cleaner Prod.*, 229, 956–963, 2019.
26. Li, C., Li, X., Zhang, Q., Li, L., Wang, S., The Alkaline Fusion-Hydrothermal Synthesis of Blast Furnace Slag-Based Zeolite (BFSZ): Effect of Crystallization Time. *Minerals*, 11, 12, 1314, 2021.
27. Koshlak, H., Synthesis of Zeolites from Coal Fly Ash Using Alkaline Fusion and Its Applications in Removing Heavy Metals. *Materials*, 16, 13, 4837, 2023.
28. Yoo, Y.S., Cheon, K.H., Lee, J.I., Kim, B.S., Shin, W.S., Seo, G.T., Zeolite Synthesis Using Sewage Sludge by Molten-Salt Method. *Mater. Sci. Forum*, 569, 329–332, 2008.
29. Choi, C.L., Park, M., Lee, D.H., Komarneni, S., Kim, Y.H., Lim, W.T., Zeolitization of mineral wastes by molten-salt method: NaOH–KNO3 system. *J. Porous Mater.*, 14, 1, 37–42, 2007.
30. Sanhueza, V., Kelm, U., Cid, R., López-Escobar, L., Synthesis of ZSM-5 from diatomite: a case of zeolite synthesis from a natural material. *J. Chem. Technol. Biotechnol.*, 79, 7, 686–690, 2004.
31. Zhang, S., Cui, M., Zhang, Y., Yu, Z., Meng, C., Synthesis of zeolite Y from diatomite and its modification by dimethylglyoxime for the removal of Ni(II) from aqueous solution. *J. Sol-Gel Sci. Technol.*, 80, 1, 215–225, 2016.

32. El Bojaddayni, I., Kuçuk, M.E., El Ouardi, Y., Jilal, I., El Barkany, S., Moradi, K., Repo, E., Laatikainen, K., Ouammou, A., A review on synthesis of zeolites from natural clay resources and waste ash: Recent approaches and progress. *Miner. Eng.*, 198, 108086, 2023.
33. Beyer, H. and Belenykaja, I., A New Method for the Dealumination of Faujasite-Type Zeolites. *Stud. Surf. Sci. Catal.*, 5, 203–210, 1980.
34. Peron, D.V., Zholobenko, V.L., de Melo, J.H.S., Capron, M., Nuns, N., de Souza, M.O., Feris, L.A., Marcilio, N.R., Ordomsky, V.V., Khodakov, A.Y., External surface phenomena in dealumination and desilication of large single crystals of ZSM-5 zeolite synthesized from a sustainable source. *Microporous Mesoporous Mater.*, 286, 57–64, 2019.
35. Ruren, X., Pang, W., Yu, J., Huo, Q., Chen, J., Chemistry of zeolites and related porous materials: synthesis and structure. *Environ. Eng. Manage.*, 7, 7, 73–75, 2008.
36. Dzwigaj, S., Peltre, M.J., Massiani, P., Davidson, A., Che, M., Dzwigaj, S., Massiani, P., Sen, T., Sivasanker, S., Incorporation of vanadium species in a dealuminated β zeolite. *Chem. Commun.*, 1, 87–88, 1998.
37. Groen, J.C., Peffer, L.A.A., Moulijn, J.A., Pérez-Ramírez, J., Mechanism of Hierarchical Porosity Development in MFI Zeolites by Desilication: The Role of Aluminium as a Pore-Directing Agent. *Chem. – Eur. J.*, 11, 17, 4983–4994, 2005.
38. Wang, L., Yao, M., Hu, X., Hu, G., Lu, J., Luo, M., Fan, M., Amine-modified ordered mesoporous silica: The effect of pore size on CO2 capture performance. *Appl. Surf. Sci.*, 324, 286–292, 2015.
39. Covarrubias, C., Gracia, F., Palza, H., Catalytic degradation of polyethylene using nanosized ZSM-2 zeolite. *Appl. Catal., A*, 384, 186–191, 2010.
40. Demir, H., Daglar, H., Gulbalkan, H.C., Aksu, G.O., Keskin, S., Recent advances in computational modeling of MOFs: From molecular simulations to machine learning. *Coord. Chem. Rev.*, 484, 215112, 2023.
41. Zhang, Z., Schott, J.A., Liu, M., Chen, H., Lu, X., Sumpter, B.G., Fu, J., Dai, S., Prediction of Carbon Dioxide Adsorption *via* Deep Learning. *Angew. Chem. Int. Ed.*, 58, 1, 259–263, 2019.
42. Zhang, X., Zhang, K., Lee, Y., Machine Learning Enabled Tailor-Made Design of Application-Specific Metal–Organic Frameworks. *ACS Appl. Mater. Interfaces*, 12, 1, 734–743, 2020.
43. Zhu, X., Tsang, D.C.W., Wang, L., Su, Z., Hou, D., Li, L., Shang, J., Machine learning exploration of the critical factors for CO2 adsorption capacity on porous carbon materials at different pressures. *J. Cleaner Prod.*, 273, 122915, 2020.
44. Namdeo, S., Srivastava, V.C., Mohanty, P., Machine learning implemented exploration of the adsorption mechanism of carbon dioxide onto porous carbons. *J. Colloid Interface Sci.*, 647, 174–187, 2023.
45. Ma, S. and Liu, Z.-P., Machine learning potential era of zeolite simulation. *Chem. Sci.*, 13, 18, 5055–5068, 2022.

46. Pophale, R., Cheeseman, P.A., Deem, M.W., A database of new zeolite-like materials. *PCCP*, 13, 27, 12407, 2011.
47. Falcioni, M. and Deem, M.W., A biased Monte Carlo scheme for zeolite structure solution. *J. Chem. Phys.*, 110, 3, 1754–1766, 1999.
48. Kim, B., Lee, S., Kim, J., Inverse design of porous materials using artificial neural networks. *Sci. Adv.*, 6, 1, 2020.
49. Schwalbe-Koda, D., Kwon, S., Paris, C., Bello-Jurado, E., Jensen, Z., Olivetti, E., Willhammar, T., Corma, A., Román-Leshkov, Y., Moliner, M., Gómez-Bombarelli, R., A priori control of zeolite phase competition and intergrowth with high-throughput simulations. *Science*, 374, 6565, 308–315, 2021.
50. Daeyaert, F., Ye, F., Deem, M.W., Machine-learning approach to the design of OSDAs for zeolite beta. *Proc. Natl. Acad. Sci.*, 116, 9, 3413–3418, 2019.
51. Jensen, Z., Kwon, S., Schwalbe-Koda, D., Paris, C., Gómez-Bombarelli, R., Román-Leshkov, Y., Corma, A., Moliner, M., Olivetti, E.A., Discovering Relationships between OSDAs and Zeolites through Data Mining and Generative Neural Networks. *ACS Cent. Sci.*, 7, 5, 858–867, 2021.
52. Ma, S., Shang, C., Wang, C.-M., Liu, Z.-P., Thermodynamic rules for zeolite formation from machine learning based global optimization. *Chem. Sci.*, 11, 37, 10113–10118, 2020.
53. Li, X., Han, H., Evangelou, N., Wichrowski, N.J., Lu, P., Xu, W., Hwang, S.-J., Zhao, W., Song, C., Guo, X., Bhan, A., Kevrekidis, I.G., Tsapatsis, M., Machine learning-assisted crystal engineering of a zeolite. *Nat. Commun.*, 14, 1, 3152, 2023.
54. Raccuglia, P., Elbert, K.C., Adler, P.D.F., Falk, C., Wenny, M.B., Mollo, A., Zeller, M., Friedler, S.A., Schrier, J., Norquist, A.J., Machine-learning-assisted materials discovery using failed experiments. *Nature*, 533, 7601, 73–76, 2016.

12

Waste-Derived Catalysts: A Way to Speed Up Environmental Remediation

Suvojit Maity[1], Soham Sarkar[1], Subhamoy Banerjee[2], Tina De[2] and Ruchira Mukherjee[2]*

[1]Department of Electronics and Communication Engineering, West Bengal Institute of Engineering Management, University of Engineering & Management, Kolkata, West Bengal, India
[2]Department of Basic Science and Humanities, Institute of Engineering Management, University of Engineering & Management, Kolkata, West Bengal, India

Abstract

Environmental remediation has become necessary due to the escalating pollution from varied sources especially industrial, agricultural and urban. In this context, waste-derived catalysts offer both a sustainable and cost-effective approach for accelerating remediation processes while at the same time addressing global waste management challenges. These catalysts are synthesized from a broad range of solid waste materials. They include agricultural biomass, plastic waste, electronic waste and industrial effluents. Their intrinsic characteristics consist of high porosity, large surface area, and diverse active sites. These features essentially make them particularly effective in various catalytic applications, including wastewater treatment, air purification and the degradation of hazardous organic and inorganic pollutants. The use of waste as raw material aligns with circular economy principles, promoting resource recovery and a sustainable environment. This chapter provides an overview of the synthesis techniques, functional properties, and environmental applications of waste-derived catalysts. It also highlights the current challenges, such as performance variability and scalability, and offers insight into future research directions. Ultimately, waste-derived catalysts represent a promising strategy to bridge the gap between effective environmental remediation and sustainable materials engineering.

**Corresponding author*: ruchira.mukherjee@iem.edu.in

Deepti Goyal, Sakshi Kabra Malpani and Renuka Jain (eds.) Catalysis from Waste: Transforming Solid Residues into Value, (321–346) © 2026 Scrivener Publishing LLC

Keywords: Waste-derived catalysts, environmental remediation, circular economy, pollution control, agricultural biomass, electronic waste, industrial effluents

12.1 Introduction

Waste-derived catalysts are green innovative materials made from industrial, agricultural, and municipal waste as shown in Figure 12.1. They provide a green chemistry alternative to traditional catalysts, minimizing waste disposal issues while maximizing resource efficiency. Their uses include energy generation, environmental remediation, chemical synthesis which are in accordance with green chemistry and the circular economy principles [1].

One of the most important categories in waste-derived catalysts is nano catalysts produced from bone waste. Due to their high content of hydroxyapatite and calcium phosphate, bone-derived nano catalysts have great potential in environmental and energy applications [2]. Bone-derived nano catalysts are applied in photocatalysis for the degradation of pollutants, water splitting for hydrogen generation, and other catalytic reactions [3]. The natural availability, low cost, and high catalytic activity of these materials render them viable options for sustainable development. This chapter discusses the synthesis, applications, and benefits of waste-derived catalysts, as well as their contribution to green chemistry and resource-efficient technologies. We have highlighted waste-derived catalysts as an important field of study in environmental cleanup. Low-cost raw materials have

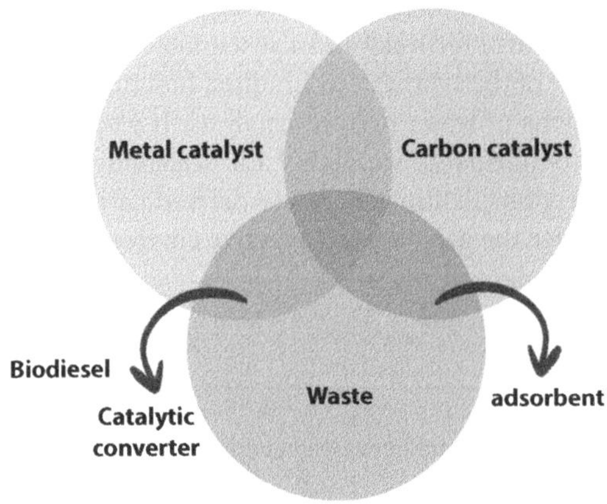

Figure 12.1 Waste-derived catalysis.

been investigated, and the environmental impact of technology has been emphasized. Synthesis of catalysts, conversion processes, adsorption and applications in advanced oxidation processes have been thoroughly investigated. Improved properties of biochar and coal fly ash (CFA) have been observed, and their potential to serve as effective decontamination agents has been suggested. The potential of sludge-derived biochar for environmental mitigation has been highlighted through an understanding of the processes, materials, and catalysts used. Sustainable production methods within the waste industry have been examined, and the techniques required for waste management have been considered. Furthermore, new waste-derived reagents have been introduced to underscore their importance in designing strategies to mitigate environmental pollution.

Although the work outlines the overall waste-derived catalyst scenario, a special focus has also been given to the topics of biochar and CFA. These catalysts are actively used in the production of green and renewable energy, which means that alongside being green and renewable themselves, they also contribute to the production of other green and renewable energies, albeit from secondary sources [1–3].

12.2 Past Explorations into the World of Waste-Derived Catalysts

Research into waste-based catalysts has led to the identification of more environmentally benign industrial processes with reduced impacts on pollution. This field of research dates back to the second half of the 20th century, when increasing public concern and the resulting government response to environmental issues provided an impetus for researchers to seek more sustainable materials and methods for catalytic applications. Research has revealed that industrial by-product wastes such as fly ash and blast furnace slag can be converted into useful materials. These by-products were shown to have properties that could make them suitable for catalytic uses, particularly in environmentally linked remediation processes [4].

12.2.1 Expansion and Diversification

The beginning of the twenty first century has typically seen a very strong expansion in the assortment of waste materials considered for catalytic applications. From agricultural residues to municipal solid wastes and even electronic wastes, all these materials have begun to find their way into the limelight as sources of catalytic materials. Methods have been developed for turning such varieties of waste streams into workable catalysts. One excellent example is that during this time, the generation of activated carbons from biomass waste made its way into catalytic applications [5].

12.2.2 Technological Advances

In the early twenty-first century, breakthroughs in material science and chemical engineering increased the drive even more. Various techniques have been developed for efficiently turning a wide variety of waste into high-performance catalysts including pyrolysis, gasification and hydrothermal processing. Furthermore, the activity, selectivity and stability of these catalysts were compared with those developed from non-renewable resources [5].

12.2.3 Current Works on Waste-Derived Catalysts

In recent years, significant progress has been made in developing waste-derived catalysts, driven by the dual goals of sustainable chemistry and effective waste management. Researchers are increasingly transforming agricultural residues, industrial by-products and electronic waste into high-performance catalytic materials. Biochar from agricultural waste, after physical or chemical activation, has shown strong potential in hydrogen production, pollutant degradation and biodiesel synthesis. Similarly, industrial slag and fly ash—wastes from the steel and coal industries, have been converted into catalysts with promising results in processes like Fischer–Tropsch synthesis and CO_2 reduction. Electronic waste, rich in precious metals like gold, silver and palladium, is processed *via* acid leaching and pyrolysis to create catalysts for organic and electrochemical reactions. These innovations align with circular economy principles, offering sustainable alternatives to conventional catalysts while minimizing environmental impact [6].

12.3 Quality of Waste-Derived Catalysts

Waste-derived catalysts possess several distinctive qualities that make them attractive for various industrial applications, emphasizing their sustainability, efficiency, and versatility.

12.3.1 Sustainability

Waste-derived catalysts are the top-notch catalysts due to their environmental sustainability. "**Waste of one is the manure of another**"-waste materials that are the by-products of industry, agricultural and municipal wastes are put to use in the form of catalysts that promote a circular economy. This principle of sustainability also matches the global aspiration to carry out the least amount of environmental damage and protect natural resources [7].

12.3.2 Cost Effectiveness

Catalysts made from waste materials typically offer economic benefits over those produced from conventional raw resources. The utilization of waste-derived catalysts, often available at minimal or no cost, can lead to a significant reduction in raw material expenditures. The effectiveness of these catalysts in saving costs is due to their supply chain optimization of production costs, while their quality remains unaffected [7].

12.3.3 Tailored Properties

Scientists can manipulate catalysts to produce specific characteristics needed for any catalytic application. Processing techniques, such as pyrolysis, hydrothermal treatment, and chemical activation, can be tuned so that maximum robustness, selectivity, and sustainability benefits can be attained [8].

12.3.4 Efficiency

The results of recent studies indicate that waste-originated catalysts can, in some cases, even outperform or match conventional catalysts. As has been shown, biodiesel synthesis and pollutant removal using waste-biomass catalysts were very efficient and effective. Moreover, these catalysts not only enable the transformation of waste matter into high-value products but also improve the reduction in the catalytic process of waste through technological advancements or increased catalytic performances [9].

12.3.5 Environmental Benefits

Compared with conventional catalyst production, waste-derived catalysis leads to a decrease in the environmental footprint. In addition to their ability to eliminate waste and reduce production costs, waste-related compounds greatly contribute to the protection of the environment and city life. This part is of the utmost importance in implementing sustainable industrial activities without severe environmental impacts from industrial processes [9].

12.4 Need for Waste-Derived Catalysts

The need for waste-derived catalysts arises from the pressing environmental and economic challenges associated with traditional catalytic processes and waste management. The development and utilization of these catalysts address several critical needs:

12.4.1 Environmental Sustainability

Traditional catalysts usually contain precious metals and other non-renewable resources whose extraction and processing produce vast amounts of environmental degradation and carbon emissions. Waste-derived catalysts use materials that would otherwise end up in landfills, causing pollution, thereby reducing the general ecological impact of waste management. It is an endeavor to reduce waste and increase resource efficiency, as prescribed by green chemistry: reduce, reuse, recycle and recover [10].

12.4.2 Waste Management

Solid waste disposal is a major issue for waste management systems. The process of converting waste into very valuable catalysts is twofold in its benefits, i.e., it minimizes the amount of waste to be disposed off and makes high-value items out of materials that are normally regarded as liabilities. This methodology not only saves landfills from being overburdened but also minimizes the environmental risks involved in waste accumulation [10].

12.4.3 Environmental Benefits

Synthesizing catalysts from wastes can be economical compared to mechanical processes. Moreover, the process of converting waste to catalysts can spawn economic opportunities such as jobs in the recycling and processing sector. The economic factor is the main benefit, true both in developed and underdeveloped countries for proper waste management and the need for resources [11].

12.4.4 Advancing the Circular Economy

Waste-derived catalysts support the concept of a circular economy, where materials are continually reused and recycled, minimizing waste and generating the need for new raw materials. By integrating waste-derived catalysts into industrial processes, industries can move towards more sustainable production models complying with the United Nations Sustainable Development Goals that reduce reliance on virgin resources and promote the reuse of existing materials [11].

12.4.5 Enhancing Catalytic Processes

Studies have proven that waste-derived solid catalysts perform equally and even more effectively than traditional catalysts in some applications. For example, catalysts made from crop residues have been used in the production of bio derived products and the elimination of toxins, both of which are selective and highly efficient. This, in turn, gives way to the development of new catalytic technologies and increases the scope of usage of these materials. The reason is that, in the first case, the catalysts are derived from a variety of agro-industrial residues and are experimentally determined to masquerade as biodiesel fuel in pollution as well as be highly selective in pollutant biosynthesis [12].

12.5 Applications of Waste-Derived Catalysts

Waste-derived catalysts, sourced from industrial and agricultural residues, have shown remarkable versatility and effectiveness across a wide range of chemical and environmental applications. These materials are used in the synthesis of fine chemicals and pharmaceuticals, for example, catalysts like biochar enhance reaction rates, yields, and selectivity in processes such as hydrogenation, oxidation, and C–C coupling. In energy production, waste-based catalysts play a key role in hydrogen generation through reforming and gasification, offering improved efficiency with lower emissions. They are also employed in converting biomass into biofuels and biochemicals, using materials like rice husk ash and sugarcane bagasse to improve yield and sustainability [12]. Environmental remediation is another major area, where catalysts like fly ash, red mud, and industrial slags aid in the degradation of pollutants in wastewater and the removal of Volatile Organic Compounds (VOCs) and oxides of Nitrogen (NOx) from industrial emissions. Additionally, waste-based catalysts are being explored in fuel cells and batteries, promoting cost-effective and sustainable energy storage.

12.6 Differences between Existing and Waste-Derived Catalysts

The difference between existing and waste-derived catalysts lies primarily in their sourcing, environmental impact and economic considerations.

12.6.1 Sourcing and Material Origin

Existing catalysts employ different materials, like precious metals, platinum and palladium, semi-precious metals, or other synthetic materials. These materials are mostly mined, refined, and processed during catalyst production. This in turn adversely impacts the environment as mining activities are energy-intensive processes. On the contrary, catalysts produced from waste often consist of mixtures of disparate waste materials which include industrial by-products, agricultural residues, and municipal solid waste. This is the case where we can sustainably utilize secondary resources, instead of destroying the environment by harsh activities such as the extraction of raw materials and waste disposal [13].

12.6.2 Environmental Impact

Refurbishing utilized catalysts with existing resources significantly solves the issues by reducing energy consumption and greenhouse gas emissions. Additionally, the disposal of used catalysts as hazardous waste may result in environmental issues caused by their chemical composition and potentially toxic properties. Waste-derived catalysts are more efficient in that they make the best possible use of resources, with lower energy consumption and by-product utilization. They also play a significant role in the reduction of greenhouse gases [12, 13].

12.6.3 Economic Considerations

Waste-derived catalysts could be a carrier of economic savings, mainly through a reduction in the material cost (waste materials are cheap or free) as well as a potential benefit from waste management savings. Further exploitation of waste-catalyzed products may become an alternative pathway of economic growth for use in the recycling and waste processing industries, which could result in better local and regional economic development.

12.6.4 Performance and Application

In contrast, the use of waste catalysts can make a noticeable difference. Studies suggest that waste-based catalysts display similar performance when compared to traditional catalysts in terms of activity, selectivity and stability, which is the case of selenium in biodiesel production and the degradation of pollutants [12]. However, the efficiency, scalability and the variability of the waste material and the process techniques used may cause the recycling of waste materials of the same efficiency and scalability. The development of necessary waste-derived catalysts will require both dedicated research and development.

12.6.5 Enhancing Catalytic Processes

Recent studies have shown that the catalysts obtained from waste are not only equal but also, in some cases, better than traditional catalysts. For example, catalysts made from agricultural residues have been effectively used in biodiesel production. A brief comparison of conventional and waste-derived catalysts has been demonstrated in Table 12.1.

12.7 Types of Waste-Derived Catalysts

Waste-derived catalysts include a variety of types; each derived from different waste materials and tailored for specific catalytic applications. These catalysts leverage the unique properties of waste materials to promote sustainable and efficient chemical processes.

Table 12.1 Comparison of conventional and waste-derived catalysts for environmental remediation.

Aspects	Waste-derived catalysts	Conventional catalysts
Source of Catalyst	Made from municipal, agricultural, and industrial waste (such as fly ash, eggshells, and biochar).	Produced using natural minerals, metals or pure chemicals.
Sustainability	Extremely environmentally friendly and sustainable (waste-to-resource approach).	Due to high resource and energy consumption, it is often less sustainable.
Cost	Because it uses waste materials, it is inexpensive or less expensive.	Comparatively costly because it requires complex synthesis pathways and high-purity precursors.
Environmental Impact	Minimize waste disposal and reduce landfill burden.	May involve mining, generation of toxic by-products, and non-biodegradable waste.
Performance and Efficiency	Similar in many situations, it may differ depending on how the material is treated and functionalized.	Usually highly effective and optimized for specific applications.
Reproducibility and Consistency	Can be variable due to heterogeneity of waste feedstock.	High reproducibility and purity; consistent catalytic properties.
Surface Area and Porosity	May need modification (e.g., activation, doping) to achieve high surface area.	Tailored surface area and porosity during synthesis.

(*Continued*)

Table 12.1 Comparison of conventional and waste-derived catalysts for environmental remediation. (*Continued*)

Aspects	Waste-derived catalysts	Conventional catalysts
Applications	Dye degradation, heavy metal removal, wastewater treatment.	CO_2 reduction, NOx/SOx removal, hydrocarbon cracking, advanced oxidation processes.
Scalability & Industrial Use	Emerging; scaling up is still under research and development.	Already established in industrial processes.
Circular Economy Contribution	Strongly supports circular economy and waste valorization.	Limited contribution unless recyclable or biodegradable.

12.7.1 Biomass-Derived Catalysts

Solid wastes that are biomass-based, such as the remaining crop straw, forest branches and others, can be transformed into electrocatalysts *via* pyrolysis, electrochemical, microwave or wet-chemical methods as shown in Figure 12.2. These catalysts are generally composed of activated carbon or biochar materials prepared in this manner and are used for the production of biodiesel, water treatment, and gas purification [13].

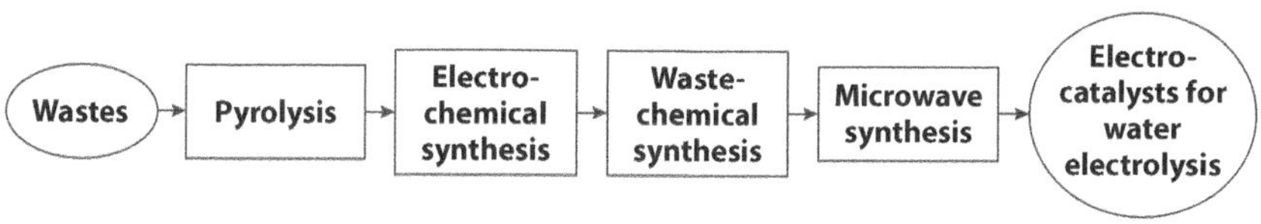

Figure 12.2 Circular economy-driven green hydrogen energy assisted by waste-derived electrocatalysts for water electrolysis.

12.7.2 Industrial Byproduct-Derived Catalysts

Residues of the industrial sector can be used as the starting material for catalysis. For example, fly ash from coal combustion, red mud from aluminum refining, and slag from steel manufacturing can be used. The metals and metal oxides associated with them can be used for catalytic applications. Studies have shown that conventional methods cannot be used here; instead, processes such as acid leaching, thermal treatment, or chemical modification are utilized.

12.7.3 Solid Waste-Derived Catalysts

Household waste materials that include plastics, food waste, and paper waste can be converted to catalysts *via* pyrolysis, gasification or chemical conversion. They can be used as catalysts for the conversion of waste into energy, for example syngas production. Another prominent usage is wastewater treatment and pollutant degradation [12].

12.7.4 Electronic Waste-Derived Catalysts

E-waste contains metals such as palladium, platinum and rare earth elements. These metals can be recycled as catalysts. E-waste-derived catalysts are quite useful since they have a high degree of reactant selectivity, for example, in hydrogenation and oxidation reactions.

There are definite benefits for e-waste-derived catalysts. These waste-derived heterogeneous catalysts serve to improve the overall energy and atom efficiency of existing and novel chemical processes. These catalysts become elements that create a circular economy by converting waste into materials of value. Furthermore, they play a vital role in sustainable development and the realization of goals [13].

12.8 Sources of Waste-Derived Catalyst

12.8.1 Biochar

Biochar is a highly carbonaceous substance resulting from the pyrolysis (thermal decomposition in a limited oxygen environment) of biomass under constrained conditions. It is defined by its high percentage of carbon and porous texture, which gives it special properties amenable to diverse environmental and agricultural uses. Biochar is usually manufactured from low-cost biomass feedstocks *via* pyrolysis, rendering it affordable compared to conventional catalysts made from precious metals or synthetic materials. The cost-effectiveness is most pronounced with large-scale industrial processes and sustainable development programs [14].

Biochar is generally made from organic biomass feedstocks like wood chips, crop residues, and animal manure. Under pyrolysis, biomass is subjected to thermal breakdown, producing biochar, bio-oil, and syngas. The process is carried out under controlled temperatures (generally between 350°C and 700°C) and in the absence or minimal presence of oxygen to avoid combustion and optimize biochar production [14]. A general synthesis procedure to produce catalysts from biochar is shown in Figure 12.3.

One of the primary peculiarities of biochar is its stability because it contains a porous carbon structure and high surface area. All these properties contribute significantly to the retention of water and nutrients, as well as carbon sequestration within the soil. This newly found porous characteristic of biochar can promote soil microorganism development and thus improve soil fertility by fostering nutrient flow and decreasing nutrient leaching [15].

Biochar has attracted considerable interest from scientists and engineers because of its potential environmental benefits. Till date, biochar has been employed in agriculture as a soil improvement agent to increase the productivity of plants. Along with this, it helps in the water retention of the soil and dampen the emission of greenhouse gases by locking carbon.

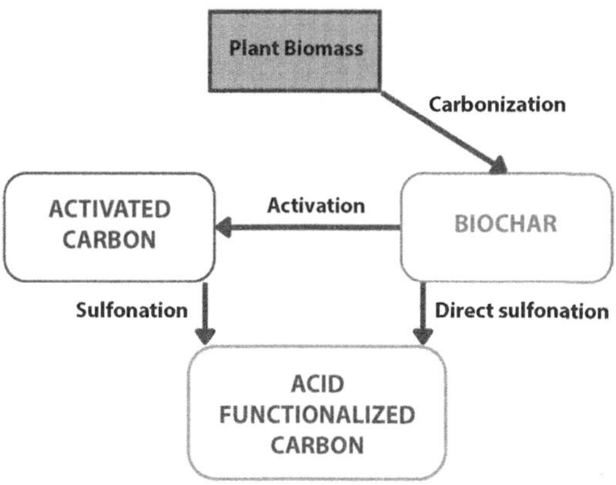

Figure 12.3 General flowchart for the synthesis of biomass-derived sulfonic acid functionalized carbonaceous materials (SAFCAMs).

On the other hand, in environmental remediation, biochar is used for soil remediation, wastewater treatment as well as a sorbent for pollutants because of its adsorption capacity and surface chemistry [16, 17].

Biochar also has catalytic properties that can be used to catalyze chemical transformations. It also provides active sites to catalyze various reactions, such as esterification, transesterification, oxidation, and hydrogenation processes. For example, the utilization of biochar-based catalysts in biodiesel production has been found to assist in the transesterification of triglycerides with alcohol to yield effective biodiesel.

Advantages

- **High surface area and porosity:** Biochar possesses a high surface area and porosity; thus, it offers active sites in adequate amounts for catalytic reactions. This characteristic improves the catalytic activity by enabling increased interactions between the reactants and the catalyst surface to enhance the reaction rates and yield [18].
- **Cost-effectiveness:** Biochar is usually manufactured from low-cost biomass feedstocks *via* pyrolysis, rendering it affordable compared to conventional catalysts made from precious metals or synthetic materials. The cost-effectiveness is most pronounced with large-scale industrial processes and sustainable development programs [19].
- **Environmental sustainability:** In contrast to catalysts from noble metals or synthetic materials, biochar is produced from renewable biomass feedstock which contributes to sustainable development [17].

Disadvantages

- **Variable Properties:** The surface chemistry and porosity of biochar may vary depending on feedstock and pyrolysis conditions and thus can change its catalytic performance; therefore, characterization and optimization of such a biochar-based catalyst must be carried out with caution in any specific application [18].
- **Limited Stability and Regeneration:** Biochar may exhibit limited stability under harsh reaction conditions or during prolonged use, affecting its catalytic longevity and performance over time. Regeneration of biochar-based catalysts

can be challenging and may require additional treatment steps to restore catalytic activity effectively [19].

- **Reaction Specificity:** The catalytic function of biochar may be specific to some types of reactions due to its surface chemistry and functional groups. Although biochar performs well in some catalytic processes, like biodiesel production or degradation of pollutants, its use in a broad spectrum of catalytic reactions might be limited relative to more general-purpose catalyst materials [17, 18].

12.8.2 Coal Fly Ash

CFA is a by-product resulting from the combustion of coal in thermal power plants. It consists of fine particles that rise with the flue gases and are then retained, typically either by electrostatic precipitators or fabric filter bag houses. Due to the large quantity of coal produced worldwide, CFA management has become a serious environmental issue. However, recent developments indicate that there are several potential applications of CFA, wherein it can now be advantageously used—i.e., the transformation of this waste into a resource.

The prime composition of CFA, however, is usually siliceous or aluminous material, although it may differ in other oxides, such as iron, calcium, magnesium, potassium, sodium, and sulfur, depending on the origin of the coal. One of the most significant uses of CFA is in construction processes, particularly as a supplementary cementitious material in concrete production. Fly ash increases concrete workability, strength, and durability while placing less demand on Portland cement—hence, reducing CO_2 emissions associated with cement production [20].

Apart from concrete, it is also used in making geopolymers—a class of synthetic aluminosilicate materials with very good mechanical properties and chemical resistance. Geopolymers derived from CFA have been tested for use in building materials, fire-resistant coatings, and hazardous waste encapsulation. The alkali activation involved in geopolymer production makes CFA a durable and environmentally benign material that can otherwise be used as a substitute for conventional materials in a sustainable manner [21].

It is also used in the manufacture of ceramics, bricks, and tiles. The fine size of the particles and pozzolanic properties of CFA render it an ideal raw material for these applications to improve the performance of products while reducing the energy consumption of the production process.

In addition, CFA can be used as a filling material in asphalt and in soil stabilization for road bases and subbases, improving their engineering properties [22].

Another vital area in which applications of CFA extends is environmental remediation. Because of its high surface area and adsorption capacity, CFA effectively protects against the heavy metals and organic pollutants present in wastewater. Modification of CFA—by chemical or physical treatment—enhances its adsorption capabilities, providing a low-cost water purification solution. Moreover, CFA can be used to reclaim mine sites and neutralize acidic soils in land restorations [23, 24].

CFA, a by-product of coal combustion, has emerged as a promising catalyst for the degradation of various environmental pollutants. Its rich composition of metal oxides—such as Fe_2O_3, Al_2O_3, SiO_2, and TiO_2—contributes to its high surface reactivity, making it suitable for catalytic applications [25, 26]. Several studies have demonstrated the effectiveness of CFA in the breakdown of organic dyes, such as methylene blue and rhodamine B, through photocatalytic and adsorption-based mechanisms. Additionally, CFA has been successfully employed in the treatment of industrial effluents, aiding in the removal of heavy metals and toxic organic compounds *via* catalytic wet oxidation and other advanced oxidation processes (AOPs) [23].

Due to its low cost, availability, and environmental relevance, fly ash continues to gain attention as a sustainable material for pollutant remediation, contributing to both waste valorization and environmental protection [24]. Modified CFA has been used for heavy metal adsorption and catalytic reduction in contaminated water [25]. Surface treatment of the CFA itself, primarily chemical, can increase its affinity for heavy metals and improve its catalytic properties, thus making it a low-cost, high-efficiency material for treating industrial wastewater [25]. CFA, a major industrial byproduct of coal combustion, has gained significant attention as a low-cost, abundant material with catalytic potential in various organic transformations. Due to its rich composition of metal oxides such as silica (SiO_2), alumina (Al_2O_3), iron oxide (Fe_2O_3), and trace alkaline earth metals, fly ash serves as an effective heterogeneous catalyst or catalyst support in multiple chemical reactions [26].

Another prominent application of CFA is in the transesterification of vegetable oils for biodiesel production. The inherent alkalinity of CFA, particularly when modified or doped with calcium oxide (CaO), enhances its basic character, making it an efficient catalyst for converting triglycerides into methyl esters [27]. Unlike conventional homogeneous alkali catalysts (e.g., NaOH, KOH), CFA-based catalysts are non-corrosive, reusable, and easily separable from the reaction mixture, offering a more sustainable and environmentally benign alternative [26, 27].

In addition, CFA has been used as a support material for metal-based catalysts, such as nickel. For example, CFA-supported nickel catalysts have demonstrated effectiveness in hydrogenation reactions, particularly in facilitating hydrogen transfer by stabilizing and dispersing nickel particles [28]. This eliminates the need to supply nickel separately in the hydrogen source, thereby reducing catalyst cost and increasing process efficiency. The surface structure and porosity of CFA aid in the even distribution of active metal sites, improving the overall catalytic performance and longevity [29, 30].

CFA as a catalyst could be utilized to gasify the biomass and derive syngas, a gas mixture of carbon monoxide and hydrogen. This catalytic ability of CFA increases the process of gasification, which in turn increases the production and quality yield of syngas. The application of CFA not only addresses disposal issues but also contributes to cleaner sources of energy [30–34].

CFA has also been utilized as a catalyst in the catalytic cracking of hydrocarbons for the production of lighter fractions, such as gasoline and olefins. The alumina and silica contents in CFA provide the acidity and stability required for the cracking process. This application of CFA as a catalyst not only aids in fly ash management but also improves the cracking process [34–37].

Zeolites are microporous aluminosilicate minerals with a high surface area, ion-exchange capacity, and potent catalytic qualities that can be made from CFA, a byproduct of burning coal [36]. Zeolites made from CFA are frequently employed in environmental remediation to remove heavy metals and other contaminants from wastewater, as well as in catalytic processes like petrochemical refining, where they act as cracking catalysts. In keeping with the ideas of the circular economy and green chemistry, this conversion not only valorizes industrial waste but also offers a sustainable substitute for zeolites that are naturally mined [37]. CFA is important for a number of green chemistry applications outside of zeolite synthesis. In hydrogenation, oxidation, and reforming reactions, it supports metal-based catalysts, and modified CFA doped with metal oxides (such as CaO or MgO) acts as a heterogeneous catalyst in acid-base reactions, transesterification, and esterification [23, 37]. Furthermore, materials based on fly ash have demonstrated promise in photocatalysis when paired with semiconductors such as TiO_2, allowing organic pollutants to degrade when exposed to light [38]. Because of their high CO_2 adsorption capacity, materials derived from CFA are also being investigated for carbon capture. Additionally, CFA is utilized in blended cement and geopolymer formulations in the construction industry, which lessens the need for conventional cement and lowers carbon emissions [38]. These various uses highlight fly ash's adaptability as a sustainable material that promotes waste reduction, pollution prevention, and environmentally friendly industrial processes [38].

Advantages

Although many potential advantages can be derived from the use of CFA, a myriad of concerns surround this by-product. There may be variability in the composition and presence of toxic elements, while various regulatory constraints may exist for their use [39]. In this respect, the comprehensive characterization and treatment of CFA have become pivotal for safe and effective applications. Continued research and development continue opening new ways and means to maximize the gainful uses of CFA with reduced environmental and health risks [39]. Some advantages and disadvantages of CFA based catalysts are summarized below:

- **Cost-effectiveness:** Being an industrial by-product, CFA is easily available and is cheaper as compared to conventional catalysts. Applying CFA curtails the use of costly raw materials, thereby reducing the overall cost of production in industrial processes [40].
- **High surface area and porosity:** CFA normally has a large surface area and porosity which is essential for good catalytic activity [23].
- **Environmental benefits:** The use of CFA as a catalyst promotes environmental sustainability through recycling waste. The material that would otherwise be disposed of in landfills causing environmental pollution. This approach aligns with the concepts of waste minimization and resource recovery, thereby creating a circular economy [23].
- **Versatility in applications:** CFA has been successfully employed as a catalyst in a wide range of applications, including the synthesis of zeolites, biodiesel production, wastewater treatment and flue gas desulfurization. Its versatility makes it a valuable resource in various industrial sectors [40].

Disadvantages

- **Variability in Composition:** The composition of CFA can vary significantly depending upon the source of coal, combustion conditions and their variability. This variation in the consistency and reliability of CFA as a catalyst poses challenges in standardizing its application in industrial processes [26].
- **Presence of Toxic Elements:** CFA can have trace levels of toxic elements, including arsenic, lead, and mercury. These elements can be detrimental to the environment and human health if not properly managed. Therefore, it is imperative that the CFA to be used as a catalyst should be treated to minimize harmful content [26].
- **Limited Activity and Selectivity:** Although CFA does exhibit catalytic activity, sometimes it does not match the efficiency of conventional catalysts in terms of both activity and selectivity. This can limit the catalytic reaction efficiency [23].
- **Regulatory and Public Perception Challenges:** There could be notable regulatory and societal issues linked with the application of CFA as a catalyst; for instance, individuals are becoming increasingly troubled over the emission of toxic elements of coal combustion residues and in general environmental pressure. Regulatory approval and public acceptability must be gained to make such CFA-based catalysts widely accepted [41].

12.9 Impedances in the Development of Waste-Derived Catalysts

- **Consistency and Quality:** Variability in waste material composition can affect the consistency and performance of catalysts [42].
- **Processing and Purification**: Transforming waste materials into effective catalysts requires advanced processing techniques, which can be technically challenging and costly.
- **Regulatory and Safety Issues:** Ensuring that waste-derived catalysts meet safety and regulatory standards is crucial for their widespread adoption [43].

12.10 Future Development and Aspects of Waste-Derived Catalysts

Future developments in waste-derived catalysts are expected to change the rules of sustainable chemistry and environmental management [44–46]. The greatest challenges for the future are improving the performance of the catalyst through material innovation and processing optimization. An exciting aspect of this line of inquiry is the incorporation of nanotechnology for the optimization of surface features and the catalytic performance of materials derived from wastes. The nanoscale structures of waste-derived materials increase the performance, surface area and maximize the interaction of catalysts and reactants [47, 48].

Furthermore, a new frontier in waste-based catalysts is being integrated with renewable sources of energy: solar and wind energy. Renewable-based hybrid systems integrate renewable energy and catalytic functionality to drive a reduced environmental footprint in catalytic operation. A recent trend in development is the focus on multipurpose catalysts that can concurrently execute more than one reaction, streamline industrial processes and reduce the need for multiple catalyst types. The scalability and economic viability of the production of waste-derived catalysts are also considered as critical factors. Advances in processing techniques, such as microwave-assisted synthesis and supercritical fluid extraction, are currently under investigation for efficiency and cost-effectiveness [49–54].

Another important step would be to perform a Life Cycle Assessment (LCA) of the catalysts, thereby ensuring that the overall process is sustainable. LCA studies also enable an understanding of the impact of waste in production through disposal, and they will guide the development of truly green catalysts [55, 56]. Further studies in this area will involve diversification in the origin of wastes for use in catalyst preparation processes, involving increased kinds of industrial and household wastes. Diversity enhances the application and effective management of various types of wastes. Waste-derived catalysts for the future involve interdisciplinary research, bridging material science, environmental engineering, and industrial chemistry to develop new solutions for sustainable development [57–61].

12.11 Conclusion

In conclusion, this chapter highlights the potential of catalysts obtained from waste for environmental remediation and catalysis. The raw materials used here are low-cost ones, which provide these catalysts with properties that make them alternatives in decontamination processes. The synthesis and conversion processes discussed here reflect the versatility and effectiveness of the materials in various chemical processes. Specifically, sludge-derived biochar has been very promising for mitigating environmental pollution because of its effective adsorption capabilities and catalytic properties. Exploration of sustainable production mechanisms for the waste industry further emphasizes and underlines the importance of environmentally friendly waste management practices. This research explains the present state of waste-derived catalyst technology but introduces new reagents that, based on their importance, identify critical strategies for curbing environmental pollution. In general, the findings call for a wider acceptance of catalysts derived from waste in providing support for a circular economy and a sustainable future for mankind.

AI disclosure statement: The author acknowledges the use of AI-based tools for minor language editing or grammatical correction. However, all intellectual contributions, analyses, and conclusions presented in this chapter are entirely the author's own.

References

1. Liu, Y., Luo, S., Sun, X., Waste-derived catalysts for biodiesel production: A review. *Renewable Sustainable Energy Rev.*, 115, 109366, 2019, https://doi.org/10.1016/j.rser.109366.
2. Gopi, D., Ramya, S., Rajeswari, D., Kavitha, L., Govindaraju, K., Bone-derived hydroxyapatite nanostructures for environmental and energy applications: A review. *Ceram. Int.*, *46*, 5, 5395–5406, 2020, https://doi.org/10.1016/j.ceramint.2019.10.154.
3. Sheikh, R.A., Mazumder, M.A.J., Begum, S., Bone-derived catalysts: Sustainable applications in photocatalysis and hydrogen production. *J. Environ. Chem. Eng.*, 9, 6, 106744, 2021, https://doi.org/10.1016/j.jece.2021.106744.
4. Chen, Y., He, Z., Zhang, L., Waste-derived catalysts for the removal of organic pollutants from wastewater: A review. *J. Hazard. Mater.*, 412, 125256, 2021, https://doi.org/10.1016/j.jhazmat.2021.125256.
5. Guo, Y., Hu, S., Li, J., Circular economy and waste-derived catalysts: A sustainable approach. *Green Chem.*, 22, 15, 4942–4953, 2020.
6. Santos, E., Sousa, S., Abreu, M.H., Mendes, A., Bone-derived hydroxyapatite: A highly efficient catalyst for the photo-oxidation of pollutants. *J. Hazard. Mater.*, 404, 124064, 2021.

7. Prathap, M.A., Suresh, M., Kumar, D., Hydroxyapatite derived from bone waste for catalytic degradation of organic pollutants. *Mater. Sci. Eng., B*, 227, 84–92, 2018.
8. Rashad, M., Yang, S., He, P., Green synthesis of nanocatalysts from bone waste for hydrogen production. *Int. J. Hydrogen Energy*, 44, 12, 5894–5902, 2019.
9. Lam, M.K., Lee, K.T., Mohamed, A.R., Homogeneous, heterogeneous and enzymatic catalysis for transesterification of high free fatty acid oil (waste cooking oil) to biodiesel: A review. *Biotechnol. Adv.*, 28, 4, 500–518, 2010.
10. Gui, M.M., Lee, K.T., Bhatia, S., Feasibility of edible oil vs. nonedible oil vs. waste edible oil as biodiesel feedstock. *Energy*, 33, 11, 1646–1653, 2008.
11. Zhang, Y., Dubé, M.A., McLean, D.D., Kates, M., Biodiesel production from waste cooking oil: Process design and technological assessment. *Bioresour. Technol.*, 89, 1, 1–16, 2003.
12. Sharma, Y.C., Singh, B., Upadhyay, S.N., Advancements in development and characterization biodiesel: A review. *Fuel*, 90, 1, 2355–2373, 2011.
13. Demirbas, A., Progress and recent trends in biodiesel fuels. *Energy Convers. Manage.*, 50, 1, 14–34, 2009.
14. Atadashi, I.M., Aroua, M.K., Aziz, A.A., High quality biodiesel and its diesel engine application: review. *Renewable Sustainable Energy Rev.*, 14, 7, 1999–2008, 2010.
15. Tiwari, J.N., Sultan, S., Lee, W.G., Yoon, T., Critical review of electronic waste (e-waste) recycling using biological agents. *Environ. Res.*, 185, 109453, 2020.
16. Kwapinski, J.W., Rodríguez-Valencia, A.B., Hayes, M., Leahy, J., Hayes, M.H.B., Physical structure and surface area characterization of biochars, in: *Biochar for Environmental Management: Science and Technology*, J. Lehmann and S. Joseph (Eds.), pp. 63–89, Routledge, Abingdon, UK, 2009, ISBN: 978-0-203-76226-4 (ebook).
17. Lehmann, J. and Joseph, S. (Eds.), *Biochar for environmental management: Science and technology*, pp. 125–138, Earthscan, London, 2009, ISBN: 9781844076581.
18. Lehmann, J., *et al.*, Biochar effects on soil biota - A review. *Soil Biol. Biochem.*, 43, 9, 1812–1836, 2011.
19. Jeffery, S., *et al.*, Biochar boosts tropical but not temperate crop yields. *Environ. Res. Lett.*, 12, 5, 053001, 2017, https://doi.org/10.1088/1748-9326/aa67bd.
20. Mohan, D., *et al.*, Organic and inorganic contaminants removal from water with biochar, a renewable, low cost and sustainable adsorbent – A critical review. *Bioresour. Technol.*, 160, 191–202, 2014.
21. Zhao, L., *et al.*, Catalytic properties and applications of biochar: A review. *Energy Procedia*, 142, 92–97, 20172017.
22. Chen, Y., *et al.*, Biochar-based catalysts for biodiesel production: A review. *Renewable Energy*, 125, 775–785, 2018.

23. Ahmaruzzaman, M., A review on the utilization of fly ash. *Prog. Energy Combust. Sci.*, 36, 3, 327–363, 2010, https://doi.org/10.1016/j.pecs.2009.11.003.
24. Yadav, V.K. and Fulekar, M.H., Green synthesis and characterization of fly ash-supported nanocomposites and their application for the removal of heavy metals from wastewater: A review. *Environ. Sci. Pollut. Res.*, 26, 35, 35401–35417, 2019, https://doi.org/10.1007/s11356-019-06544-w.
25. Van Riessen, A., Chen-Tan, N., Rees, C.A., Geopolymers: Structures, processing, properties and industrial applications, in: *Advances in Ceramic Matrix Composites*, I.M. Low (Ed.), pp. 230–250, Woodhead Publishing, Cambridge, UK, 2013, https://doi.org/10.1533/9780857097334.2.230.
26. Mishra, D.P., Patel, R.K., Mishra, S., Fly ash as a catalyst and catalyst support for industrial applications: A review. *Environ. Chem. Lett.*, 19, 2351–2367, 2021, https://doi.org/10.1007/s10311-020-01145-z.
27. Sharma, Y.C., Singh, B., Korstad, J., Application of an efficient CaO impregnated fly ash catalyst for biodiesel synthesis from the transesterification of soybean oil. *Energy Fuels*, 25, 7, 3370–3376, 2021, https://doi.org/10.1021/ef200543e.
28. Wu, C., Wang, Z., Williams, P.T., Huang, J., Pyrolysis of waste plastics using Ni/coal fly ash catalysts to produce gases and carbon nanotubes. *Waste Manage. (Oxford)*, 33, 3, 650–656, 2013, https://doi.org/10.1016/j.wasman.2012.09.012.
29. Zhou, H., Jin, Y., Yang, J., Xu, Q., Coal fly ash-supported Ni catalysts for hydrogenation reactions: Enhanced dispersion and catalytic performance. *Appl. Catal., B*, 180, 483–492, 2016.
30. Yao, Z.T., Xia, M.S., Sarker, P.K., Chen, T., A review of the utilization of fly ash worldwide in producing geopolymer-based products and other applications. *J. Hazard. Mater.*, 286, 34–52, 2015.
31. Wang, S. and Wu, H., Environmental-benign utilization of fly ash as low-cost adsorbents. *J. Hazard. Mater.*, 136, 3, 482–50, 2006, https://doi.org/10.1016/j.jhazmat.2006.01.064.
32. Gupta, V.K. and Kumar, R., Fly ash-supported catalysts for environmental applications. *Environ. Sci. Pollut. Res.*, 26, 3, 2065–2076, 2019.
33. Wang, S., Boyjoo, Y., Choueib, A., A comparative study of dye removal using fly ash treated by different methods. *Chemosphere*, 60, 10, 1401–1407, 2008.
34. Bhagiyalakshmi, M., Yun, L.M., Anuradha, R., Synthesis of nickel fly ash catalysts and their application in hydrogenation reactions. *Chem. Eng. J.*, 160, 2, 673–68, 2010.
35. Gohain, M., Bora, A.J., Das, P., Utilization of waste fly ash as a heterogeneous catalyst in biodiesel production. *Renewable Energy*, 132, 68–79, 2019.
36. Zhao, B., Zhang, Q., Liu, X., Catalytic gasification of biomass using fly ash for syngas production. *Energy Convers. Manage.*, 124, 252–259, 2016.

37. Ghorpade, R.P., Parale, V.G., Gaikwad, S.P., Catalytic cracking of hydrocarbons using fly ash-based catalysts. *Fuel Process. Technol.*, 124, 52–61, 20142014.
38. Querol, X., Moreno, N., Umaña, J.C., Synthesis of zeolites from coal fly ash: An overview. *Int. J. Coal Geol.*, 50, 1-4, 413–423, 2002.
39. Matos, A.M., Veloso, C.M., Ribeiro, M.J., Zeolites from coal fly ash for adsorption and catalysis: A review. *Waste Manage. (Oxford)*, 95, 513–528, 2019.
40. Kumar, S., Pandey, A.K., Shukla, S.K., Photocatalytic degradation of organic pollutants using flyash–TiO_2 composite under UV irradiation. *J. Environ. Chem. Eng.*, 7, 2, 102925, 20192019.
41. Lima, E.C., Hosseini-Bandegharaei, A., Moreno-Piraján, J.C., Application of biochar for the removal of pollutants from aqueous solutions. *Chemosphere*, 224, 261–286, 2019.
42. Blissett, R.S. and Rowson, N.A., A review of the multicomponent utilization of coal fly ash. *Fuel*, 97, 123, 2012.
43. Gupta, M., Agarwal, M., Rathi, A., Utilization of waste materials in the production of biodiesel. *Renewable Sustainable Energy Rev.*, 15, 5, 1779–1787, 2011.
44. Hossain, A.B.M.S., Salleh, A., Boyce, A.N., Chowdhury, P., Naqiuddin, M., Biodiesel fuel production from algae as renewable energy. *Am. J. Biochem. Biotechnol.*, 4, 3, 250–254, 2011.
45. Izquierdo, M. and Querol, X., Leaching behaviour of elements from coal combustion fly ash: An overview. *Int. J. Coal Geol.*, 94, 54–66, 2012.
46. Li, H., Guo, S., Dai, H., Catalytic gasification of biomass using waste-derived catalysts for hydrogen production. *Renewable Energy*, 85, 30–36, 2016.
47. Zhu, X., Liu, S., Liu, S., Recent advances in the applications of biochar-based catalysts: A review. *Chem. Eng. J.*, 375, 121929, 2019.
48. Kiddee, P., Naidu, R., Wong, M.H., Electronic waste management approaches: An overview. *Waste Manage. (Oxford)*, 33, 5, 1237–1250, 2013.
49. El-Fadel, M., Findikakis, A.N., Leckie, J.O., Environmental impacts of solid waste landfilling. *J. Environ. Manage.*, 50, 1, 1–25, 2019.
50. Chen, W.-Q., Graedel, T.E., Mulligan, S.A., Sustainable recycling of rare earths. *Nat. Mater.*, 17, 5, 394–395, 2018.
51. Geyer, R., Jambeck, J.R., Law, K.L., Production, use, and fate of all plastics ever made. *Sci. Adv.*, 3, 7, 1700782, 2017.
52. Ragaert, K., Delva, L., Van Geem, K., Mechanical and chemical recycling of solid plastic waste. *Waste Manage. (Oxford)*, 69, 24–58, 2017.
53. Muller, C., Anaerobic digestion of biodegradable solid waste in low- and middle-income countries. *J. Waste Manage.*, 27, 1, 116–118, 2007.
54. Baby, R., Hussein, M.Z., Abdullah, A.H., Zainal, Z., Nanomaterials for the treatment of heavy metal contaminated water. *Polymers (Basel)*, 14, 3, 583, 2022, doi: 10.3390/polym14030583.

55. Zhao, Y., Qiu, F., Yang, D., Renewable energy-driven advanced oxidation processes (AOPs) for environmental remediation: A review of recent developments. *Chem. Eng. J.*, 388, 124275, 2020.
56. Singh, R., Shukla, A., Tiwari, S., Srivastava, M., Sharma, Y.C., Recent progress in microwave assisted synthesis of heterogeneous catalysts for biodiesel production. *Renewable Sustainable Energy Rev.*, 119, 109560, 2020.
57. Sharma, G., Van Dael, M., Van Passel, S., You, F., How to assess the potential of emerging green technologies? Towards a prospective environmental and techno-economic assessment framework. *Green Chem.*, 21, 3, 486–513, 2019, https://doi.org/10.1039/C8GC03066E.
58. Zhang, Z., Wang, J., Chen, Y., Zhao, Y., Waste-derived catalysts for sustainable technologies: Bridging materials science and environmental engineering. *Renewable Sustainable Energy Rev.*, 161, 112393, 2020.
59. Pires, A., Martinho, G., Goncalves, D., Sustainable solid waste management in the 21st century: A perspective for the Mediterranean region. *Waste Manage. Res.*, 29, 10, 1074–1080, 2011.
60. Rautela, R., Sharma, A., Prakash Ranjan, V., Rathika, K., Pratap, V., Ram Yadav, Ram Yadav, B., Kumar, S., Waste bone derived catalysts for biofuel production: A review. *Environ. Res.*, 200, 111767, 20212021.
61. Tahir, M., Pan, L., Idrees, F., Zhang, X., Wang, L., Zou, J.-J., Wang, Z.L., Turning Solid Waste into Catalysts: A Path for Environmental Solutions. *ChemPlusChem*, 89, 12, 202300293, 2024, https://doi.org/10.1002/cplu.202300293.

13

Waste-Derived Catalysts Paving the Way to a Circular Economy: Advancing Sustainable Solutions

Amita Somya[1]*, Anish Khan[2] and Amit Prakash Varshney[3]

[1]Department of Chemistry, Amity School of Applied Sciences, Amity University, Bengaluru, Karnataka, India
[2]Center of Excellence for Advanced Materials Research, King Abdulaziz University, Jeddah, Saudi Arabia
[3]Bengaluru, Karnataka, India

Abstract
The process of wastewater treatment is crucial to preserving both human health and the environment. Conventional wastewater treatment techniques frequently use costly and energy-intensive procedures, which raise operating costs and have an adverse effect on the environment. Utilizing catalysts made from waste is a viable substitute that tackles these issues and is consistent with sustainable development ideal. It is a feasible way to minimize waste output, reduce resource consumption, and lessen environmental effects in the context of a fast-changing industrial landscape and growing environmental concerns. The current chapter explores the fundamental ideas behind the invention and production of waste-derived catalysts, emphasizing their distinct benefits over traditional catalysts, such as affordability, energy efficiency, and adaptability. The chapter highlights the potential of waste-derived catalysts to propel the shift to a circular economy by elucidating their varied uses across a range of chemical research sectors through an extensive assessment of the latest research works.

***Keywords*:** Waste, waste-derived catalysts, wastewater treatment, sustainable development, environment, SDG-6, circular economy

**Corresponding author*: somya.amita@gmail.com

Deepti Goyal, Sakshi Kabra Malpani and Renuka Jain (eds.) Catalysis from Waste: Transforming Solid Residues into Value, (347–368) © 2026 Scrivener Publishing LLC

13.1 Introduction

Allowing easy access to pure water resources is one of the many challenges that society faces if we want to sustain growth. The ongoing problem of water pollution resulting from fast industrial and urban expansion is a serious concern for the health of the ecosystem and humanity [1]. There are certain pollutants like heavy metals, microorganisms, microplastics, and antibiotics that are life threatening; thus, it is essential to remove them from water sources immediately. This problem has emerged as a focus for many researchers, and due to this, many advanced and efficient methods for treating wastewater have been developed.

Some of the most effective methods for treating wastewater include electrochemical treatment, adsorption, photocatalytic degradation, advanced oxidation processes (AOPs), and ion exchange [2]. Each of these methods functions through different mechanisms and processes. For example, electrochemical treatment uses electric currents to drive redox reactions that break down pollutants, while adsorption involves the binding of contaminants to the surface of adsorbent materials. Photocatalytic degradation utilizes light-activated catalysts to break down harmful substances [3], and AOPs harness highly reactive species to oxidize and eliminate pollutants. In contrast, ion exchange replaces unwanted ions in the water with more benign ions from a solid resin, successfully removing contaminants such as heavy metals [4], and nitrates.

All these techniques, despite their variations, rely on functional micro- and nanomaterials to accomplish effective pollutant removal and degradation. These resources are crucial for improving the efficiency and efficacy of treatment procedures. Therefore, the advancement of water remediation technologies depends on the development of sophisticated materials such as metal oxides, metal sulfides, and carbon-based compounds.

For instance, carbon compounds are effective adsorbents because of their large porosity and surface area. AOPs and photocatalysis both make extensive use of metal oxides due to their diverse catalytic capabilities. Ion exchange resins are highly effective at eliminating charged pollutants because they are designed to selectively exchange particular ions. They are frequently made of synthetic polymers.

As a result, maintaining clean water is crucial for sustainable growth, and removing dangerous pollutants from water is a top priority. To effectively remediate wastewater and protect the environment and public health, new and effective wastewater treatment techniques must be developed, backed by the creation and use of cutting-edge micro- and nanomaterials.

One of the most actively researched areas today is the efficient conversion of waste materials into functional substances. The paradigm of going from waste to value is very important to both the economy and the environment because it encourages using waste as a raw material and turning it into valuable products, which in turn promotes sustainability and resource efficiency. A growing number of carbon and transition metal containing compounds, which are made from waste materials available in the earth's crust, have recently shown great promise for water purification.

In this paradigm, practical methods like ball milling, pyrolysis, and precipitation are used to transform troublesome solid waste from municipalities into substances that are beneficial for water conservation [5]. In order to process a wide variety of waste materials and turn them into useful materials for wastewater treatment, these techniques are essential.

Numerous biomass wastes have been effectively converted into beneficial adsorbents and catalysts, such as animal feces [6], lotus seedpod [7], tea waste [8], and soy waste [9]. Furthermore, electronic waste like spent batteries [10], electrical waste from packaging [11], and waste liquid crystal displays [12], as well as commercial waste like red mud [13] and sewage sludge [14] have been transformed into useful products. In this context, waste from construction, such as brick waste, may also find use. The many waste kinds that could be transformed into useful materials for use in wastewater treatment are shown in Figure 13.1. The suitability of these materials for various water treatment applications is influenced by their differences in dimensions, densities, chemical compositions, morphologies, and moisture content [15].

One essential component of wastewater purification is the conversion of these various waste materials into useful substances like adsorbents, photocatalysts, and electrocatalysts. The numerous cutting-edge water remediation methods previously covered [3, 4], such as ion exchange, adsorption, electrochemical treatment, AOPs, and photocatalytic degradation, depend on these functional materials. For instance, biochar made from soy waste has demonstrated excellent efficacy in removing heavy metals from tainted water. Likewise, when exposed to light, photocatalysts made from red mud could indeed efficiently break down organic contaminants. These examples underscore the potential of waste-derived materials in addressing the pressing issue of water pollution.

Hence, the conversion of waste materials into valuable products for water purification not only mitigates environmental pollution but also contributes to the sustainable management of resources. The ongoing development and optimization of these materials are vital for advancing water treatment technologies and promoting a circular economy.

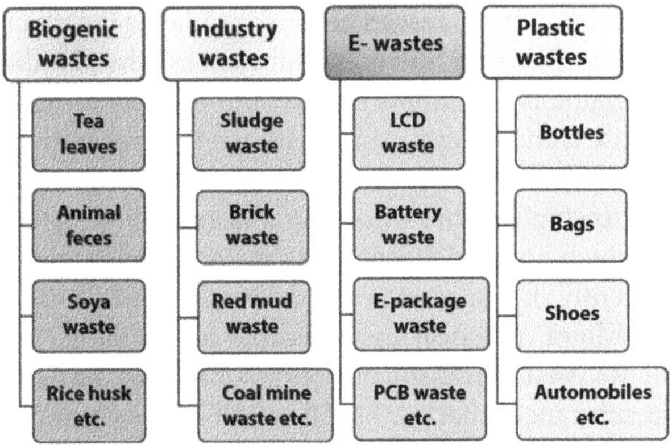

Figure 13.1 Variety of wastes, converted to functional materials for wastewater treatment applications.

For instance, Liu *et al.* [16], derived biochar from the agricultural waste through a pyrolytic process, which demonstrated a higher adsorption capacity i.e., 49.70 mg/g for Pb^{2+}. The "waste-to-value" principle holds significant environmental and economic importance as it greatly reduces the environmental impact of solid/liquid waste and lowers the production costs of functional materials such as adsorbents and catalysts. Beyond adsorption, waste-derived functional materials are extensively used in photocatalysis, AOPs, and electrochemical treatment for water purification [17].

Improving the performance of waste-derived materials significantly relies on the careful design of their nanostructures and components. This involves controlling the precursors and experimental conditions during synthesis. Consequently, it is vital to develop effective strategies to convert solid waste into functional materials with desirable properties, such as high surface area and stability. Numerous cutting-edge processes, including ball milling, pyrolysis, combustion, and wet-chemical procedures, are efficient in turning various kinds of solid waste into useful products. The properties of the waste constituents have a major role in choosing a synthesis technique. For instance, pyrolysis is frequently used to create carbon compounds [18], from biomass waste and sewage wastes, whereas combustion and sol-gel techniques are usually used to create metal oxides [19]. It is encouraging to note that the use of these methods has been successful in turning waste into high-performance materials, which has advanced water treatment technology and helped manage resources sustainably.

While some researchers have focused on the environmental uses of biogenic waste-derived biochar in AOPs and adsorption in recent years, there is still a big lack of systematic summaries of the wider use of different waste-derived materials, like biomass, industrial waste, and electronic waste, in adsorption, electrochemical treatments, photocatalytic breakdown, and AOPs for wastewater remediation. This chapter fills this knowledge gap by giving a summary of recent developments in the conversion of various waste into cutting-edge materials appropriate for wastewater treatment. It discusses efficient strategies for synthesis, explores recent applications in key treatment processes, and emphasizes the design optimization of waste-derived materials to enhance their effectiveness and sustainability in water purification. Future perspectives on the evolution of next-generation waste-derived materials are also proposed to guide further research and innovation in this critical area.

13.2 Strategies for Transforming Waste into Functional Materials

Waste is typically not directly utilized for wastewater purification; instead, they require adequate procedures to be transformed into expected functional materials. Recently, a wide range of biomass, electronic waste, and industrial waste—varying in size, shape, densities, chemical compositions, and moisture content—have been successfully converted into functional materials for wastewater treatment. This transformation is achieved through methods such as hydrothermal synthesis, pyrolysis, calcination, sol–gel processes, co-precipitation, and ball milling [15], which are explained in the next section. Furthermore, the chapter also analyzes a few experimental parameters which significantly affect properties and effectiveness of waste-derived materials in wastewater treatment.

13.2.1 Thermal Decomposition Techniques

Thermal decomposition techniques are essential processes that use heat to transform waste materials into simpler, functional components for various applications, including wastewater purification. Diverse organic and inorganic waste can be efficiently transformed into useful materials with improved qualities using these techniques. Pyrolysis, combustion, calcination, and hydrothermal carbonization (HTC) are some of the main methods of thermal decomposition.

- **Pyrolysis**

Pyrolysis is the process of heating organic materials to high temperatures, usually between 300 and 900°C, without the presence of oxygen. Through the breakdown of the waste's intricate molecular structure, biochar, liquid oils, and syngas—a blend of hydrogen (H_2), carbon monoxide (CO), and few other gases are produced. Because of its high specific surface area and high porosity, which makes it an efficient adsorbent for pollutants, biochar is especially useful in wastewater treatment [15]. Pyrolysis also produces valuable byproducts and lowers the amount of waste.

- **Calcination**

Calcination is a technique that produces thermal breakdown by roasting inorganic substances to extremely high temperatures, often above 600°C, with little to no oxygen present. Oxides and gases like CO_2 are usually produced as a result of this procedure. For example, lime (CaO) is created when limestone ($CaCO_3$) is calcined. The resultant calcined materials, like metal oxides, are frequently employed as adsorbents or catalysts in water purification procedures because of their great thermal stability and reactivity [15].

- **Combustion**

In contrast to pyrolysis and calcination, combustion is the process of igniting organic matter with the aid of excess oxygen, producing ash, heat, water vapors, and carbon dioxide. While combustion ash is not commonly used directly to produce adsorbents, it could be processed more thoroughly to extract valuable materials, and the heat produced during combustion can be used for energy recovery, which contributes to sustainability for all [15].

- **Hydrothermal Carbonization**

Wet biomass is treated by HTC at high water pressures and moderate temperatures (from 180 to 250°C). Through this process, the biomass is transformed into a liquid phase with different dissolved organic components and hydrochar, a solid rich in carbon. HTC is a useful waste conversion method because of its large carbon content and advantageous structural characteristics, which makes hydrochar an efficient adsorbent for water contaminants.

A variety of studies have been documented that integrate different techniques to create effective functional materials from assorted waste. For instance, fallen leaves [20], from the *Platanus orientalis* tree were transformed into activated carbon microspheres *via* hydrothermal synthesis followed by calcination. In a similar fashion, red mud was processed into porous γ-Al_2O_3 microspheres [21], using a combined hydrothermal and thermal method. Furthermore, wastewater sludge was converted into mesoporous biochar [22], through a process involving coagulation, hydrothermal reaction, and pyrolysis. These combined techniques effectively transform various waste into valuable materials.

13.2.2 Sol-Gel Technique

A chemical procedure called the sol-gel method is used to create materials, usually glass and ceramics from tiny molecules. This method uses hydrolysis and polymerization reactions to change a sol (a suspension that consists of particles), into a gel, which is a network of interconnecting particles. First, a solution is created by dissolving metal salts or alkoxides in a solvent. The solution becomes a sol by regulated hydrolysis and condensation, and then it develops into a network that resembles gel. To create a dense, solid substance, this gel can thereafter be dried and heated to eliminate any leftover solvent and organic materials. The capacity of the sol-gel process to yield materials with great homogeneity and purity at comparatively low temperatures makes it noteworthy. It makes it possible to precisely regulate the composition, structure, and characteristics of the material, which makes it popular for usage in coatings, fibers, catalysts, and porous water purification systems. Nowadays, a variety of waste are transformed into liquid precursors by the sol-gel technique [15], which produces functional materials. For instance, a mixed oxide with strong photocatalytic degradation capability for textile dye was created from the leaching solution of used Ni-Cd batteries [23].

13.2.3 Complex Co-Precipitation Technique

Co-precipitation is the process of precipitating numerous substances from a solution at the same time to create new materials. To create a uniform solution, metal salts must be dissolved in a solvent. The metal ions are then turned into insoluble compounds that precipitate out of the solution collectively by changing the pH or introducing a precipitating agent. At the molecular level, the different components are evenly distributed in the precipitate. The precipitate can then be further treated, usually by calcination, to create the desired end material after filtering, washing, and drying. Because co-precipitation allows exact control over composition and homogeneity, it is frequently utilized to create catalysts, as well as magnetic materials, and other sophisticated materials.

13.2.4 Enhanced Ball Milling Technique

A high-energy ball mill is used in enhanced ball milling which is a sophisticated mechanical alloying method. It refines particles and causes structural alterations in materials. This process involves placing materials into a rotating drum with heavy balls, which exert severe plastic deformation on the particles through repeated collisions. The high kinetic energy input promotes alloying, microstructural refinement, and the formation of metastable phases. Widely used for synthesizing nanomaterials and modifying bulk materials, enhanced ball milling offers precise control over particle size distribution and material properties, making it crucial in various fields, including electronics, catalysis, and composite production. Ball milling is extensively used to prepare functional materials from solid waste, particularly biochar-based materials [5], by reducing particle size and increasing surface area to enhance reactions with pollutants.

13.3 Applications of Waste-Derived Catalysts in Wastewater Treatment

Adsorption is an economically viable method for removing pollutant species from wastewater. The efficiency of the process largely depends on the adsorbent's surface area and functional groups. Waste-derived economical materials are well-known for their versatile physical and chemical properties which have garnered significant attention for adsorbing metal/anion ions, dyes, and antibiotics. The following section gives details of the applications of various waste-derived catalysts in removing different types of pollutants from wastewater through different processes.

13.3.1 Degradation of Organic Pollutants

The extensive use of dyes, herbicides, plastics, pesticides, surfactants, fungicides, and pharmaceuticals in various industries poses threats to aquatic environments and human health [1, 24]. Waste-derived catalysts are particularly effective in treating organic pollutants in wastewater through adsorption. These catalysts, often created from agricultural residues, industrial by-products, or electronic waste, possess high surface areas and abundant functional groups that enhance their adsorption capabilities. For instance, activated carbon produced from coconut shells or sawdust can efficiently adsorb dyes, phenols, and other organic contaminants, and remove them from wastewater. The large surface area allows for more interaction sites with pollutants, while functional groups on the catalyst surface can form bonds with organic molecules, trapping them effectively. This method is not only cost-effective but also sustainable, as it repurposes waste materials into valuable resources for environmental remediation. Carbon-based materials are extensively researched because of their large surface area and versatile surface chemistry. Bahamon and associates [25], used molecular simulations to show that activated carbon has a variety of adsorption qualities, emphasizing its potential for successfully extracting pharmaceuticals from wastewater depending on their atom affinities and molecular structures. Given the diverse structures of organic contaminants, waste-derived carbon materials have been modified structurally and functionally. The texture and surface chemistry of adsorbents are critical in the determination of their efficiency to adsorb organic pollutants, with porous structures and high surface areas generally proven more effective. In contrast, the influence of surface functional groups on adsorption varies, depending on the specific case. For the removal of organic pollutants, including antibiotics like amoxicillin and metronidazole, Alivand *et al.* [26], showed that adding melamine to create mesopores and immobilizing $-NH_2$ groups greatly increases the adsorption capacity of nanoporous carbons made from asphaltene residue. This improvement is mainly due to improved charge-transfer effects and electrostatic interactions.

The addition of functional groups and active sites to carbon-based composites containing metal oxides, zeolites, clay minerals, silica, and nano zero-valent iron (nZVI) results in improved adsorption capabilities [27, 28]. Fe-N co-modified biochar (Fe-N-RSBC) was created by Mei *et al.* [29], through pyrolysis, which combines magnetic components (Fe_3O_4, γ-Fe_2O_3, and Fe_3C), functional groups, and a graphitized carbon structure. This sophisticated material used a variety of processes, such as surface complexation, electrostatic interactions, hydrogen bonding, pore filling, and π-π interactions, to produce a high adsorption capacity for tetracycline (TC) (156 mg/g). The potential of these composites to effectively remove organic contaminants from aqueous environments is demonstrated. Layered clay minerals are noted for their large surface areas and interaction potential with pollutants *via* exchangeable cations within mineral layers. Basaleh *et al.*, synthesized a steel slag-acrylamide acrylic acid (SSAA) copolymer that effectively removes cationic and anionic dyes like methylene blue (MB), methyl orange (MO) with maximum efficiencies i.e., 94% and 97%, respectively, attributing physical adsorption for MB and chemisorption for MO [30]. These low-cost inorganic/organic composites offer high performance and are pivotal in developing efficient multifunctional adsorbents.

13.3.2 Degradation of Inorganic Pollutants

Inorganic ions present in wastewater are increasing concerns owing to their toxicity and cumulative biological effects [31, 32]. The treatment of inorganic pollutants using waste-derived catalysts involves innovative approaches that capitalize on the unique properties of materials sourced from various waste streams. These catalysts, often derived from industrial by-products, agricultural residues, or electronic waste, are engineered to effectively remove heavy metals, metal ions, and other inorganic contaminants from wastewater.

For example, waste-derived materials such as modified clays, zeolites, and metal oxides have shown significant promise in adsorbing and catalytically transforming inorganic pollutants. These materials are tailored to have high surface areas, ion exchange capacities, and specific chemical functionalities that enable efficient binding and removal of contaminants like arsenic, lead, cadmium, and mercury. Processes like ion exchange, surface complexation, and redox reactions are employed to facilitate the removal and transformation of these pollutants into less harmful forms.

Carbon compounds originating from bio-waste, including activated carbon (AC), are widely used for adsorbing ions of heavy metals due to their huge surface area, cost-effectiveness, and abundance of oxygen-based functional groups, such as -OH, -COOH, and -C-O [33]. Through treatments with acids, bases, oxidizing agents, organic compounds, and salts, these materials' surface changes and nanostructure modulation improve their adsorption efficiency. Bases (KOH, NaOH) and chloride ($ZnCl_2$) activations during pyrolysis optimize the porosity of activated biochar, increasing surface area and facilitating mass/charge transfer for improved inorganic ion adsorption. Acid treatments (e.g., H_2SO_4) focused on introduction of the functional groups (e.g., $-SO_3H$) that enhance metal ion attraction and adsorption capacity. Similarly, oxidizing agents (e.g., NaClO) increase active binding sites by converting biochar functional groups to carboxylic groups, enhancing metal ion adsorption and removal [34].

Additionally, waste-derived carbon materials effectively remove anions (e.g., F^-, Cl^-, NO_3^-, SO_4^{2-}, PO_4^{3-}) through adsorption mechanisms. Introducing positively charged segments onto carbon surfaces *via* chemical modifications ensures high removal efficiency. In metal ion regulation, Yang *et al.* [35], demonstrated that Fe-modified biochar derived from waste-activated sludge, with amorphous hematite and hydroxides as predominant Fe species, exhibited superior PO_4^{3-} adsorption compared to untreated biochar, highlighting the role of ligand exchange in the adsorption process.

Apart from biowaste-based materials, adsorbents derived from industrial waste (like, spent limestone, slag, fly ash, mine waste, and electronic waste) also demonstrate effective conduction in removing inorganic ions [12, 36]. Typically composed of (mixed) metal hydroxides, these adsorbents provide ample active sites for adsorbing metal ions and anions. The nanostructure of some of the industrial waste-derived adsorbents further increases their adsorption capabilities. For instance, Zou *et al.*, utilized hydrothermal treatment to convert spent $LiFePO_4$ battery waste into zeolite with 3-D, 12-member-ring channels, resulting in high structural porosity and strong Pb^{2+} adsorption capacity (maximum up to 723.8 mg/g) [12].

The effectiveness of ion adsorption on waste-derived oxides is affected significantly by reaction conditions, like pH, reaction time, and temperature. Qiu *et al.*, demonstrated that the adsorption characteristics of Cd^{2+} by hydrothermally modified circulating fluidized bed fly ash (HM-CFB-FA) are pH-dependent. At lower pH values (<6.2), adsorption dominates due to inhibition of chemical precipitation by H^+ ions, whereas at higher pH values (>6.2), hydrolysis and ionization reactions of oxides of Ca and Al lead to Cd^{2+} chemical precipitation [37]. Optimizing these conditions is crucial for enhancing adsorption performance and understanding the underlying mechanisms.

Moreover, the time-dependent adsorption mechanism is also noteworthy. Perez *et al.*, observed that waste-derived Ca $(OH)_2/CaCO_3$ materials effectively adsorb orthophosphate ions, with Ca $(OH)_2$ playing a primary role within the initial 15 minutes, followed by $CaCO_3$ facilitating further removal through mechanisms like direct ligand exchange with CO_3^{2-} under high orthophosphate concentrations [37]. These findings highlight the dynamic nature of adsorption processes and underscore the importance of optimizing conditions for efficient pollutant removal.

To enhance pollutant removal, researchers used surfactants as media and combined different adsorbents, leveraging their respective advantages such as large surface area, cost-effectiveness, and high adsorption capacity. Carbon/zeolite-based composites have garnered significant interest. For example, Han *et al.* [38], developed SCZ composites by loading nanoscale zerovalent iron (nZVI) onto microspherical carbon (SC) derived from waste cartons (Figure 13.2) [15]. They effectively adsorb Cr (VI) ions and subsequently reduce them to Cr (III) *via* chemical reactions involving nZVI and hydroxyl groups on SC. This integrated approach accelerates the removal of hazardous metal ions. Additionally, metal hydroxides, sulfides, and metal-organic frameworks (MOFs) are commonly combined with carbon/zeolite materials to form composites with enhanced adsorption capacities. Ngambia *et al.* [39], created an SDBC-Mg (II) composite from

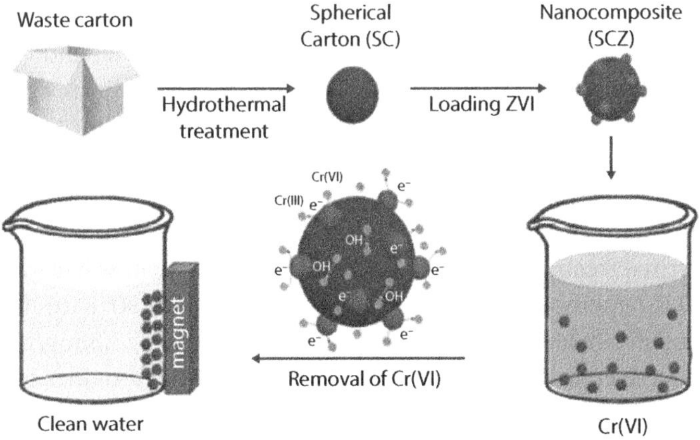

Figure 13.2 Removal of Cr (VI) ions by new nanocomposite derived from waste carton [19].

sewage sludge-derived biochar using a precipitation-calcination method, achieving over 99% removal of Ag^+, Pb^{2+}, Cu^{2+}, and Cd^{2+} ions from pond water, owing to its high surface area (91.57 m^2/g) and abundant sorption sites.

13.3.3 Photocatalytic Degradation of Pollutants

Photocatalysis harnesses solar energy to degrade pollutants, where photocatalysts under light generate excited holes (h^+) and electrons (e^-) [3]. These species possess strong oxidation capabilities, directly oxidizing different pollutants. Additionally, radicals like ·OH formed from h^+ and e^- react with OH^-, H_2O, and O_2, further degrade pollutants into less toxic compounds such as CO_2 and H_2O [40]. To optimize sunlight utilization and reduce semiconductor photocatalyst costs, interest has grown in waste-derived metal oxide/carbon-based materials for photocatalytic applications.

Metal oxide-based semiconductor catalysts are extensively used for pollutant photodegradation, often derived from metal-containing solid waste. For instance, Abdo *et al.*, demonstrated that SnO_2 nanoparticles recovered from waste printed circuit boards (WPCBs) leachate exhibited efficient UV light-driven MB dye degradation [41]. The preparation's success relied on selectively leaching Sn from WPCBs using Na_2-EDTA. Eggshell, a common biowaste, was also utilized to produce CaO nanopowders *via* calcination, which effectively degraded Lanasyn Rez F5B and MB dyes through photodegradation processes [42].

To enhance the photocatalytic efficiency of semiconductor photocatalysts, developing waste-derived composite photocatalysts with low electron-hole recombination, broad light absorption, strong photo stability, and high electronic conductivity is crucial. Carbon-based composites, such as e-waste-derived g-C_3N_4-Fe_2O_3, exemplify this approach [43]. As compared to individual g-C_3N_4 and Fe_2O_3, the composite exhibits superior photocatalytic performance in degrading MO and textile effluents attributed to enhanced visible light absorption, heterojunction formation, and improved electron-hole pair separation efficiency. Additionally, the iron oxide component facilitates easy recovery *via* magnetic separation from the solution.

Carbon-based, metal-free photocatalysts derived from biowaste, such as carbon dots and graphitic carbon nitride (g-C_3N_4), demonstrate effective performance in pollutant mineralization. g-C_3N_4, synthesized by Yang *et al.* [44], from dicyandiamide and mushroom waste *via* thermal condensation and exfoliation, exhibited high activity in MB degradation (>90% removal in 270 min) due to its high graphitic degree and pyrrolic-N units. Similarly, carbon dots and graphene synthesized from biowaste also show potential in photodegradation and photoreduction applications [45]. For instance, cellulose-derived carbon dots efficiently removed Cr (VI) from wastewater under sunlight, demonstrating their dual capability for organic and inorganic pollutants co-removal.

Biochar-based composites derived from waste biomass have also extensively been used in photodegradation of organic compounds and heavy metals [5]. Biochar serves as an excellent support for photoactive materials, leveraging its large surface area to enhance catalyst durability and shape nanoparticle morphology. Additionally, biochar's strong adsorption capacity, abundant surface functional groups, and good electrical conductivity contribute significantly to improving the photocatalytic efficiency of composite catalysts [46].

13.3.4 Electrical Degradation of Pollutants

Because of its high efficacy, energy efficiency, and simplicity of use, electrocatalysis has made considerable strides in the treatment of wastewater as a powerful method for pollutant degradation [2]. Waste-derived catalysts are essential for increasing efficiency in the electrochemical breakdown of contaminants. These catalysts, which act as electrodes or electrocatalysts in procedures like electrooxidation and electrocoagulation, are made from biomass leftovers, industrial byproducts, and electronic waste. By encouraging either reduction or oxidation reactions *via* specific surface characteristics and active spots, they enhance electron transport processes and make pollution removal easier. They hold promises for developing environmentally friendly wastewater treatment methods because of their affordability and sustainability.

Biomass waste, sludge, and industrial waste are examples of waste-derived materials that are widely used in the electrochemical degradation of pollutants [47]. Because of its great chemical resistance, huge specific surface area, superior electrical conductivity, and adaptable functional groups, biochar is frequently used as an electrocatalyst or as a substrate for metal-based electroactive materials. Liu *et al.* [48] showed that adding Fe and Zn to biochar enhanced its specific surface area and functional groups, which in turn improved nitrobenzene adsorption and electrolysis. In the light of crucial role that electronic characteristics and nanostructure play in electrocatalytic performance, the latest study has concentrated on loading electroactive chemicals onto porous biochar to create composites for efficient electrocatalytic applications [49, 50].

13.3.5 Advanced Oxidation Processes

AOPs use highly reactive hydroxyl (·OH) or sulfate (SO_4^-) radicals to decompose organic contaminants in water. Waste-derived catalysts have found significant applications in enhancing these processes due to their unique properties such as high surface area, diverse surface functionalities, and cost-effectiveness. Nanomaterials often play a crucial role in activating oxidizers and facilitating the oxidation process. Among these waste-derived materials, particularly carbon-based ones have been extensively studied for their potential in enhancing AOPs.

Waste-derived metal-bearing Fenton-like catalysts (such as those containing Fe, Cu, Cr, Co, Ce, and Ru) are highly efficient in decomposing H_2O_2 to produce hydroxyl radicals (·OH). For instance, Wang *et al.* [52], developed an iron-doped biochar (Fe@BC) Fenton-like catalyst from sawdust, which achieved over 92.7% degradation of RhB [51] within 140 minutes. The biochar support itself was ineffective in activating H_2O_2, highlighting the importance of the iron species as the central active sites for RhB degradation. Similarly, various Fe-based Fenton catalysts have been designed from waste such as lithium-ion batteries [52], iron ore tailings [53], sludge [51], and biomass [54], for organic contaminant degradation.

Metal leakage is a common issue with metal-bearing Fenton-like catalysts, leading to reduced performance and the formation of metallic sludge. This has spurred interest in metal-free Fenton-like catalysts derived from biowaste. For example, Zhuang and co-workers [55], developed a metal-free, 3D graphene-based Fenton-like catalyst from biowaste (OG). The 3D graphene assembly, with its confinement effect, resulted in a lower energy barrier (1.1 vs. 1.6 eV) for H_2O_2 decomposition compared to the traditional 2D graphene structure. Additionally, electron migration *via* the C–O–C bridge in OG increased electron loss from the C–F bonds, enhancing the mineralization of perfluorooctanoic acid (PFOA). These unique structural and electronic properties of OG contributed to a high degradation performance, achieving 93.4% PFOA removal in 150 minutes.

13.4 Challenges

Waste-derived catalysts have a bright future, but there are still a number of obstacles to overcome. Variability in the make-up processes and characteristics of waste materials are the major problems since they might impact the catalysts' performance and consistency. Furthermore, there is a risk of metal leakage, especially in Fenton-like catalysts containing metals, which could have negative effects on the environment, operations and result in metallic sludge and decreased performance. The other important challenge lies in transferring these catalysts from laboratory to industrial production levels, since it might be challenging to retain efficiency and cost-effectiveness at greater scale. To increase the catalysts' stability and activity, it is difficult to optimize their surface characteristics and nanostructure. To increase waste-derived catalysts' environmental safety, scalability, and dependability in pollutant degrading applications, more research and development is needed to address these issues.

13.5 Future Scope

Waste-derived catalysts have a wide range of possible applications in pollution degradation in the future. It is expected that developments in the fields of material science and engineering will create more stable and effective catalysts from a variety of waste materials. Advances in nanotechnology may improve these catalysts' reactivity and surface characteristics, which would enhance their performance in a variety of environmental applications. It may be possible to develop multifunctional systems that can handle several pollutants at the same time by exploring hybrid and composite substances that blend waste-derived catalysts with other useful materials. Furthermore, industrial applications will require initiatives for scaling up manufacturing procedures while preserving environmental sustainability and cost-effectiveness. The incorporation of waste-derived catalysts into circular economy models, which support resource recovery and sustainable waste management, is also gaining traction. To fully realize the promise of waste-derived catalysts in environmental remediation, interdisciplinary research and cooperation will be essential.

13.6 Conclusion

Waste-derived catalysts successfully solve waste management and environmental pollution problems by offering a practical and long-lasting solution for pollutant degradation. These catalysts exhibit exceptional efficacy in adsorption, photocatalysis, electrocatalysis, and AOPs. They are derived from a variety of waste materials, including biomass, industrial by-products, and electronic waste. They are great substitutes for traditional materials because of their large surface area, variety of functional groups, and affordability. However, ongoing research and innovation are required to overcome issues like metal leakage and long-term stability. These catalysts have a bright future ahead of them—thanks to developments in material science and nanotechnology, which should increase their scalability and efficiency and greatly aid in the development of a tool for circular economy and sustainable environmental remediation.

Acknowledgements

The Chancellor, Vice-Chancellor, Vice-President (RBEF) and Director, Amity School of Applied Sciences, Amity University in Bengaluru are all acknowledged by the author for the aforementioned project work. Additionally, the author is grateful to Mr. Anirudh and Miss Anushi Varshney for their assistance in compiling this work.

AI disclosure statement: The author acknowledges the use of AI-based tools for minor language editing or grammatical correction. However, all intellectual contributions, analyses, and conclusions presented in this chapter are entirely the author's own.

References

1. Somya, A., An investigation study on water quality of few lakes of Bengaluru, in: *Industry 4.0 with Modern Technology*, pp. 200–204, CRC Press, 2024, https://doi.org/10.1201/9781003450924-37.
2. Chen, Z., Liu, Y., Wei, W., Ni, B.-J., Recent advances in electrocatalysts for halogenated organic pollutant degradation. *Environ. Sci.: Nano*, Royal Society of Chemistry (RSC), 6, 8, 2332–2366, 2019, https://doi.org/10.1039/c9en00411d.
3. Al-Zahrani, S.A., Patil, M.B., Mathad, S.N., Patil, A.Y., Al Otaibi, A., Masood, N., Mansour, D., Khan, A., Gupta, V., Topare, N.S., Somya, A., Ayyar, M., Photocatalytic Azo Dye Degradation Using Graphite Carbon Nitride Photocatalyst and UV-A Irradiation. *Crystals*, MDPI, AG, 13, 4, 577, 2023, https://doi.org/10.3390/cryst13040577.
4. Somya, A., Metal Phosphates: Their Role as Ion Exchangers in Water Purification, in: *Metal Phosphates and Phosphonates*, pp. 341–356, Springer International Publishing, 2023, https://doi.org/10.1007/978-3-031-27062-8_19.
5. Amusat, S.O., Kebede, T.G., Dube, S., Nindi, M.M., Ball-milling synthesis of biochar and biochar–based nanocomposites and prospects for removal of emerging contaminants: A review. *J. Water Process Eng.*, 41, 101993, 2021, https://doi.org/10.1016/j.jwpe.2021.101993.
6. Jothi Ramalingam, R., Sivachidambaram, M., Vijaya, J.J., Al-Lohedan, H.A., Muthumareeswaran, M.R., Synthesis of porous activated carbon powder formation from fruit peel and cow dung waste for modified electrode fabrication and application. *Biomass Bioenergy*, 142, 105800, 2020, https://doi.org/10.1016/j.biombioe.2020.105800.
7. Liu, N., Liu, Y., Zeng, G., Gong, J., Tan, X., JunWen, Liu, S., Jiang, L., Li, M., Yin, Z., Adsorption of 17β-estradiol from aqueous solution by raw and direct/pre/post-KOH treated lotus seedpod biochar. *J. Environ. Sci.*, 87, 10–23, 2020, https://doi.org/10.1016/j.jes.2019.05.026.
8. Çelebi, H., Recovery of detox tea wastes: Usage as a lignocellulosic adsorbent in Cr6+ adsorption. *J. Environ. Chem. Eng.*, 8, 5, 104310, 2020, https://doi.org/10.1016/j.jece.2020.104310.
9. Batool, A. and Valiyaveettil, S., Chemical transformation of soya waste into stable adsorbent for enhanced removal of methylene blue and neutral red from water. *J. Environ. Chem. Eng.*, 9, 1, 104902, 2021, https://doi.org/10.1016/j.jece.2020.104902.
10. Zou, W., Feng, X., Wei, W., Zhou, Y., Wang, R., Zheng, R., Li, J., Luo, S., Mi, H., Chen, H., Converting Spent LiFePO4 Battery into Zeolitic Phosphate for Highly Efficient Heavy Metal Adsorption. *Inorg. Chem.*, 60, 13, 9496–9503, 2021, https://doi.org/10.1021/acs.inorgchem.1c00614.
11. Liou, T.-H., Liu, S.-M., Chen, G.-W., Utilization of e-wastes as a sustainable silica source in synthesis of ordered mesostructured titania nanocomposites

with high adsorption and photoactivity. *J. Environ. Chem. Eng.*, 10, 2, 107283, 2022, https://doi.org/10.1016/j.jece.2022.107283.
12. Tsai, C.-K., Doong, R., Hung, H.-Y., Sustainable valorization of mesoporous aluminosilicate composite from display panel glasses waste for adsorption of heavy metal ions. *Sci. Total Environ.*, 673, 337–346, 2019, https://doi.org/10.1016/j.scitotenv.2019.04.056.
13. Wu, W., Chen, Z., Huang, Y., Li, J., Chen, D., Chen, N., Su, M., Red mud for the efficient adsorption of U(VI) from aqueous solution: Influence of calcination on performance and mechanism. *J. Hazard. Mater.*, 409, 124925, 2021, https://doi.org/10.1016/j.jhazmat.2020.124925.
14. Jiang, M., Yang, Y., Lei, T., Ye, Z., Huang, S., Fu, X., Liu, P., Li, H., Removal of phosphate by a novel activated sewage sludge biochar: Equilibrium, kinetic and mechanism studies. *Appl. Energy Combust. Sci.*, 9, 100056, 2022, https://doi.org/10.1016/j.jaecs.2022.100056.
15. Chen, Z., Wei, W., Chen, H., Ni, B.-J., Recent advances in waste-derived functional materials for wastewater remediation. *Eco-Environ. Health*, 1, 2, 86–104, 2022, https://doi.org/10.1016/j.eehl.2022.05.001.
16. Liu, L., Huang, Y., Zhang, S., Gong, Y., Su, Y., Cao, J., Hu, H., Adsorption characteristics and mechanism of Pb(II) by agricultural waste-derived biochars produced from a pilot-scale pyrolysis system. *Waste Manage. (Oxford)*, 100, 287–295, 2019, https://doi.org/10.1016/j.wasman.2019.08.021.
17. Zhao, C., Shao, B., Yan, M., Liu, Z., Liang, Q., He, Q., Wu, T., Liu, Y., Pan, Y., Huang, J., Wang, J., Liang, J., Tang, L., Activation of peroxymonosulfate by biochar-based catalysts and applications in the degradation of organic contaminants: A review. *Chem. Eng. J.*, 416, 128829, 2021, https://doi.org/10.1016/j.cej.2021.128829.
18. Solangi, N.H., Kumar, J., Mazari, S.A., Ahmed, S., Fatima, N., Mubarak, N.M., Development of fruit waste derived bio-adsorbents for wastewater treatment: A review. *J. Hazard. Mater.*, 416, 125848, 2021, https://doi.org/10.1016/j.jhazmat.2021.125848.
19. Chen, Z., Wei, W., Chen, H., & Ni, B.-J., Recent advances in waste-derived functional materials for wastewater remediation, in: *Eco-Environment & Health*, 1, 2, 86–104, Elsevier BV, 2022. https://doi.org/10.1016/j.eehl.2022.05.001
20. Jiang, X., Sun, P., Xu, L., Xue, Y., Zhang, H., Zhu, W., Platanus orientalis leaves based hierarchical porous carbon microspheres as high efficiency adsorbents for organic dyes removal. *Chin. J. Chem. Eng.*, 28, 1, 254–265, 2020, https://doi.org/10.1016/j.cjche.2019.03.030.
21. Li, J., Xu, L., Sun, P., Zhai, P., Chen, X., Zhang, H., Zhang, Z., Zhu, W., Novel application of red mud: Facile hydrothermal-thermal conversion synthesis of hierarchical porous AlOOH and Al2O3 microspheres as adsorbents for dye removal. *Chem. Eng. J.*, 321, 622–634, 2017, https://doi.org/10.1016/j.cej.2017.03.135.

22. Cai, L., Sun, J., Cui, L., Jiang, Y., Huang, Z., Stabilization of heavy metals in piggery wastewater sludge through coagulation-hydrothermal reaction–pyrolysis process and sludge biochar for tylosin removal. *J. Cleaner Prod.*, 260, 121165, 2020, https://doi.org/10.1016/j.jclepro.2020.121165.
23. Ribeiro, J.S., Moreira, T.F.M., Santana, I.L., Ferreira, S.A.D., Lelis, M.F.F., Freitas, M.B.J.G., Sol-gel synthesis, characterization, and catalytic properties of Ni, Cd, Co, and Fe oxides recycled from spent Ni-Cd batteries using citric acid as a leaching agent. *Mater. Chem. Phys.*, 205, 186–194, 2018, https://doi.org/10.1016/j.matchemphys.2017.11.025.
24. Bao, T., Damtie, M.M., Hosseinzadeh, A., Wei, W., Jin, J., Phong Vo, H.N., Ye, J.S., Liu, Y., Wang, X.F., Yu, Z.M., Chen, Z.J., Wu, K., Frost, R.L., Ni, B.-J., Bentonite-supported nano zero-valent iron composite as a green catalyst for bisphenol A degradation: Preparation, performance, and mechanism of action. *J. Environ. Manage.*, 260, 110105, 2020, https://doi.org/10.1016/j.jenvman.2020.110105.
25. Bahamon, D. and Vega, L.F., Pharmaceutical Removal from Water Effluents by Adsorption on Activated Carbons: A Monte Carlo Simulation Study. *Langmuir*, 33, 42, 11146–11155, 2017, https://doi.org/10.1021/acs.langmuir.7b01967.
26. Alivand, M.S., Najmi, M., Tehrani, N.H.M.H., Kamali, A., Tavakoli, O., Rashidi, A., Esrafili, M.D., Ghasemy, E., Mazaheri, O., Tuning the surface chemistry and porosity of waste-derived nanoporous materials toward exceptional performance in antibiotic adsorption: Experimental and DFT studies. *Chem. Eng. J.*, 374, 274–291, 2019, https://doi.org/10.1016/j.cej.2019.05.188.
27. Ashiq, A., Sarkar, B., Adassooriya, N., Walpita, J., Rajapaksha, A.U., Ok, Y.S., Vithanage, M., Sorption process of municipal solid waste biochar-montmorillonite composite for ciprofloxacin removal in aqueous media. *Chemosphere*, 236, 124384, 2019, https://doi.org/10.1016/j.chemosphere.2019.124384.
28. Murphin Kumar, P.S., Ganesan, S., Al-Muhtaseb, A.H., Al-Haj, L., Elancheziyan, M., Shobana, S., Kumar, G., Tropical fruit waste-derived mesoporous rock-like Fe2O3/C composite fabricated with amphiphilic surfactant-templating approach showing massive potential for high-tech applications. *Int. J. Energy Res.*, 45, 12, 17417–17430, 2021, https://doi.org/10.1002/er.6798.
29. Mei, Y., Xu, J., Zhang, Y., Li, B., Fan, S., Xu, H., Effect of Fe–N modification on the properties of biochars and their adsorption behavior on tetracycline removal from aqueous solution. *Bioresour. Technol.*, 325, 124732, 2021, https://doi.org/10.1016/j.biortech.2021.124732.
30. Basaleh, A.A., Al-Malack, M.H., Saleh, T.A., Poly(acrylamide acrylic acid) grafted on steel slag as an efficient magnetic adsorbent for cationic and anionic dyes. *J. Environ. Chem. Eng.*, 9, 2, 105126, 2021, https://doi.org/10.1016/j.jece.2021.105126.
31. Somya, A., Khan, A., Hussein, M.A., Al-Romaizan, A.N., Malash, A.M., Topare, N.S., Asiri, A.M., Biological and environmental studies on

pectin-zirconium (IV) phosphate, a biopolymer doped hybrid cation exchanger. *Int. J. Biol. Macromol.*, 250, 125803, 2023, https://doi.org/10.1016/j.ijbiomac.2023.125803.

32. Somya, A. and Maheria, K., Studies on polyoxyethylene octyl phenyl ether supported thorium(IV) phosphates: a new cation exchange material. *Bull. Mater. Sci.*, 43, 1, 2020, https://doi.org/10.1007/s12034-020-02153-z.
33. Ramrakhiani, L., Ghosh, S., Majumdar, S., Heavy metal recovery from electroplating effluent using adsorption by jute waste-derived biochar for soil amendment and plant micro-fertilizer. *Clean Technol. Environ. Policy*, 24, 4, 1261–1284, 2022, https://doi.org/10.1007/s10098-021-02243-4.
34. Wan Ngah, W.S. and Hanafiah, M.A.K.M., Removal of heavy metal ions from wastewater by chemically modified plant wastes as adsorbents: A review. *Bioresour. Technol.*, 99, 10, 3935–3948, 2008, https://doi.org/10.1016/j.biortech.2007.06.011.
35. Yang, Q., Wang, X., Luo, W., Sun, J., Xu, Q., Chen, F., Zhao, J., Wang, S., Yao, F., Wang, D., Li, X., Zeng, G., Effectiveness and mechanisms of phosphate adsorption on iron-modified biochars derived from waste activated sludge. *Bioresour. Technol.*, 247, 537–544, 2018, https://doi.org/10.1016/j.biortech.2017.09.136.
36. Qiu, R., Cheng, F., Huang, H., Removal of Cd2+ from aqueous solution using hydrothermally modified circulating fluidized bed fly ash resulting from coal gangue power plant. *J. Cleaner Prod.*, 172, 1918–1927, 2018, https://doi.org/10.1016/j.jclepro.2017.11.236.
37. Pérez, S., Muñoz-Saldaña, J., Garcia-Nunez, J.A., Acelas, N., Flórez, E., Unraveling the Ca–P species produced over the time during phosphorus removal from aqueous solution using biocomposite of eggshell-palm mesocarp fiber. *Chemosphere*, 287, 132333, 2022, https://doi.org/10.1016/j.chemosphere.2021.132333.
38. Han, J., Zhang, G., Zhou, L., Zhan, F., Cai, D., Wu, Z., Waste Carton-Derived Nanocomposites for Efficient Removal of Hexavalent Chromium. *Langmuir*, 34, 21, 5955–5963, 2018, https://doi.org/10.1021/acs.langmuir.8b00225.
39. Ngambia, A., Ifthikar, J., Shahib, I.I., Jawad, A., Shahzad, A., Zhao, M., Wang, J., Chen, Z., Chen, Z., Adsorptive purification of heavy metal contaminated wastewater with sewage sludge derived carbon-supported Mg(II) composite. *Sci. Total Environ.*, 691, 306–321, 2019, https://doi.org/10.1016/j.scitotenv.2019.07.003.
40. Ma, M., Liu, Y., Wei, Y., Hao, D., Wei, W., Ni, B.-J., A facile oxygen vacancy and bandgap control of Bi(OH)SO4·H2O for achieving enhanced photocatalytic remediation. *J. Environ. Manage.*, 294, 113046, 2021, https://doi.org/10.1016/j.jenvman.2021.113046.
41. Abdo, D.M., Abdelbasir, S.M., El-Sheltawy, S.T., Ibrahim, I.A., Recovery of Tin as Tin oxide nanoparticles from waste printed circuit boards for photocatalytic dye degradation. *Korean J. Chem. Eng.*, 38, 9, 1934–1945, 2021, https://doi.org/10.1007/s11814-021-0838-9.

42. Mohamed, F., Shaban, M., Aljohani, G., Ahmed, A.M., Synthesis of novel eco-friendly CaO/C photocatalyst from coffee and eggshell wastes for dye degradation. *J. Mater. Res. Technol.*, 14, 3140–3149, 2021, https://doi.org/10.1016/j.jmrt.2021.08.055.
43. Babar, S., Gavade, N., Shinde, H., Gore, A., Mahajan, P., Lee, K.H., Bhuse, V., Garadkar, K., An innovative transformation of waste toner powder into magnetic g-C3N4-Fe2O3 photocatalyst: Sustainable e-waste management. *J. Environ. Chem. Eng.*, 7, 2, 103041, 2019, https://doi.org/10.1016/j.jece.2019.103041.
44. Yang, H.-C., Liu, M.-T., Chao, M.-W., Wang, K.-H., Hu, C., Polymeric g-C3N4 Derived from the Mixture of Dicyandiamide and Mushroom Waste for Photocatalytic Degradation of Methyl Blue. *Top. Catal.*, 63, 11–14, 1182–1192, 2020, https://doi.org/10.1007/s11244-020-01237-8.
45. Khare, P., Singh, A., Verma, S., Bhati, A., Sonker, A.K., Tripathi, K.M., Sonkar, S.K., Sunlight-Induced Selective Photocatalytic Degradation of Methylene Blue in Bacterial Culture by Pollutant Soot Derived Nontoxic Graphene Nanosheets. *ACS Sustainable Chem. Eng.*, 6, 1, 579–589, 2017, https://doi.org/10.1021/acssuschemeng.7b02929.
46. Men, Q., Wang, T., Ma, C., Yang, L., Liu, Y., Huo, P., Yan, Y., In-suit preparation of CdSe quantum dots/porous channel biochar for improving photocatalytic activity for degradation of tetracycline. *J. Taiwan Inst. Chem. Eng.*, 99, 180–192, 2019, https://doi.org/10.1016/j.jtice.2019.03.019.
47. Petala, A., Bampos, G., Frontistis, Z., Using Sawdust Derived Biochar as a Novel 3D Particle Electrode for Micropollutants Degradation. *Water*, 14, 3, 357, 2022, https://doi.org/10.3390/w14030357.
48. Liu, Q., Jiang, S., Su, X., Zhang, X., Cao, W., Xu, Y., Role of the biochar modified with ZnCl2 and FeCl3 on the electrochemical degradation of nitrobenzene. *Chemosphere*, 275, 129966, 2021, https://doi.org/10.1016/j.chemosphere.2021.129966.
49. Wang, Y., Tian, X., Wang, S., Cui, C., Xin, Y., Zhang, G., Zhou, C., *In situ*-synthesized amorphous Pd/N-C microspheres derived from shrimp shells as a three-dimensional electrocatalyst for hydrodechlorination of diclofenac. *Chem. Eng. J.*, 428, 131231, 2022, https://doi.org/10.1016/j.cej.2021.131231.
50. Gong, Y., Wan, J., Zhou, P., Wang, X., Chen, J., Xu, K., Oxygen and nitrogen-enriched hierarchical MoS2 nanospheres decorated cornstalk-derived activated carbon for electrocatalytic degradation and supercapacitors. *Mater. Sci. Semicond. Process.*, 123, 105533, 2021, https://doi.org/10.1016/j.mssp.2020.105533.
51. Wang, C., Sun, R., Huang, R., Highly dispersed iron-doped biochar derived from sawdust for Fenton-like degradation of toxic dyes. *J. Cleaner Prod.*, 297, 126681, 2021, https://doi.org/10.1016/j.jclepro.2021.126681.
52. Xu, L., Chen, C., Huo, J.-B., Chen, X., Yang, J.-C.E., Fu, M.-L., Iron hydroxyphosphate composites derived from waste lithium-ion batteries for lead adsorption and Fenton-like catalytic degradation of methylene blue.

Environ. Technol. Innovation, 16, 100504, 2019, https://doi.org/10.1016/j.eti.2019.100504.

53. Gao, L., Wang, L., Li, S., Cao, Y., Highly active Fenton-like catalyst derived from solid waste-iron ore tailings using wheat straw pyrolysis. *Environ. Sci. Pollut. Res.*, 29, 21, 31567–31576, 2022, https://doi.org/10.1007/s11356-021-17168-0.
54. Ahsan, M.A., Puente Santiago, A.R., Rodriguez, A., Maturano-Rojas, V., Alvarado-Tenorio, B., Bernal, R., Noveron, J.C., Biomass-derived ultrathin carbon-shell coated iron nanoparticles as high-performance tri-functional HER, ORR and Fenton-like catalysts. *J. Cleaner Prod.*, 275, 124141, 2020, https://doi.org/10.1016/j.jclepro.2020.124141.
55. Zhuang, Y., Wang, X., Zhang, L., Kou, Z., Shi, B., Confinement Fenton-like degradation of perfluorooctanoic acid by a three dimensional metal-free catalyst derived from waste. *Appl. Catal., B*, 275, 119101, 2020, https://doi.org/10.1016/j.apcatb.2020.119101.

14

Conventional and Data-Driven Solid Waste Management Strategies

M. Shabeena Begam[1], Mohammad Faiz Afzal[2], Pawan Devidas Meshram[3], Kuldeep Agnihotri[4], Manu Vasudevan Unni[5], Amar Chipade[6] and V. Bhoopathy[7]*

[1]*Department of MBA, SNS Institutions, Coimbatore, Tamilnadu, India*
[2]*School of Interdisciplinary and Transdisciplinary Studies, Indira Gandhi National Open University, Delhi, New Delhi, India*
[3]*Institute of Chemical Technology, Kavayitri Bahinabai Chaudhari North Maharashtra University, Jalgaon, Maharashtra, India*
[4]*Department of Commerce & Management, ISBA Group of Institutes, Indore, Madhya Pradesh, India*
[5]*Department of Management, St. Claret College Autonomous, Bangalore, Karnataka, India*
[6]*Department of Civil Engineering, Dr. D. Y. Patil Institute of Technology Pimpri, Pune, Maharashtra, Indian*
[7]*Department of CSE, Sree Rama Engineering College, Tirupati, Andhra Pradesh, India*

Abstract
An integral part of sustainable development is solid waste management, which calls for efficient solutions to the problems of waste production and their negative environmental effect. Strategies for managing solid waste, including garbage reduction, recycling, composting, and energy recovery, are discussed in this chapter. It highlights the importance of improving waste management systems by combining modern technologies such as artificial intelligence, blockchain, and Internet of Things with current practices. Waste management efficiency can be enhanced through public-private partnerships, community engagement, and regulatory frameworks, which are also examined in this chapter. By using case studies and best practices, the chapter insights into the implementation of sustainable

**Corresponding author*: v.bhoopathy@gmail.com

Deepti Goyal, Sakshi Kabra Malpani and Renuka Jain (eds.) Catalysis from Waste: Transforming Solid Residues into Value, (369–392) © 2026 Scrivener Publishing LLC

management waste approaches that improve resource efficiency and environmental preservation.

Keywords: Solid waste management, recycling, waste-to-energy, composting, blockchain, artificial intelligence, sanitary landfills

14.1 Introduction

The Reduce, Recycle and Reuse (3R) project is being advocated to concurrently promote economic development and environmental preservation, aiming to foster a sound material-cycle society that effectively utilizes resources for 3R activities [1]. Moreover, it is universally acknowledged that numerous 3R efforts aid in the mitigation of greenhouse gas (GHG) emissions [2]. As a result, people have started talking about ways to make resource utilization more sustainable and to create systems economic that can encourage the efficient circulation of resources. Additionally, integrating waste management regulations, such as procedures for the correct control of chemicals is vital due to the fact that there are significant unresolved challenges of waste management [3]. The terms like waste prevention, reuse, and reduce have different meanings in different parts of the world. Nevertheless, it is widely acknowledged that when a community effectively applies the concepts of reduction and reuse, waste will be effectively minimized [4]. The formulation and suggestion of quantitative indicators, particularly material flow indicators, have facilitated the assessment and oversight of the implementation of the 3Rs. Recycling, composting, waste-to-energy technology, waste reduction, and governmental interventions are some of the solid waste management (SWM) strategies discussed in this chapter.

This chapter also looks at contemporary SWM systems with an eye on Internet of Things (IoT), blockchain, and artificial intelligence (AI) to improve monitoring, processing, and garbage collection. They are highlighted as crucial drivers in sustainable waste management, public-private cooperation and regulatory structures. Furthermore, relevant case studies have also been discussed to show the real use of these strategies in various worldwide contexts.

14.2 Management Strategies for Solid Waste

Current and future approaches to SWM should be taken into account for establishing goals for technological innovations and regulations that helps in reducing waste, as well as setting up institutional intents to support such efforts. Cutting down on trash production is a crucial step in diminishing garbage problems in the future. Recycling, reusing, and decreasing trash is an important responsibility of industry, businesses, and common people. A few ways to cut down consumption are to reuse containers and bags, be more disciplined while shopping, use fewer disposable products, and reduce, reuse, and recycle packaging. Alternative technologies can also play a role in resource production [5]. For example, rapid advancements in information technology will enhance information and communication distribution through electronic media rather than printed materials, hence significantly reducing paper use and its associated waste percentage.

14.3 Types of Waste

14.3.1 Personal Protective Equipment (PPE) Waste

PPE is regarded as the most dependable and cost-effective safeguard against viral transmission and infection. At the start of the COVID-19 pandemic, practitioners, healthcare professionals mostly used PPE, like as medical gloves, masks, and protective clothes to reduce transmission of virus. However, due to the virus's rising spread and the forecast of scientific communities about the outbreak of various new infections in the coming future; many countries now mandate everyone to have PPE kits [6].

Industries were manufacturing and people were using millions of PPE every day to combat the pandemic. For example, health care providers on the tiny island nation of Ireland used around 9 million masks every week, despite the fact that their population is just around 4.5 million [7]. But a sizable portion of that sum comes from regular people's use of PPE. Wearing a facemask is now lawful for all residents in over fifty nations and territories. The average daily disposal volume of masks for Italy is 40 million or around 44,000 tons. The inclusion of 40 million pairs of single-use gloves would elevate the proportion of utilized PPE by the populace to 0.59%, accounting for 0.14% of the total national MSW [8]. Even after the pandemic, healthcare professionals and the common public are using PPE more often. People have started using facemasks while suffering from common coughs and colds as well. Apart from medical gloves and masks, helmets, special protective jackets and clothes are used by construction site workers. In addition, lab coats, safety goggles are used in chemical industries.

14.3.2 Medical Waste

According to Rupani and co-workers [9], medical waste encompasses a wide variety of materials generated by several sources, including research and biomedical organizations, laboratories, and hospitals. Although hospitals are major producers of medical waste, they are just one of several trash generators. On average, non-hazardous general garbage, including plastic, paper, and office trash, accounts for around 75% to 90% of the overall waste, as compared to waste produced by healthcare activities [10]. Under typical circumstances, this trash can be treated using standard techniques without requiring any extra management.

14.3.3 Municipal Solid Waste (MSW)

- **Food Waste**

Higher levels of need for cognitive closure (Need for closure scale, NFC) are associated with greater food waste in scenarios of uncertainty like the pandemic COVID-19. People with high NFC waste more food than those with low NFC. This issue stems from the feeling of food shortages, which leads to unnecessary food acquisitions [11]. Spanish consumption statistics reveal that food purchases surged in the first two weeks of the shutdown, increasing by 29.8 and 10.9%, respectively, before falling by 20.3% in the third week. The overall quantity of food loss and waste (FLW) in Spain was moderately influenced by the pandemic COVID-19; as reported by [8], FLW production in homes rose by 12% [12]. The overall amount of FLW is similar to what it was in 2019, assuming that households absorbed non-household consumption during the outbreak [13].

Certain areas, particularly those that receive a lot of tourists, saw a dramatic decrease in garbage production as the news of the travel restrictions and lockdowns spread [14]. Waste output in Barcelona decreased by 25% due to the commercial, tourism activities, and concentration of labor in areas that were then closed or had diminished operations. Generation of MSW in Macao, a quintessential tourism destination in China, declined by 17% to 25% between February and May 2020 [15]. Within a span of 10–14 weeks in 2020, the severe system blockade in Milan, Italy, reduced overall garbage production by 28%. This included a 24% drop in residual waste, a 14% drop in household food waste, and an almost 80% drop in business food waste [16].

The reduction in recycling efforts is another factor leading to the decline in waste production. Individuals are opting to store certain MSW at home rather than at collection facilities. The amount of MSW that ends up in scrapyards and landfills has decreased by 20% to 40% in India since recycling activities have stopped [17]. There has been a roughly 17% drop in MSW generation in Catalonia, Spain during COVID affected month, as compared to a normal month.

- **Plastic Waste**

Our habits have shifted due to our worries about the virus's ability to persist. Consumers' concern about the environmental implications of plastic trash have changed during the pandemic. Initially they preferred plastic packaging due to health reasons and hygiene point of view which was shifted to an increase in plastic use driven by lifestyle changes that emphasized reassurance and convenience [18].

Fear of the virus and the unexpected lockdown caused an initial uptick in disposable goods and panic buying [19]. A significant increase in packaging demands was seen in different sectors including food, online shopping, medicines and many more [20]. In the USA, the 9.5% increase in demand for consumer-packaged goods could be possibly due to the food hoarding by people during pandemic times [21].

14.4 Recycling and Material Recovery

A Material Recovery Facility (MRF) sorts, cleans, processes, and stores materials for reprocessing and remanufacturing. These components may be amalgamated or kept distinct at the source [22]. Maximizing the amount of processed recyclables, while producing materials with the potential highest market value, is the main goal of the MRF. In addition to recycling materials, MRFs can turn trash into fuel for power plants or biological precursors for feedstock conversion [23]. This study did not consider the biological transformations that can occur during anaerobic and aerobic composting, nor the chemical transformations that might arise from incineration coupled with energy recovery, although they are effective and validated alternatives of waste management.

14.4.1 Collection

A major factor taken into account while designing the MRF plant is the method for collecting the recyclables that are brought to the MRF [24]. In order to process, transport, and store the obtained resources, the layout buildings and machinery must be constructed to ensure secured external as well as internal flow and access. The collection method of trash will significantly impact the MRF's expenses and resource utilization. Residential garbage and recycling collection choices include:

- ❖ **Collection of Dry/Wet Waste:** It includes separate or combined wet/dry collection of waste, with or without recyclables. Composting is a process that uses damp items, such as grass clippings, leftover food, vegetable and fruit peels, and converts them into manure.
- ❖ **Collection of Recyclables:** The process of collecting mixed recyclables in a vehicle with two sections or in two distinct vehicles, with one section reserved for paper items and the other for non-paper materials.
- ❖ **Collection of Mixed Refuse:** Gathering trash and recyclables in one truck without sorting them at the source point.
- ❖ **Co-Collection:** Sorting recyclables and trash into various colored bags for loading onto trash trucks or other vehicles with many compartments.

The proposed MRF in New York City will process source-separated commingled recyclables collected weekly or biweekly. It is engineered to integrate with the current collection system [25]. Recyclables papers go into the green bags or garbage bins and metal, glass, and plastic (MGP) recyclables into the blue bags or garbage bins; these two streams of co-collection are shown in Table 14.1.

Optimizing collection is the primary factor determining the total cost-effectiveness of any SWM; nevertheless, this study does not address

Table 14.1 Materials for recycling separated at source.

Eco-friendly green bags	Blue bags
• Catalogues and magazines • Food items • Bioplastics, compostable plastics • Alternative cardboard (e.g., egg cartons, smooth, tubes) • Business stationery • Technical reports • Newspapers • Directory of phone numbers • Protective covers • Papers of a different mixed type	• Carton for beverages • Metals for domestic use • High density polyethylene (HDPE) • Polyethylene terephthalate (PET) • Bottles and jars made of glass • Tin cans • Trays and aluminum foil • Metal products (e.g., empty aerosol cans, hangers)

changing New York City's system current collection [26]. Most of the money spent on a SWM goes towards collection, making it the costliest part of the overall system.

14.4.2 Specification of Market

A successful recycling program depends on identifying and developing stable, reliable markets for targeted recyclables, as unsold commodities will require storage space and ultimately lead to landfilling [27]. The market of a certain material is influenced by the comparison between the prices and quality of raw materials recovered from the MRF and the virgin raw materials extracted from nature. Materials should not be targeted for recycling unless there are stable and reliable markets, as processing and collection can be quite expensive. After deciding which materials should be recycled, the next step is to find out their market specs so that the unit separation procedures can be tailored according to their needs [28].

14.4.3 Floor for Tipping

The tipping floor is a big, recessed space where materials are dropped upon arrival at the facility. The tipping level can accommodate an additional second operational shift for the materials and a minimum of two days estimated volume of material [29]. The floor is engineered to support substantial weight, endure the abrasion from material movement, and facilitate effective liquid drainage.

The transfer of materials from vehicles collection to the floor tipping must be efficient while safeguarding the commodities [30]. The deposited material is typically transferred onto conveyors which were then carried to the systems separated by front-end loaders on tipping floors. The floor tipping method necessitates more machinery and results in substantial glass breakage, making it one of the most inefficient parts of the MRF. As a result of this, a lot of facilities across the nation can only recover mixed glass because the floor tipping method causes too much breakage.

14.5 Composting and Organic Waste Management

Segregation of waste materials into moderately biodegradable, and non-biodegradable categories is based on their nature of decomposition. Biodegradable waste can get decomposed by biological processes while the non-biodegradable waste cannot be decomposed in nature.

The degradability rate of biodegradable wastes is increased by aerobic and anaerobic organisms [31]. Examples of biodegradable wastes include products of agriculture such as cow dung and chicken droppings [32]. Municipal domestic waste (MDW) is gradually decomposing material. The components provide a robust texture. Examples of such waste include wood and cardboard [33]. The chapter's latter portion delves deeply into this. The opposite is true for non-biodegradable waste: it cannot be broken down by biological processes [34]. Materials such as leather, plastics, mineral waste, and polythene bags are not biodegradable.

14.6 Conventional Waste Disposal Methods

Composting or more traditional forms of waste disposal could be considered. The composting and conventional methods of garbage disposal are categorized below.

14.6.1 Waste Disposal through River and Ocean Dumping

The discharge of garbage containing different chemicals into water bodies poses a substantial risk, potentially making the water toxic for both human and aquatic life [35]. Toxic pollutants can enter the human body *via* the ingestion of aquatic animals [36]. Many industries located near rivers or oceans discharge their effluents directly into these water bodies without any proper pre-treatment. This is done to reduce garbage disposal costs. In certain under-developed or developing countries, homes are built on top of water, allowing trash to fall directly into bodies of water and be carried away with the water flow. The water from this source may receive water from another area, potentially acting as a conduit for water-borne pathogens, which could lead to diseases in the people who use this water for their various domestic activities [37].

14.6.2 Open Dump Waste Disposal

This method involves the careless dumping of waste in any open area. It includes disposing waste on the street or highway. Policy makers and environmental regulations recommended that this form of trash disposal should be discouraged as this method of garbage disposal promotes the growth of pests like rats and houseflies, which can spread diseases like plague, Lassa and cholera fever. Roadside litter also spread foul smell in their vicinity [36].

14.6.3 Animal Consumption

Domestic animals such as dogs, sheep, and goats fed cassava leaves, yam peels, and other food waste. Animals or humans that ingest the contaminated food scraps are both at risk of contacting infectious diseases [38]. Meat from animals that have been exposed to tainted food can cause illness in humans. One of the examples is trichinosis disease in people caused by consuming undercooked or raw meat infected with Trichinella spiralis worm [39].

14.6.4 Sanitary Landfills

The utilization of garbage as landfills is the core principle of this waste disposal approach. Controlled tipping is another name for this technique [40]. To implement this technique, garbage is spread out in thin layers and then compressed with a heavy bulldozer when each layer is added. The trash is covered with a thin layer of clean soil and pressed again when it reaches a height of around 3 meters. When the hole is half full, the procedure is repeated. Soil creatures and soil, as well as humans (via breathing polluted air or drinking polluted water), could be harmed by pathogenic bacteria or hazardous chemical components of the waste that is disposed of in this way. As a result of these negative effects, the member states of the European Union (EU) have discouraged this technique of garbage disposal, favoring alternatives including anaerobic waste treatment and energy recovery with incineration [41]. In 2001, a landfill regulation was developed, according to which by 2030, all member nations of the EU will remove sanitary landfills and reduce landfill waste disposal 10% by 2025.

14.6.5 Compost

Biodegradable Organic Waste (BOW) can be eliminated through composting. It is also known as biodegradable waste [42]. Composting is an effective method for recycling organic waste into biofertilizers and other useful goods. It is not possible to compost certain materials, such as polythene bags and plastics. Composting is a risk-free method of organic waste management; but it does produce GHG and create odors.

14.6.6 Incineration

Combustible garbage is burned at high temperatures in this procedure. The amount of these wastes is cut in half with this technique. A sanitary landfill is the last destination for the by-products of combustion, including ash, glass, and metals [43]. This approach is linked to fire hazards and the emission of GHG; it is not a comprehensive trash disposal strategy; and it merely decreases the size of wastes. Incineration has the potential to generate energy. The energy produced by this method is less expensive as compared to that produced by coal. This strategy could save around 2-2.6 million tons of CO_2 equivalents per year [44].

14.6.7 The Deep-Well Injection Method

The deep-well injection technique involves burying waste materials in layers of impermeable rock below the surface. The treatment of hazardous industrial fluid waste is facilitated by this approach [45]. This procedure is used for the disposal of waste materials that contain acidic or caustic chemicals, as well as brine from oil fields and uranium processing plants from radioactive waste. The local geological subsurface area should be addressed before disposing garbage by this method to prevent underground water pollution. One possible drawback of this approach is that it has the potential to contaminate groundwater.

14.7 Waste-to-Energy (WTE) Technology

Besides waste management strategies, at this level, it is our responsibility to transform non-recyclable garbage into usable energy in the form of heat, electricity, or transportation fuels [46]. Waste energy technology processes a wide range of waste types, including liquid, semi-solid, solid, and gaseous. Thanks to WTE technology, waste—which was formerly thought of as useless—is now used as a feedstock for these processes and generates heat. By repurposing trash as fuel, fertilizer, and electricity, WTE facilities not only reduce landfill space and associated expenses but also make good use of previously discarded materials [47]. There are two categories of WTE technologies: thermochemical and biochemical technologies.

14.7.1 Gasification

Gasification is a thermochemical process that converts organic material (biomass or municipal solid waste) into alternative energy forms *via* a chemical reaction rather than combustion. It transforms liquid and solid substances into clean and valuable syngas [48]. The primary chemical events responsible for this transformation essentially involve the combination of carbon-based materials (feedstock) with a tiny quantity of oxygen/air or steam (insufficient for combustion). The syngas is a by-product of the organic materials breakdown into their component molecules. Gaseous by-products with a high yield (up to 85%) are usually produced immediately from the process, which takes place at high temperatures (about 800°C).

14.7.2 Pyrolysis

It is the thermochemical breakdown of organic materials in an inert environment without using oxygen. The composition of pyrolytic end-products can be changed by adjusting the heating rate, waste particle size, pyrolysis temperature, and vapor resident time [49]. Pyrolytic oils are liquid by-products of a pyrolysis process that occurs at relatively low temperatures, whereas syngas is formed at higher temperatures. The three distinct reactions of pyrolysis—slow, flash, and fast pyrolysis are defined by the feedstock's temperature, heating rate, and residence time.

Syngas, a mixture of carbon monoxide, hydrogen gas, carbon dioxide gas, and methane, and a mixture of carbonates and hydrocarbons, are the by-products of pyrolysis, as illustrated in Figure 14.1.

14.7.3 Incineration

Incineration is a tried and reliable method for recovering energy from MSW. Complete combustion of waste organic materials allows for heat recovery and volume reduction of MSW. Also, it lessens the infectiousness and possible toxicity of hazardous waste and medical waste, respectively.

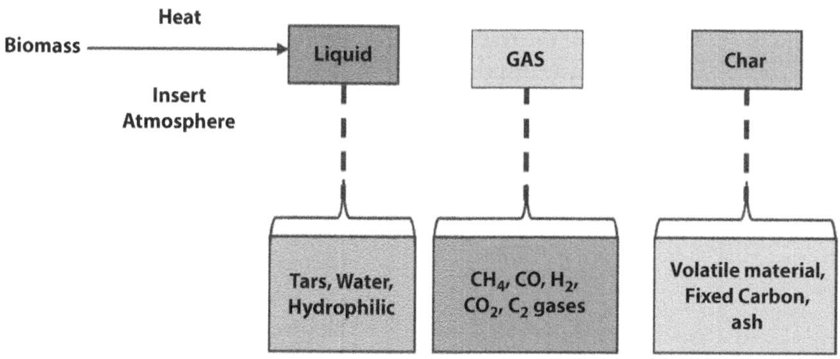

Figure 14.1 Product of biomass pyrolysis.

Table 14.2 Comparison of conventional and composting waste management methods.

S. no.	Conventional	Composting
1.	The pests, diseases, and insects generated due to river and ocean open dump, animal feeding and incineration, have negative effects on both humans and animals.	Composting improves soil quality and aids in the suppression of plant diseases.
2.	The volume of garbage typically increases due to the piling of various types of waste (open dump, river and ocean dumping, etc.).	Significantly decrease in waste volume
3.	The burning of garbage *via* incineration has substantial impact on the greenhouse effect	The generation of hydrocarbons such as methane, the amount of carbon dioxide gas released during composting is lower than that of other waste treatment methods that involve combustion.
4.	Incineration is a versatile process that can treat various types of polymers and polythene bags.	Composting is not an option for materials that are resistant to the process, such as polythene bags and plastics.
5.	Open dumps, sanitary landfills, ocean and river dumping, and incineration are some of the traditional ways of waste disposal that damage the soil, water, and air bodies. They produce unpleasant smell and make unsightly sights.	By reducing soil erosion and keeping soil particles together, composting contributes to environmental sustainability. It aids in the regulated storage and subsequent recycling of waste materials. In the process of bioremediation, they aid degraded soil. Additionally, they attract other beneficial bacteria, insects, fungi, etc., which increases soil biodiversity and ultimately benefits the crop.

Incineration cannot be guaranteed to achieve full oxidation of waste without an adequate quantity of oxygen. In an incineration factory, the typical combustion temperature is around 850°C, with water and carbon dioxide being the primary by-products. Incinerator bottom ash is a solid by-product of incineration that is formed when non-combustible materials like glass and metal are removed from MSW. This requires additional energy to power the electromagnetic belt used for sorting the MSW. The four primary components of an incinerator are the feed system, the combustion chamber, the exhaust gas system, and the residue disposal chamber. The redesigned incinerators, on the other hand, feature a continuous feed system and a primary combustion chamber walled with heat-resistant materials that contain moving grates.

Burning MSW mostly produces two by-products: heat and ash. Steam turbines typically employ the heat produced to create steam, which then generates power [50]. As an alternative, you can get fly ash or bottom ash from the process. Produced ash is a substantial by-product since it accounts for 20–25% of the entire feedstock weight. About 75 to 90% of the ash weight is bottom ash, also called slag [51]. Approximately 10–25% of the ash weight is fly ash, while the majority is non-combustible and partially burned materials like metals and glass. The tiny particles of fly ash are more poisonous than bottom ash because of the heavy metals and associated dioxins that condense on them. Consequently, it is stated that in regulating the generated fly ash, specialized equipment is required, which in turn increases the expenses of construction and maintenance.

A comparison between conventional and composting waste management methods is summarized in Table 14.2.

14.8 Policy and Regulatory Framework

When it comes to managing solid waste, government rules and laws are quite important. Successful interventions in policymaking encompass:

- **Public-Private Partnerships (PPPs):** Working together with private companies to create effective waste management strategies.
- **Strict Waste Disposal Regulations:** Implementing rubbish disposal standards can help to lessen the harmful effects on the environment.

- **International Cooperation:** Integrating waste management strategies with international initiatives to combat climate change and promote sustainability.
- **Zero Waste Policies:** Urging businesses and municipalities to reduce their trash output as much as possible.

14.9 Role of Modern Technologies

- **Artificial Intelligence (AI)**

Systems driven by AI improve route optimization, predictive analytics, and waste sorting for waste collecting companies. By better detecting and grouping waste products, ML models help to lower contamination in recycling sources.

- **Block-chain Technology**

Block-chain technology makes traceable and open waste management techniques more convenient. It guarantees correct documentation for trash transactions, therefore lowering fraud rates and improving responsibility in garbage disposal and recycling.

- **Internet of Things (IoT)**

By real-time bin fill level monitoring, IoT-enabled smart bins and sensors maximize waste collecting. This method powered by data lowers running expenses and improves service quality.

14.10 Case Studies

14.10.1 Study Case 1: Tokyo's Waste Management Model: Public-Private Partnership (PPP) Power

In Japan, PPPs green is very popular. Various programs run by the government encourage partnerships between the public and commercial sectors. The efficient use of recycling and resources of materials are given greater importance in Japan as a means to lessen its reliance on foreign suppliers. A recycling-based society is established by the country's basic law, which in turn defines the notion of the circular economy. In Japan, it is the responsibility of the individual towns to offer services in accordance with the country's laws and to make any new laws that are required for their implementation; this will benefit the country's producers, retailers, and shoppers.

The Tokyo Model, which combines technology, service provisioning, and community involvement, is eloquently described in research by Hongo. It is now common practice in Japan to convert waste into power, especially municipal waste. Numerous nations in the Middle East, South America, and Asia have embraced its best practices. Electricity generated from renewable trash has been provided at premium prices in Japan under the Feed-in Tariff (FiT) program since 2011. The combined electrical output of the world's more than 300 garbage incineration facilities is over 17 times that of a nuclear power station, which is about 1700 megahertz. Municipal trash from Tokyo's 23 cities, with a moisture level of 40%, has a calorific value of around 2,200 kcal/kg. The calorific value of lignite coal ranges from 4,500 to 5,500 kcal/kg whereas the moisture percentage is 30%, but peat has a value of 2,300 kcal/kg or more, with the moisture content 50% or higher. Currently, incineration technology can generate power from garbage with calorific values ranging from 1,700 to 3,400 kcal/kg, completely without requiring additional fuels. Since the characteristics of waste materials changes daily, waste-to-power conversion necessitates not just technology but also an effective management system and operators with the necessary expertise.

14.10.2 Study Case 2: Sustainable Energy Initiative in Serbia: Kladovo's First Solar Power Plants (SPP)

With a population of about 7.1 million, Serbia generates most of power on its own. Coal accounts for over 70% of Serbia's electricity output, with hydropower units contributing the remaining 30%. Elektroprivreda Srbije (Power Industry of Serbia), Serbia's national power utility, controls most of the consumer's power and owns about all the big power generation facilities across the nation.

The national Serbian energy balance for 2020 states that the following amounts of power were generated in 2019 in Gigawatt hours (GWh): hydropower - 10,172 GWh, district heating - 346 GWh, coal - 26,295 GWh, industrial generation - 363 GWh, solar - 14 GWh, wind - 848 GWh, biogas - 116 GWh, gas cogeneration - 108 GWh, and other sources - 587 GWh.

Serbia is making efforts to diversify its energy mix and increase its focus on energy efficiency in response to the unreliability of its old generation methods brought about by climate change. Renewable energy sources hold great promise for this nation, which has set a goal of increasing energy efficiency by 20% till 2020 in accordance with the Energy Community Treaty (ECT).

14.10.3 Study Case 3: P4G: Global Goals 2030 and Partnering for Green Growth

The 2018 initiative P4G—Partnering for Green Growth and the Global Goals 2030—is a great illustration of how worldwide efforts can be united to speed up PPPs while also addressing potential solutions for growth and sustainable development [52]. The Sustainable Development Goals (SDGs) and the Paris Agreement on Climate Change may be effectively implemented through the development of large-scale PPPs and this conference is at the forefront of this effort. In its mission to discover and foster creative partnerships for sustainable development and growth, four important organizational partners, P4G collaborates with nine partner countries, and assistance from prominent organizations in the commercial sector and civil society. P4G encompass both public and private entities collaborating to promote innovative and economically feasible projects in one of the five SDG sector's P4G objectives.

14.11 Challenges and Future Direction

Possible avenues for further research have been uncovered—thanks to the studies that have been conducted so far. SWM has come a long way; however, there are still a lot of obstacles:

- **Unique issues with trash:** - E-waste or electronic waste, along with other waste such as hazardous trash and used vehicle tyres, cause several environmental issues. Using developing technology to manage various waste types can significantly reduce their environmental and social implications; needing more focused attention and improved recycling procedures. Research should focus on creating more efficient collection and recycling systems for these specialized wastes. Research should be directed towards developing more recycling systems and efficient collection for this type of garbage.
- **Recovery of energy and environmentally friendly remedies** - Innovative solutions such as geopolymer concrete made from energy recovery waste, waste materials, and waste-to-energy conversions should be further investigated because of the significant influence they could have on the sustainability and efficiency of metropolitan areas.

- **Environmental consciousness and involvement in the community** - Although there have been encouraging developments, there are still obstacles to overcome, such as a lack of public involvement, environmental consciousness, enough technology, and effective institutional cooperation [53]. Living Labs and other AI-driven systems, as well as community involvement and basic tools and frameworks, can help in achieving success. A lot of research is required to find out ways to raise environmental consciousness and community involvement. Strategies that encourage people to gather their own trash and use basic, resource-friendly methods look promising. Countries can plan to enforce regulations, laws around this.
- **Concerns regarding confidentiality and safety** - To better safeguard data and the waste management system (WMS) integrity, future iterations of IoT-enabled solutions should include stronger security features. Investigations could focus on developing strong security protocols to forestall any possible intrusions into the ever-expanding web of interconnected WMS.
- **Progress in technology** - AI, Intelligent Transportation System (ITS), the IoT, and industry 4.0 are just a few of the many new technological possibilities that smart waste management may take advantage of. Waste may be efficiently separated, collected, transported, and recycled with the help of these technologies, which are turning trash into useful materials. How various technologies might be more effectively combined to form a cohesive waste management system requires additional investigation. Cognitive computing methods for trash recovery, energy generation prediction, and energy management should be studied. Illegal dumping should also be addressed.
- **Formulation of policies and standardisation** - To ensure the seamless implementation and integration of WMS enabled by the IoT, future research should be centred on developing universal standards and legislative norms. To stay up to date with technology developments and conquer institutional obstacles, research also needs to think about the policy and regulatory environment.
- **Moving trash around for better management** - Waste collection, transportation, and sustainable urban planning are

all parts of a complex web that is about to undergo some changes. The use of algorithms for optimal path planning, like the IoT, Vehicle Routing Problem (VRP), Long Range Wide Area Network (LoRaWAN), Geographic Information Systems (GIS), and garbage collection routes optimization is reducing harmful environmental consequences. Better, more secure, and longer-lasting transportation systems for WM should be the goal of future research with the integration of the IoT, ITS, and blockchain for building smart cities [54].

- **Economic circularity and digitization** - Reduced usage of virgin materials and GHG emissions can only be achieved through the digitalization of waste management systems. Using ML, the IoT, and other technologies, researchers can better understand digitalization's revolutionary potential to move us closer to a circular economy.
- **International ramifications** - Within the urban landscape and the global sustainability context, WM plays a crucial role in smart cities. Aligning with the larger objectives of sustainable development, the study's findings provide a holistic viewpoint that different metropolitan centers across the world might embrace and modify.
- **Financial and ecological impact evaluation** - To assess the real advantages and find places for improvement, thorough evaluations of the technological possibilities and ecological consequences are required. The environmental effects of waste-to-energy conversions, recycling, and other long-term waste management solutions require more research.
- **Fresh conceptualizations and commercial strategies** - To create a WMS that is thorough, effective, and long-lasting, smart cities can investigate new frameworks and business models that use current and future technology.
- **Reducing waste** - It is critical that we investigate new waste avoidance and 'zero waste' paradigm research pathways as the world's waste problem worsens. Existing approaches are substantial, but they might not be able to handle the complexity of customers' ever-evolving behaviors [55]. Concurrently, there is an increasing need to investigate efficient methods of public education to guarantee that people fully grasp the issue and actively participate in efforts to reduce waste.

- **Disruption to existing systems** - The collection and processing of waste is still inadequate in many areas.
- **Limitations in technology and funding** - The importance of funding cutting-edge waste management solutions is increasing. Incorporating digital breakthroughs like AI-driven trash sorting systems, SWM solutions, and concepts of the circular economy should be the emphasis of future plans.
- **Raising awareness and encouraging engagement** - Make sure that trash management projects incorporate the community.

14.12 Conclusion

A zero-waste plan, aiming to transform trash into resources has been implemented by EU legislation with the goal of benefiting both society and the environment. As a framework for waste management, the solid waste hierarchy idea lends credence to such laws. Sorting and recycling facilities, composting/incineration units, and landfill prisms for residual bulk disposal should be installed at MSW programs under this approach. But there are by-products requiring treatment that are produced by each of those operations. Leachates from landfill prisms are the primary focus of this project. These include newer types of prisms that comply with EU regulations, as well as older types that store permitted biodegradable waste and the often-overlooked technological wastewaters from sorting units and composting units. Public health, resource preservation, and environmental sustainability all depend on well-managed solid waste. Communities can lessen their impact on the environment by implementing integrated waste management techniques. These strategies include reducing garbage, recycling, composting, and energy recovery. To create a better, more sustainable future, it is crucial for governments, companies, and individuals to work together.

AI disclosure statement: The author acknowledges the use of AI-based tools for minor language editing or grammatical correction. However, all intellectual contributions, analyses, and conclusions presented in this chapter are entirely the author's own.

References

1. Adhikari, S., Dangi, M.B., Cohen, R.R.H., Dangi, S.J., Rijal, S., Neupane, M., Ashooh, S., Solid waste management in rural touristic areas in the Himalaya – A case of Ghandruk, Nepal. *Habitat Int.*, 143, June 2023, 102994, 2024, doi: 10.1016/j.habitatint.2023.102994.
2. Hotta, Y., Visvanathan, C., Kojima, M., Pariatamby, A., Developing 3R policy indicators for Asia and the Pacific region: experience from Regional 3R Forum in Asia and the Pacific. *J. Mater. Cycles Waste Manage.*, 18, 1, 22–37, 2016, doi: 10.1007/s10163-015-0442-3.
3. Sakai, S., Yano, J., Hirai, Y., Asari, M., Yanagawa, R., Matsuda, T., Yoshida, H., Yamada, T., Kajiwara, N., Suzuki, G., Kunisue, T., Takahashi, S., Tomoda, K., Wuttke, J., Mählitz, P., Rotter, V.S., Grosso, M., Astrup, T.F., Cleary, J., Oh, G.-J., Liu, L., Li, J., Ma, H.-W., Chi, N.K., Moore, S., Waste prevention for sustainable resource and waste management. *J. Mater. Cycles Waste Manage.*, 19, 4, 1295–1313, 2017, doi: 10.1007/s10163-017-0586-4.
4. Shabani, T., Mutekwa, T.V., Shabani, T., Solid waste characteristics and management strategies at ST Theresa (STT) and Holy Cross (HC) hospitals in Chirumanzu rural District, Zimbabwe. *Environ. Sci. Eur.*, 36, 1, 2024, doi: 10.1186/s12302-024-00882-0.
5. Das, S., Lee, S.H., Kumar, P., Kim, K.H., Lee, S.S., Bhattacharya, S.S., Solid waste management: Scope and the challenge of sustainability. *J. Cleaner Prod.*, 228, 658–678, 2019, doi: 10.1016/j.jclepro.2019.04.323.
6. Liang, Y., Song, Q., Wu, N., Li, J., Zhong, Y., Zeng, W., Repercussions of COVID-19 pandemic on solid waste generation and management strategies. *Front. Environ. Sci. Eng.*, 15, 6, 2021, doi: 10.1007/s11783-021-1407-5.
7. Kennelly, B., *et al.*, The COVID-19 pandemic in Ireland: An overview of the health service and economic policy response. *Health Policy Technol.*, 9, 4, 419–429, 2020, doi: 10.1016/j.hlpt.2020.08.021.
8. Ike, C.C., Ezeibe, C.C., Anijiofor, S.C., Nik Daud, N.N., Solid waste management in Nigeria: Problems, prospects, and policies. *J. Solid Waste Technol. Manag.*, 44, 2, 163–172, 2018, doi: 10.5276/jswtm.2018.163.
9. Rupani, P.F., Nilashi, M., Abumalloh, R.A., Asadi, S., Samad, S., Wang, S., Coronavirus pandemic (COVID-19) and its natural environmental impacts. *Int. J. Environ. Sci. Technol.*, 17, 11, 4655–4666, 2020, doi: 10.1007/s13762-020-02910-x.
10. Sarkodie, S.A. and Owusu, P.A., Impact of COVID-19 pandemic on waste management. *Environ. Dev. Sustain.*, 23, 5, 7951–7960, 2021, doi: 10.1007/s10668-020-00956-y.
11. Mian, M.M., Zeng, X., al N. Bin Nasry, A., Al-Hamadani, S.M.Z.F., Municipal solid waste management in China: a comparative analysis. *J. Mater. Cycles Waste Manage.*, 19, 3, 1127–1135, 2017, doi: 10.1007/s10163-016-0509-9.
12. Ragazzi, M., Cristina, E., Schiavon, M., Since January 2020 Elsevier has created a COVID-19 resource centre with free information in English and

Mandarin on the novel coronavirus COVID- 19. The COVID-19 resource centre is hosted on Elsevier Connect, the company's public news and information, no. January, 2020.

13. Jerie, S., Shabani, T., Mudyazhezha, O.C., Shabani, T., A review towards developing a hierarchical model for sustainable hospital solid waste management in rural areas of Zimbabwe. *Environ. Monit. Assess.*, 196, 3, 1–21, 2024, doi: 10.1007/s10661-024-12488-3.
14. Mujtaba, M.A., *et al.*, Evaluating sustainable municipal solid waste management scenarios: A multicriteria decision making approach. *Heliyon*, 10, 4, e25788, 2024, doi: 10.1016/j.heliyon.2024.e25788.
15. Zhang, Z., *et al.*, Municipal solid waste management challenges in developing regions: A comprehensive review and future perspectives for Asia and Africa. *Sci. Total Environ.*, 930, February, 172794, 2024, doi: 10.1016/j.scitotenv.2024.172794.
16. Kwakye, S.O., Amuah, E.E.Y., Ankoma, K.A., Agyemang, E.B., Owusu, B.G., Understanding the performance and challenges of solid waste management in an emerging megacity: Insights from the developing world. *Environ. Challenges*, 14, October 2023, 100805, 2024, doi: 10.1016/j.envc.2023.100805.
17. Joshi, R. and Ahmed, S., Status and challenges of municipal solid waste management in India: A review. *Cogent Environ. Sci.*, 2, 1, 2016, doi: 10.1080/23311843.2016.1139434.
18. Somani, M., Srivastava, A.N., Gummadivalli, S.K., Sharma, A., Since January 2020 Elsevier has created a COVID-19 resource centre with free information in English and Mandarin on the novel coronavirus COVID-19. The COVID-19 resource centre is hosted on Elsevier Connect, the company's public news and information, no. January, 2020.
19. Turcott Cervantes, D.E., López Martínez, A., Cuartas Hernández, M., Lobo García de Cortázar, A., Using indicators as a tool to evaluate municipal solid waste management: A critical review. *Waste Manage.*, 80, 51–63, 2018, doi: 10.1016/j.wasman.2018.08.046.
20. Banerjee, P., Hazra, A., Ghosh, P., Ganguly, A., Murmu, N. C., & Chatterjee, P. K. (2018). Solid waste management in India: a brief review. Waste management and resource efficiency: Proceedings of 6th IconSWM 2016, 1027-1049.
21. Bhat, R., Ahmad Dar, S., Ahmad Dar, D., Hamid Dar, G., Municipal Solid Waste Generation and current Scenario of its Management in India. *Int. J. Adv. Res. Sci. Eng.*, 7, 2, 419–431, 2018, [Online]. Available: https://www.researchgate.net/publication/324756753_Municipal_Solid_Waste_Generation_and_current_Scenario_of_its_Management_in_India.
22. Adeniran, A.E., Nubi, A.T., Adelopo, A.O., Solid waste generation and characterization in the University of Lagos for a sustainable waste management. *Waste Manage.*, 67, 3–10, 2017, doi: 10.1016/j.wasman.2017.05.002.
23. Dubanowitz, A. and Themelis, N.J., Design of a Materials Recovery Facility (MRF) For Processing the Recyclable Materials of New York City's Municipal Solid Waste Design of a Materials Recovery Facility (MRF) For Processing

the Recyclable Materials of New York City's Municipal Solid Waste. *Design*, May, 43, 2000.
24. Rai, S., Gurung, A., Bhakta Sharma, H., Prakash Ranjan, V., Ravi Sankar Cheela, V., Sustainable solid waste management challenges in hill cities of developing Countries: Insights from eastern Himalayan smart cities of Sikkim, India. *Waste Manage. Bull.*, 2, 2, 1–18, 2024, doi: 10.1016/j.wmb.2024.02.009.
25. Awino, F.B. and Apitz, S.E., Solid waste management in the context of the waste hierarchy and circular economy frameworks: An international critical review. *Integr. Environ. Assess. Manag.*, 20, 1, 9–35, 2024, doi: 10.1002/ieam.4774.
26. Janeeshma, E., *et al.*, Enzymes-mediated solid waste management: A sustainable practice for recycling. *Waste Manage. Bull.*, 1, 4, 104–113, 2024, doi: 10.1016/j.wmb.2023.10.007.
27. Hirbod, F., Karimi, T., Mohammadnazari, Z., Rabbani, M., Aghsami, A., Municipal solid waste management using multiple disposal location-arc routing and waste segregation approach: a real-life case study in England. *J. Ind. Prod. Eng.*, 41, 1, 81–100, 2024, doi: 10.1080/21681015.2023.2265925.
28. Taye, A., Assefa, E., Simane, B., Analysis of practices and factors of solid waste management among urban households of Addis Ababa city, Ethiopia. *Environ. Challenges*, 14, November 2023, 100811, 2024, doi: 10.1016/j.envc.2023.100811.
29. Vargas, N.G., *et al.*, Environmental Perception of Solid Waste Management in the Hurtado Balneary, Colombia. *Water Air Soil Pollut.*, 235, 2, 1–22, 2024, doi: 10.1007/s11270-023-06877-6.
30. Ojuri, O.O., Olowoselu, A.S., Akinrele, J., Ayodele, F.O., Jayejeje, O.O., Sustainable integrated solid waste management for a university campus – A case study of the Federal University of Technology Akure (FUTA), Nigeria. *Waste Manage. Bull.*, 2, 2, 161–170, 2024, doi: 10.1016/j.wmb.2024.04.004.
31. Lawal, I.M., *et al.*, Proximate analysis of waste-to-energy potential of municipal solid waste for sustainable renewable energy generation. *Ain Shams Eng. J.*, 15, 1, 102357, 2024, doi: 10.1016/j.asej.2023.102357.
32. Ayilara, M.S., Olanrewaju, O.S., Babalola, O.O., Odeyemi, O., Waste management through composting: Challenges and potentials. *Sustain.*, 12, 11, 1–23, 2020, doi: 10.3390/su12114456.
33. Xie, Y., Qiu, N., Wang, G., Toward a better guard of coastal water safety—Microbial distribution in coastal water and their facile detection. *Mar. Pollut. Bull.*, 118, 1–2, 5–16, 2017, doi: 10.1016/j.marpolbul.2017.02.029.
34. Debrah, J.K., Vidal, D.G., Dinis, M.A.P., Raising awareness on solid waste management through formal education for sustainability: A developing countries evidence review. *Recycling*, 6, 1, 1–21, 2021, doi: 10.3390/recycling6010006.
35. Oyebode, O.J., Promoting Integrated and Sustainable Solid Waste Management System in A Developing City for Public Health and Cleaner Environment

Promoting Integrated and Sustainable Solid Waste Management System in A Developing City for Public Health and Cleaner Environ. *J. Harbin Eng. Univ.*, 45, 01, 2024.

36. Kubanza, N.S., Analysing the challenges of solid waste management in low-income communities in South Africa: a case study of Alexandra, Johannesburg. *S. Afr. Geogr. J.*, 1–21, 2024, doi: 10.1080/03736245.2024.2356563.
37. Tahiru, A.-W., Cobbina, S.J., Asare, W., Takal, S.U., Unlocking Energy from Waste: A Comprehensive Analysis of Municipal Solid Waste Recovery Potential in Ghana. *World*, 5, 2, 192–218, 2024, doi: 10.3390/world5020011.
38. Wang, D., Tang, Y.T., Long, G., Higgitt, D., He, J., Robinson, D., Future improvements on performance of an EU landfill directive driven municipal solid waste management for a city in England. *Waste Manage.*, 102, 452–463, 2020, doi: 10.1016/j.wasman.2019.11.009.
39. Nabavi-Pelesaraei, A., Bayat, R., Hosseinzadeh-Bandbafha, H., Afrasyabi, H., Chau, K., Modeling of energy consumption and environmental life cycle assessment for incineration and landfill systems of municipal solid waste management - A case study in Tehran Metropolis of Iran. *J. Cleaner Prod.*, 148, 427–440, 2017, doi: 10.1016/j.jclepro.2017.01.172.
40. Barbieri, F., Rajakaruna, S., Ghosh, A., Very Short-Term Photovoltaic Power Forecasting with Cloud Modeling: A Review. *Renewable Sustainable Energy Rev.*, 75, 242–263, 2017, [Online]. Available: http://scioteca.caf.com/bitstream/handle/123456789/1091/RED2017-Eng-8ene.pdf?sequence=12&isAllowed=y%0Ahttp://dx.doi.org/10.1016/j.regsciurbeco.2008.06.005%0Ahttps://www.researchgate.net/publication/305320484_SISTEM_PEMBETUNGAN_TERPUSAT_STRATEGI_MELESTARI.
41. Chikukula, A.A., Omokaro, G.O., Godswill, O.O., Cassim, S.Y., Mabangwe, H.S., Kaisi, I., Problems and Possible Solutions to Municipal Solid Waste Management in Malawi Urban Areas – An Overview. *Asian J. Environ. Ecol.*, 23, 6, 42–52, 2024, doi: 10.9734/ajee/2024/v23i6553.
42. Jeswani, H.K. and Azapagic, A., Assessing the environmental sustainability of energy recovery from municipal solid waste in the UK. *Waste Manage.*, 50, 346–363, 2016, doi: 10.1016/j.wasman.2016.02.010.
43. Sagar, N.A., Pareek, S., Sharma, S., Yahia, E.M., Lobo, M.G., Fruit and Vegetable Waste: Bioactive Compounds, Their Extraction, and Possible Utilization. *Compr. Rev. Food Sci. Food Saf.*, 17, 3, 512–531, 2018, doi: 10.1111/1541-4337.12330.
44. Hosseinpoor, S., Dadashi, T., Mohammadi, A., Comprehensive analysis of hospital solid waste levels and HSE risks using FMEA technique: A case study in northwest Iran. *Case Stud. Chem. Environ. Eng.*, 9, February, 100646, 2024, doi: 10.1016/j.cscee.2024.100646.
45. Qazi, W.A., Abushammala, M.F.M., Azam, M.H., Multi-criteria decision analysis of waste-to-energy technologies for municipal solid waste

management in Sultanate of Oman. *Waste Manage. Res.*, 36, 7, 594–605, 2018, doi: 10.1177/0734242x18777800.
46. Dimani Tharuka Hapuarachchi, H.A., Community participation in municipal solid waste management: Special reference to Gampaha municipality. *Int. J. Sci. Res. Arch.*, 11, 2, 1153–1158, 2024, doi: 10.30574/ijsra.2024.11.2.0574.
47. Olawade, D.B., *et al.*, Trends of solid waste generation during COVID-19 Pandemic: A review. *Waste Manage. Bull.*, 1, 4, 93–103, 2024, doi: 10.1016/j.wmb.2023.10.002.
48. Mdukaza, S., Isong, B., Dladlu, N., Abu-Mahfouz, A.M., Analysis of IoT-enabled solutions in smart waste management. *Proc. IECON 2018 - 44th Annu. Conf. IEEE Ind. Electron. Soc.*, pp. 4639–4644, 2018, doi: 10.1109/IECON.2018.8591236.
49. Campuzano, R. and González-Martínez, S., Characteristics of the organic fraction of municipal solid waste and methane production: A review. *Waste Manage.*, 54, 3–12, 2016, doi: 10.1016/j.wasman.2016.05.016.
50. Liu, L., Barlaz, M.A., Johnson, J.X., Economic and environmental comparison of emerging plastic waste management technologies. *Resour. Conserv. Recycl.*, 205, 107531, 2024, https://doi.org/10.1016/j.resconrec.2024.10753.
51. Chen, Y., Wang, H., Ghofrani-Isfahani, P., Gu, L., Liu, X., Dai, X., Electronic regulation to achieve efficient anaerobic digestion of organic fraction of municipal solid waste (OFMSW): strategies, challenges and potential solutions. *Front. Environ. Sci. Eng.*, 18, 4, 1–7, 2024, doi: 10.1007/s11783-024-1812-7.
52. Njonge, T., Influence of Psychological Well-Being and School Factors on Delinquency, During the Covid-19 Period Among Secondary School Students in Selected Schools in Nakuru County: Kenya, vol. VII, no. 2454, pp. 1175–1189, 2023, doi: 10.47772/IJRISS.
53. Vassileva, A., Green Public-Private Partnerships (PPPs) as an Instrument for Sustainable Development. *J. World Econ. Transform. Transitions*, 2, 05, 1–18, 2022, doi: 10.52459/jowett25221122.
54. Fayomi, G.U., *et al.*, Smart Waste Management for Smart City: Impact on Industrialization. *IOP Conf. Ser. Earth Environ. Sci.*, 655, 1, 012040, Feb. 2021, doi: 10.1088/1755-1315/655/1/012040.
55. Fudala-Ksiazek, S., Pierpaoli, M., Kulbat, E., Luczkiewicz, A., A modern solid waste management strategy – the generation of new by-products. *Waste Manage. (Oxford)*, 49, 516–529, 2016, doi: https://doi.org/10.1016/j.wasman.2016.01.022.

15

Unlocking the Value of Solid Waste: Challenges and Opportunities

Niharika Shringi

Department of Chemistry, School of Sciences, JECRC University, Jaipur, Rajasthan, India

Abstract

Waste is a complex challenge that requires collaboration among multiple stakeholders to achieve sustainable development. Improper disposal worldwide has caused severe environmental damage and is impacting human health. Lack of enforcement of rules and their implementation, minimal participation by citizens, and less involvement of technology are challenges in waste management. In this chapter, to counter these problems, the circular economy approach and zero waste initiatives, and new government initiatives and policies such as Swachh Bharat Abhiyan, extended producer responsibility (EPR), etc. are discussed. The contribution of technological innovations such as AI, IoT, and machine learning applications in waste tracking and processing, the role of startups and the private sector are also discussed in detail. Success stories from global and Indian cities that have successfully implemented sustainable waste management strategies are included. The future approach for managing waste, policy recommendations, and better and wider use of technological advancements are also considered.

Keywords: Solid waste management, circular economy, Internet of Things (IoT), Swachh Bharat Abhiyan, policy frameworks, sustainable business models, extended producer responsibility (EPR), machine learning

15.1 Introduction

The anthropocentric issue of uncontrolled solid waste generation is one of the most serious threats to the long-term territorial development of cities,

Email: shringi.niharika@gmail.com

Deepti Goyal, Sakshi Kabra Malpani and Renuka Jain (eds.) Catalysis from Waste: Transforming Solid Residues into Value, (393–414) © 2026 Scrivener Publishing LLC

countries, regions, and, on a larger scale, terrestrial ecosystems. It relates to productive processes, consumption, urban space formation, and the extraction, production, and disposal of Earth's natural resources [1]. The international scientific community is seeking solutions from states, governments, and public and private organizations to address the challenge of unlimited economic growth based on a linear economic model [2, 3]. Over 200 years after the industrial revolution, institutions are recognizing the urgency of shifting to a sustainable model with minimal entropy due to the irreversible effects on Earth's natural systems and planetary boundaries [1]. Today, we have a technologically advanced economy and globalization that offers better longevity and encourage sustainability. However, along with these changes, factors such as rapid socio-economic changes and infrastructure development to cater rising population are now also appearing in the scenario. Countries worldwide are experiencing significant urban agglomeration as a consequence of adopting industrial-focused economic strategies [4]. Consequently, by 2050, two-thirds of the global population is expected to reside in cities [5], producing 3.40 billion tons of municipal solid waste (MSW) annually, a 70% increase from 2.01 billion tons [6] generated every day.

Hence, municipal solid waste management is a public policy developed in reaction to the economic system's negative externality. To begin, optimize the management of this negative externality, particularly with solutions at the forefront of the phenomenon [7], whether initially with direct release into the environment, and later with landfilling of these resources in landfills, controlled landfills, burning, and/or transformation into energy (WtE) through combustion or plasma [8].

In this chapter, we have discussed the current scenario of waste generated in India and all over the globe and environmental and health hazards due to its improper disposal. The role of governments and the lack of strict enforcement and fragmented policies are also discussed. The significance of public awareness and behavior is a key factor in sustainable management. To achieve the goals of the sustainable development goals (SDGs), the role of the circular economy and other waste valorization techniques is also included. Involvement of AI, IoT, and deep learning in waste tracking and processing is also discussed. The initiatives taken at the national and international levels, and the contribution of start-ups and the private sector, are also documented in this chapter. The successfully implemented sustainable waste management strategies, as case studies, have also been taken into account. In conclusion, to achieve sustainable waste management in the long run, future road development through technological advancements and research directions is discussed.

15.2 Solid Waste Generation Scenario (Global and Indian)

One of the most significant challenges arising from contemporary socio-economic development is the steady increase in MSW generation, which exacerbates the accompanying environmental and economic issues [9]. Both population growth and a significant rise in the global gross domestic product (GDP) will increase solid waste generation, imposing a serious threat to sustainability [10, 11]. According to the World Bank, global garbage generation is expected to rise by nearly 70%, from 2.01 billion tons in 2016 to 3.40 billion tons by 2050 [12]. According to an estimate, household waste accounts for 55-80% of the net MSW produced from developing countries [13]. Due to worldwide abrupt urbanization, an increase in solid waste generation has been observed. Disposal of solid waste varies from region to region all over the world. Only 40% of the waste is directly disposed off in landfills; on the other hand, only 19% undergoes recycling. In low-income countries, the percentage of openly dumped waste is almost 93%, including burning and dumping by the side of the road, whereas only 2% waste is dumped in high-income countries [12].

On average, each person generates about 0.74 kg of waste every year; on the other hand, North Americans are at the top of this list by producing 800 kg of waste per capita annually [14]. The countries with high income generate 34% of the total global waste, while low-income countries contribute only 5% of the total waste generation worldwide [15]. In India, rapid economic development, increasing population, swift urbanization and rising living standards have accelerated the MSW generation. India has diverse geographic and climatic conditions. The people living in these zones have varying usage habits; they produce distinct types of waste, resulting in a variety of waste. However, no serious initiatives have been taken to counter this garbage menace. According to the Central Pollution Control Board (CPCB) 2021 report, in India, MSW generation rate is 160,038.9 tons per day (TPD), which is approximately 0.119 kg per capita per day. 95.40 % (approx. 152,749.5 TPD) of the total waste was collected and disposed off. Nevertheless, 50% (approx. 79,956.3 TPD) was processed and underwent all types of waste treatment processes before disposal. 18.40% (29,427.2 TPD) was directly disposed off into the landfill, and the remaining 31.70% (50.655.4 TPD) was unaccounted for [16]. The comparative data of total waste generated for the years 2000, 2009, 2011 and 2021 in the different states of India were collected from various publications [16–18]. In the context of MSW generation, the year 2021, as compared to 2000, shows that Tripura has secured the highest percentage growth, whereas Andhra Pradesh acquired the last position. In the last decade, Manipur has achieved the highest percentage growth in MSW generation, while five states highlighted in green color have reduced their waste production [17]. Regardless of the alternatives to waste management that have been proposed, many countries are refusing them, and the most common treatment methods are open dumping [19, 20] and landfilling [21, 22]. Day by day, the conventional waste management approaches, such as landfilling, are becoming increasingly ineffective [23, 24].

Waste pollution is a major environmental concern affecting ecosystems, public health, and sustainable development worldwide. Rapid waste production and poor waste management practices lead to numerous ecological and health hazards. Improper disposal of pharmaceuticals causes disruptions in aquatic ecosystems and promotes antimicrobial resistance [25]. Incorrect disposal of medical waste, particularly radioactive materials, presents serious public health threats due to environmental radiation. This issue stems from mishandling and improper disposal of waste created by medical procedures involving radioactive substances, such as certain diagnostic scans and cancer treatments. Careless handling of radioactive waste can pollute land, water, and air, creating health and environmental dangers. Radiation exposure increases the risk of various diseases, including leukaemia and thyroid cancer, especially for communities near inadequately managed waste sites. Long-term exposure to even low levels of radiation can cause cumulative DNA damage, leading to mutations and health problems over time. Contaminated water sources pose a serious threat, as radioactive materials can seep into groundwater and surface water from improper disposal sites. Ingesting or inhaling radioactive particles from polluted water or air can cause internal radiation exposure, raising health risks related to environmental contamination. Additionally, radioactive waste can degrade soil quality, impacting agriculture and food safety. Radioactive isotopes in soil can enter the food chain, contaminating crops and livestock. Consuming contaminated food may introduce new health risks, increasing the public health burden. To mitigate these risks, proper disposal techniques and strict regulatory measures must be enforced to prevent environmental contamination.

15.3 Challenges in Solid Waste Management

Conventional solid waste management practices are inadequate to solve the problem of managing huge masses of solid waste generated every year around the globe. The challenges of this process start with basic sorting of solid waste and public awareness, and get more complex when social, infrastructure, legal, environmental, economic and wellness-related aspects are added to this equation. The following sub-sections represent a clear overview of the challenges associated with solid waste management:

15.3.1 Policy and Regulatory Gaps: Lack of Strict Enforcement and Fragmented Policies

Waste management has become a critical challenge in urban areas, involving environmental, socio-economic, and governance issues. This critical situation comes into play due to gaps in policies and regulations, inadequate infrastructure, dependence on the informal sector, poor public awareness and participation, financial limitations, and insensitivity towards adopting advanced recycling and waste-to-energy solutions. Adeborode *et al.* [26], discussed the challenges of food waste management in urban areas, such as the lack of systematic food waste segregation at the source, inadequate infrastructure for recycling and composting, poor policy execution, and fragmented governance, finally causing environmental and public health crises. Food wastage is also greatly influenced by public perceptions and behaviors. Misreading expiry dates, over-purchasing, improper storage practices, and cultural customs are some of the major reasons for food being cast away by customers. Another factor is that many people are unaware of how to properly store perishable items, resulting in early spoilage and unnecessary waste.

15.3.2 Infrastructure Deficiencies: Insufficient Waste Collection, Segregation, and Recycling Facilities

In 2007, to control the "quadruple crisis" involving climate change, biodiversity loss, pollution, and public health and economic impacts due to the uncontrolled use of plastic, the Federal Democratic Republic of Ethiopia (FDRE) established policies for the management of solid waste (FDRE 2007) [27]. Proclamation No. 62/1999 and 2007 directives include the ban on plastic bag production and their import (thickness less than 0.03 mm), plastic waste reduction, and compliance and enforcement of the ban. Nevertheless, due to a lack of details on plastic waste management pathways and implementation strategy, this policy couldn't be effectively implemented to achieve the goal of zero plastic waste. No regional environmental authorities have been established in all parts of the country

[27, 28]. For effective implementation of this policy, the United Nations Development Program (UNDP) supplied 13 baling machines (to compress plastic 1/3rd of its volume to make transportation easy for waste collection) to the Ministry of Urban and Infrastructure of Ethiopia to equip major cities, including Adama, Bahir Dar, Bishoftu, Dire Dawa, and Hawassa, in managing waste. But it was not continued or implemented to get the maximum advantage.

15.3.3 Public Awareness and Behavior: Low Participation in Waste Segregation and Sustainable Disposal

Biodegradable plastics (BPs) have been widely promoted as environmentally friendly alternatives to conventional plastics, but they also face challenges from a solid waste management perspective. In a recent review article [29], a comprehensive summary of 13 existing European Union (EU) laws related to waste management and the circular economy, along with national legislation and source separation guidelines from 13 countries, is provided to verify the optimal utilization of resources in the future. The various types of BPs create a significant hurdle to establishing profitable waste management infrastructure. BPs are used in food waste collection, but due to the lack of knowledge about waste, consumers remain unaware of proper disposal methods, and they are dependent on manufacturers for green washing. According to the reports given by industry studies, the lack of full-scale mechanical recycling processes is also another reason. The current EU legislation only broadly addresses issues related to BPs. Therefore, a new proposal on plastic packaging waste and the EU policy framework for BPs, also focusing on their disposal, is required. Therefore, the development of efficient waste processing and material utilization infrastructure, coupled with increased public awareness, has become essential for the effective management of BP waste.

15.4 Opportunities and Potential Solutions

Utilization of solid waste through the circular economy, upcycling, and promotion of zero-waste initiatives has emerged as sustainable and greener approaches to prevent its hazardous environmental impact on our ecology. The circular economy encourages recycling and upcycling of products. The zero waste initiatives promote pathways to avoid waste incineration and discharge, leading to maximizing the efficiency of natural resource utilization.

15.4.1 Circular Economy Approach: Promoting Recycling, Upcycling, and Zero-Waste Initiatives

The term "circularity" refers to numerous approaches aimed at maintaining resources and materials within extended virtuous loops of production and consumption, such as reducing, reusing, recycling, and renting rather than purchasing goods [30, 31]. The circular economy (CE) emphasizes the retention and optimization of products for reusability, redesign, and renovation to enhance the recyclability of products, as illustrated in Figure 15.1 The purpose of introducing CE concept through the natural cycle model is to highlight the shift away from the traditional 'extraction-production-disposal' pattern of the linear economy (LE), which aligns with the current industrial landscape. This approach has gained worldwide attention to emphasize sustainability issues in public policies, with governments, as the European Union and China adopting it as part of their development strategies. Estimates indicate that by 2050, the global population will reach 9 billion [32], leading to serious anthropogenic and environmental impacts, such as decreased biodiversity, large-scale ecological imbalance, scarcity of natural resources, and resulting fluctuations in market prices and instability in the global economic system [33, 34]. Globally, the circular plastics economy has emerged as a flagship eco-friendly shift from the conventional "take-make-dispose" model to combat plastics pollution [35]. Single-use hygienic products, drinking straws, plates, etc., contribute significantly to plastic pollution; over twenty billion such products are used in less than a year [36]. Therefore, implementing bans or prohibitions on single-use plastics could improve the situation. Using biodegradable plastics

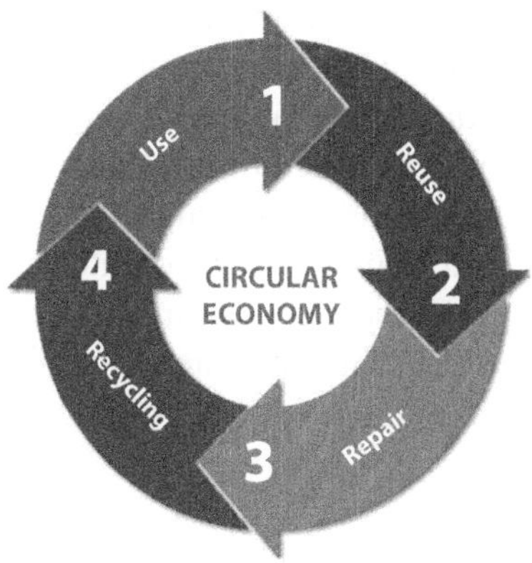

Figure 15.1 Core steps of circular economy. Reprinted from ref. [42].

as a reusable alternative to disposables can be a sustainable solution to this problem [37]. Furthermore, developing infrastructure for plastic recycling and adopting a systematic approach to plastic waste management can turn this waste into a valuable resource [38]. Achieving zero waste requires a comprehensive strategy that encompasses product design, alternative materials, advanced recycling infrastructure, and improved technologies. Enhancing recycling requires better collection and sorting systems, including more recycling bins, a deposit-refund scheme, and improved sorting technologies, which will help increase recycling rates and reduce reliance on landfills [39]. Accountability measures, such as extended producer responsibility (EPR), rely on binding recycling programs, quotas, and levies for producers [40]. Awareness efforts and incentive schemes promote safe disposal, whereas single-use plastic prohibitions in environmentally sensitive locations reduce littering [41].

The zero waste initiative or strategy was encouraged to reduce the environmental impact and depletion caused by solid waste [43]. The zero-waste is primarily responsible for production and consumption approaches, resource reuse and efficient recycling, exploring pathways to avoid waste incineration and discharge, and finally maximizing the efficiency of natural resource utilization [44].

A zero-waste city is an urban paradigm that strives to eliminate trash in landfills or incineration by implementing circular economy principles, resource efficiency (product and service production), and environmental responsibility. The model assumes a closed-loop system in cities, where garbage is constantly recovered, repurposed, reused, or not generated at all. The zero-waste city promotes a culture of shared responsibility among the government, commercial sector, and civil society, with a focus on technical innovation and integrated urban planning [45]. Qian *et al.* [46], studied the implementation of the Zero Waste pilot policy in China, which significantly improved the green transformation in the pilot sector. Mechanism tests demonstrated that this policy facilitated corporate green transformation by promoting enterprises' green innovations. Heterogeneity test confirmed significant changes found in non-high-tech industries, non-heavily polluting enterprises, and non-state-owned enterprises, which also benefited from the policy. Nevertheless, future research is still required to employ extended time series data and international comparative analyses to assess the effectiveness of interventions [46].

15.4.2 Government Initiatives and Policies (Global and Indian): Swachh Bharat Abhiyan, Extended Producer Responsibility (EPR), Waste-to-Wealth Mission

For generations, Bharat has struggled to provide adequate sanitation to its citizens. To address this issue, in 2014, the union government launched the Swachh Bharat Mission (SBM), a statewide drive to address the serious concern of poor sanitation. SBM has been regarded as the world's largest sanitation initiative, with the lofty objective of guaranteeing that Bharat is free of open defecation by October 2019 [47]. The SBM has achieved substantial success by helping approximately 100 million rural households to get access to toilets, equating to 500 million citizens across approximately 630,000 villages in the nation [48]. This has resulted in major economic, environmental, and health benefits across the country. In the context of financial gains, it is projected that SBM has saved around INR 4.9 trillion (roughly USD 76 billion) [49]. This economic benefit has been experienced across all levels of society, with financial returns on household spending over a decade estimated at 1.7 times the cost, returns to society on overall spending over a decade at 4.3 times the cost, the poorest in society seeing a financial return of 2.6 times the cost, and the societal benefit being 5.7 times the cost [50]. Nevertheless, continuous media coverage is needed to raise more awareness about cleanliness, change in mindsets, and invest considerably to ensure that waste disposal and segregation are consistently and efficiently implemented throughout the country. Additionally, both qualitative and quantitative research are required for a better understanding of perceptions of SBM and continuous efforts to improve its implementation approach [51].

Another promising waste management approach known as Extended Producer Responsibility (EPR) has become an organizational and economic tool [52]. Here, manufacturers should be responsible for every stage, from manufacturing to the disposal of their products. This initiative aims to encourage companies to minimize the adverse environmental effects of their packaging and products through effective recycling, reuse, and disposal [53]. EPR initiatives require genuine engagement from stakeholders such as producers, retailers, waste management companies, consumers, and government agencies for successful implementation. In 2020, India adopted an EPR program through the Plastic Waste Management (PWM) for plastic packaging materials using an integrated mixed-method case study, known as the Uniform Framework of EPR [54]. By 2024, the adoption of an EPR scheme for packaging will have been widely accepted in many parts of the world. In the 1970s, the US introduced EPR for various materials, including hazardous waste, advanced disposal costs, electronic waste, and beverage containers. States such as Oregon and Maine have implemented fee-based mandatory EPR systems for packaging waste. In 2021, Oregon enacted the Plastic Pollution and Recycling Modernization Act, while Maine passed LD 1541 to improve municipal recycling systems and reduce taxpayers' costs (Maine Department of Environmental Protection, Oregon Department of Environmental Quality). Following the EU Directive 94/62/EC on Packaging and Packaging Waste in 1994, this law sets current net packaging recycling targets of 55% by 2025, with a goal of 25% by weight plastic recycling [55]. Additionally, by 31 December 2030, the target will be increased to ensure that at least 70% by weight of all packaging waste is recycled, with a 55% recycling target specifically for plastic [53]. In 1995, Japan established its packaging EPR program under the Act on the Promotion of Sorted Collection and Recycling of Containers and Packaging. Between 1997 and 2010, packaging recycling rates in Japan rose by 27%, marking their success. In 2020, India introduced the Uniform Framework of EPR, a program for plastic packaging materials, under the Rules for Plastic Waste Management (India Ministry of Environment). The success of this EPR plan is largely attributed to the voluntary actions of stakeholders [54].

15.4.3 Technological Innovations: Applications of AI, IoT, and Machine Learning in Waste Management

The introduction of artificial intelligence (AI) and the Internet of Things (IoT) in waste management strategies has emerged as a cost-effective and reliable solution to shift from traditional methods. AI helps in analyzing complex data to optimize waste collection schedules, reducing the number of pickups, checking the status of bins before overflowing, and preventing the emptying of unused bins [56]. Resource optimization enhances the reliability of the collection process while reducing fuel costs and emissions. IoT involves interconnected devices and sensors that enable real-time monitoring, data collection, and analysis [57]. It allows waste bins and collection vehicles to communicate their status, location, and capacity levels to central management systems. Sensors in waste bins monitor fill levels, while GPS-enabled vehicles track routes and operational activities. These technologies communicate *via* advanced systems such as LoRaWAN and NB-IoT, ensuring seamless data transmission to centralized platforms.

IoT-powered platforms then analyze the data received and determine the most appropriate response, resulting in fewer unnecessary trips, lower fuel consumption, and reduced operational costs [58]. When AI algorithms process IoT-generated data to forecast waste generation patterns in urban areas, they also assist in planning collection schedules and optimizing resource allocation. For instance, ML models can predict peak waste periods during events and seasons, enabling cities to respond more effectively. Additionally, AI-powered automated sorting systems can significantly improve recycling efficiency by sorting waste with high accuracy. The integration of AI and IoT enhances waste collection and sorting processes and promotes the circular economy, contributing to greater sustainability. Using AI and IoT, urban centers can apply SMART (self-monitoring, analysis, and reporting technology) and adaptive management solutions that address solid waste generation challenges, reduce environmental impacts, support optimal utilization of available resources, and assist cities in achieving sustainability goals [58].

With the introduction of deep learning (DL) techniques, there has been a renewed emphasis on addressing waste management issues by creating image recognition systems for waste categorization and automated waste sorting [59, 60]. They are increasingly used for garbage classification and detection, particularly in household waste sorting. Sirimewan *et al.* [61], used cutting-edge deep learning-based semantic segmentation models, DeepLabv3+ and U-Net, to recognize the composition of CRD (construction, renovation, and demolition) debris in the wild. The study discovered that DeepLabv3+ and U-Net models, with a ResNet-101 backbone network, achieved the maximum segmentation accuracy. Additionally, the models performed satisfactorily with RegNetX-1.6 and MobileOne-S3, which are lightweight backbones. These findings highlight the usefulness of DL models in identifying CRD waste in the field, which can be used to automate waste sorting in the CRD context. This study can play a significant role in mitigating the waste sent to landfills through improved recycling and recovery efforts, encouraging better environmental management. However, time and resource constraints hindered the data collection process. The authors also manually created ground truth labels for the whole dataset. Therefore, future research should focus on increasing the dataset to maintain the category imbalance and automating the labelling process, thereby reducing the need for time-consuming annotations.

15.4.4 Role of Startups and Private Sector: Waste Management Startups and Business Models for Sustainability

In today's data-driven world, growing concern about environmental issues encourages more entrepreneurs to pursue business model innovation that supports eco-friendly solutions. These firms aim to develop technological products and services that promote environmental sustainability. In this context, AI becomes a crucial tool for creating, capturing, and delivering value. The business model offers economic advantages to start-ups, along with additional economic, social, and environmental benefits for both their organizations and the environment. The concept of sustainable innovation is based on broader normative principles such as environmental sustainability, sustainable development [62], and start-up ventures [62]. The basic idea of sustainable innovation as a value provided by businesses refers to a single-sided market in a transaction between the supplier and the customer within the value chain of the Business-to-Business (B2B) or Business-to-Citizens (B2C) segments. Multiple customer groups and transacting parties from various sectors of the economy, including Business-to-Government (B2G), Government-to-Business (G2B), Government-to-Citizens (G2C), Citizens-to-Government (C2G), and Government-to-Government (G2G), maintain multiple relationships with each other, with numerous value propositions considered [63]. Here, social impact accelerators (SIAs) are organizations specialized in boosting the growth of sustainability-focused start-ups. The European Commission (EU) promotes the development of accelerators concentrating on specific technology sectors or areas as part of its economic development initiatives. These accelerators are funded by the European Institute of Innovation & Technology (EIT), which has established numerous Knowledge and Innovation Communities (KICs) to tackle significant societal issues and evolving environments [64]. KICs aim to foster a supportive environment for innovation, including start-ups developing relevant solutions, and to accelerate their growth through funding and appropriate entrepreneurial training. The EIT Climate-KIC RIS Accelerator was launched in 2016, working with a total of 83 start-ups across 12 European countries from the EIT RIS (Bulgaria, Croatia, Cyprus, Estonia, Greece, Latvia, Lithuania, Malta, Portugal, Romania, Serbia, and Slovenia). Bergmann also analyzed that out of these 83 start-ups, only 63 programs participated in the study.

15.4.5 Community Engagement: Citizen Participation, Awareness Campaigns, and Behavioral Shifts

Municipal Solid Waste Management (MSWM) greatly impacts urban livability and public health. Unfortunately, MSWM is often seen as only a government duty. However, all key stakeholders agree that residents must actively take part in this process because they are the primary producers of MSW. Nonetheless, a lack of understanding among individuals and poor communication between the government and citizens pose significant barriers to progress. Therefore, various legislative measures are needed to close this communication gap. Public awareness boosts entrepreneurial activity by encouraging community engagement, proper waste disposal, recycling programs, and advocating sustainable behavioral changes [65]. Furthermore, greater public awareness, acceptance, and engagement lead to successful businesses, job creation, and community development, as MSWM firms demand trained labor throughout the value chain [66]. Along with economic benefits, promoting waste management is vital for reducing pollution and conserving resources [66]. Effective MSWM provides significant public health and environmental benefits [67]. As a result, raising public awareness about proper waste disposal and recycling will improve waste segregation and decrease landfill use.

15.5 Case Studies and Success Stories

Segregation of waste is a crucial step in the waste management process, as it facilitates successful reuse, recycling, and recovery (RRR). Sadly, it has received limited attention and is mostly practised informally in most developing countries (DC). Through effective policies, South Korea has achieved remarkable progress in food waste recycling. Food waste makes up about 30% of total rubbish, with over 90% being successfully sorted and collected [68]. Optimizing biogas production and converting food waste into animal feed remain challenging tasks. The Volume-Based Waste Free system introduced in 1995, reduced waste while promoting recycling. The 2005 ban on direct food waste landfilling encouraged more sorting and careful disposal. The Master Plan for Reducing Food Waste, launched in 2010, led to the nationwide adoption of the Weight-Based Food Waste Fee (WBFWF) system in 2013. South Korea has enhanced its food waste management by drawing inspiration from Japan's policies and refining them with the WBFWF system. It also adopted European measures to strengthen its regulations. Still, the efficient utilization of waste materials necessitates a multifaceted approach rather than the creation of a single technology. Optimization of approaches for dealing with food waste based on the characteristics of each region is required, emphasizing societal and economic benefits through participation programs. Continuous policy effectiveness monitoring is required, along with critical technical improvements that take into account energy and carbon considerations. Further research into more precise waste management systems would help to promote sustainable resource rotation in local communities.

Singapore, which has a population of 5.9 million, generated an average of 759 tons of food waste during the last five years, worsening food security concerns and putting significant strain on the country's food and land resources. The Singapore government's Zero Waste Plan and 30-by-30 food security target aim at waste reduction and resource efficiency, while also encouraging local agri-food firms to collectively meet 30% of Singapore's nutritional requirements by 2030. In recent years, valorizing food waste streams into higher-value goods has been a widely accepted approach for reducing food waste and providing a new supply of food ingredients. Local research and development capabilities have effectively utilized food wastes such as okara, spent barley grains, and fruit and vegetable wastes to create a variety of prototypes for culinary and agricultural purposes. Nevertheless, food waste valorization has numerous hurdles, including infrastructure development, economic feasibility, consumer awareness, collaborative partnerships, regulatory backing, and data management. This assessment can be used as an instrument for other countries seeking to ensure food security and achieve the Sustainable Development Goals [69].

Hamdan *et al.* [4], comprehensively analyzed the present waste management strategies against the demographics of ten Indian smart cities. The study also focused on the current state of MSW functional elements, including financial and administrative frameworks. The gathered information was collected from secondary sources, government and local government databases, and reports from important stakeholders. The article gives recommendations and offers insights into the feasibility, constraints, and promise of implementing waste-to-energy (WtE) systems in India. Among all ten Indian smart cities, Indore's success in achieving high segregation rates is due to good governance, public participation initiatives like "Do Bin Har Din," and the use of ICT tools such as vehicle tracking and real-time data systems. This signifies the importance of community interaction and technology innovation in improving trash management [70]. On the contrary, cities such as Ranchi, Srinagar, and Varanasi suffer hurdles due to ineffective policy enforcement, poor literacy, and inadequate budgets. Larger cities, such as Pune and Surat, while unaffected by these factors, face challenges in trash processing and collecting due to poor resource utilization, limited public knowledge and engagement, and a trained labor scarcity. This emphasizes the importance of customized interventions and strong multi-stakeholder collaborations. Replicating the Indore approach calls for raising public awareness, enforcing strict source segregation rules, integrating the informal waste sector to increase recycling, incentivizing public engagement, and developing ICT infrastructure. The financial limits for such commitments can be overcome by soliciting financial contributions from the private sector in exchange for offering incentives in energy, by-products, and the supply chain from WtE systems [4]. Additionally, India's heterogeneous MSW necessitates the use of essential sorting techniques for effective WtE treatment. This includes a financial and physical investment by MNCs, preventing them from incorporating WtE into their MSWM structure.

15.6 Future Roadmap for Sustainable Waste Management

This chapter aims to critically assess the effectiveness of waste management with the SDGs and their capacity to meet the socio-economic needs of diverse populations worldwide. To address waste-related challenges, a radical change towards the circular economy approach is required so that resources can be efficiently utilized with minimal waste generation. Government strategies to promote the CR should include effective waste management plans, proper implementation of EPR schemes, and encouraging recycling and upcycling programs among the people and manufacturers. Das *et al.* [71], highlighted that to combat transboundary plastic pollution across countries, fostering cross-border collaboration and partnerships in the sector is essential. Such collaborative initiatives and opportunities strengthen regional cooperation among South Asian countries and international organizations.

Policymakers can play a significant role in encouraging a supportive environment for adopting digital waste management using AI and ML to make the whole process faster and easily trackable. Proper implementation of laws and maintaining environmental guidelines by industries, public-private partnerships in the waste management sector can play a pivotal role in addressing challenges such as resource scarcity, technological hurdles, and system integration [72]. Environment-related regulations, such as the European Union's Waste Framework Directive and Extended

Producer Responsibility regulations, also give essential guidance for promoting sustainable waste practices to achieve minimal waste generation. To advance in the waste management sector in the future, emerging technologies such as blockchain, robotics, ML and DL models should be implemented according to the type of waste and region, which could improve the transparency, accountability, and resource recovery in a brief period. DL can create image recognition systems for waste categorization and an automated waste sorting system. Blockchain can contribute by improving traceability across supply chains, while robotics can automate waste sorting and handling. Hence, the adoption of these digital technologies and waste recycling strategies would significantly improve lifestyles in cities and business cultures, leading to a greener, cleaner, and more sustainable future.

AI disclosure statement: The author acknowledges the use of AI-based tools for minor language editing or grammatical correction. However, all intellectual contributions, analyses, and conclusions presented in this chapter are entirely the author's own.

References

1. Rittl, L.G.F., Zaman, A., de Oliveira, F.H., Digital Transformation in Waste Management: Disruptive Innovation and Digital Governance for Zero-Waste Cities in the Global South as Keys to Future Sustainable Development. *Sustainability*, 17, 1608, 2025.
2. Herbert, E., Giraud, G., Louis-Napoléon, A., Goupil, C., Macroeconomic dynamics in a finite world based on thermodynamic potential. *Sci. Rep.*, 13, 1–27, 2023.
3. Borrello, M., Cembalo, L., D'Amico, V., Narratives to revert overconsumption: human-nature interdependence and Circular Economy. *Agric. Food Econ.*, 11, 1–8, 2023.
4. Hamdan, A., Panda, S., Jain, M.S., Raj, V., Mathew, S., Assessing municipal solid waste in Indian smart cities: A path towards Waste-to-Energy. *Heliyon*, 11, e42770, 2025.
5. Dutta, A. and Jinsart, W., Waste generation and management status in the fast-expanding Indian cities: A review. *J. Air Waste Manage. Assoc.*, 70, 491–503, 2020.
6. Singh, M., Solid Waste Management in Urban India: Imperatives for Improvement. *J. Contemp. Issues Bus. Gov.*, 25, 328, 2019.
7. Velis, C.A., Wilson, D.C., Gavish, Y., Grimes, S.M., Whiteman, A., Socio-economic development drives solid waste management performance in

cities: A global analysis using machine learning. *Sci. Total Environ.*, 872, 161913, 2023.

8. Bartl, A., Moving from recycling to waste prevention: A review of barriers and enables. *Waste Manage. Res.*, 32, 3–18, 2014.
9. Nurzhan, A., Ruan, X., Chen, D., A Review of Life Cycle Assessment Application in Municipal Waste Management: Recent Advances, Limitations, and Solutions. *Sustainability (Switzerland)*, 17, 302, 2025.
10. Zamri, M.F.M.A., Bahru, R., Amin, R., Khan, M.U.A., Abd Razak, S.I., Hassan, S.A., Kadir, M.R.A., Nayan, N.H.M., Waste to health: A review of waste derived materials for tissue engineering. *J. Cleaner Prod.*, 290, 125792, 2021.
11. Ho, H.J., Iizuka, A., Shibata, E., Chemical recycling and use of various types of concrete waste: A review. *J. Cleaner Prod.*, 284, 124785, 2021.
12. Kaza, S., Yao, L., Bhada-Tata, P., Van Woerden, F., *What a waste 2.0: a global snapshot of solid waste management to 2050*, World Bank Publications, 2018.
13. Zhang, A., Venkatesh, V.G., Liu, Y., Wan, M., Qu, T., Huisingh, D., Barriers to smart waste management for a circular economy in China. *J. Cleaner Prod.*, 240, 118198, 2019.
14. Wowrzeczka, B., City of Waste-Importance of Scale. *Sustainability*, 13, 3909, 2021.
15. Abdullah, N., Al-Wesabi, O.A., Mohammed, B.A., Al-Mekhlafi, Z.G., Alazmi, M., Alsaffar, M.M., Alsaffar, M., Anbar, M., Sumari, P., Integrated Approach to Achieve a Sustainable Organic Waste Management System in Saudi Arabia. *Foods*, 11, 1214, 2022.
16. MSW_AnnualReport_2020-21 https://cpcb.nic.in/uploads/MSW/MSW_AnnualReport_2020-21.pdf
17. Shahab, S. and Anjum, M., Solid Waste Management Scenario in India and Illegal Dump Detection Using Deep Learning: An AI Approach towards the Sustainable Waste Management. *Sustainability*, 14, 15896, Nov. 2022.
18. Joshi, R. and Ahmed, S., Status and challenges of municipal solid waste management in India: A review. *Cogent Environ. Sci.*, 2, 1139434, 2016.
19. Shrestha, P.P., Ghimire, A., Dangi, M.B., Urynowicz, M.A., Development of a Municipal Solid Waste Management Life Cycle Assessment Tool for Banepa Municipality. Nepal. *Sustainability (Switzerland)*, 15, 9954, 2023.
20. Silva, V., Contreras, F., Bortoleto, A.P., Life-cycle assessment of municipal solid waste management options: A case study of refuse derived fuel production in the city of Brasilia, Brazil. *J. Cleaner Prod.*, 279, 123696, 2021.
21. Zaikova, A., Vinitskaia, N., Deviatkin, I., Havukainen, J., Horttanainen, M., Life Cycle Assessment of Existing and Alternative Options for Municipal Solid Waste Management in Saint Petersburg and the Leningrad Region, Russia. *Recycling*, 7, 19, 2022.
22. Zhang, J., Qin, Q., Li, G., Tseng, C.H., Fang, G., Assessing the impact of waste separation on system transition and environmental performance through a city-scale life cycle assessment. *Ecol. Econ.*, 211, 107886, 2023.

23. Mpofu, A.B., Oyekola, O.O., Welz, P.J., Anaerobic treatment of tannery wastewater in the context of a circular bioeconomy for developing countries. *J. Cleaner Prod.*, 296, 126490, 2021.
24. Amobonye, A., Bhagwat, P., Singh, S., Pillai, S., Plastic biodegradation: Frontline microbes and their enzymes. *Sci. Total Environ.*, 759, 143536, 2021.
25. Arke, M., Massoud, M.A., Mourad, Y.F., Jaffa, M.A., Habib, R.R., Environmental and Health Consequences of Pharmaceutical Disposal Methods: A Scoping Review. *Environ. Manage.*, 75, 1388–1400, 2025.
26. Olushola Adeborode, D. and Victor Owoigbe, K., Food Waste Management in Urban Areas: Challenges and Sustainable Solutions. *IRE J.*, 8, 2025.
27. Aragaw, T.A., Plastic waste management strategies toward zero waste: Status, perspectives and recommendations for Ethiopia. *Cambridge Prisms: Plast.*, 3, e1, 2025.
28. Inclusive Green Economy Policy Review for Ethiopia, EfD - Initiative, Accessed: Jun. 17, 2025. [Online]. Available: https://www.efdinitiative.org/publications/inclusive-green-economy-policy-review-ethiopia.
29. Mhaddolkar, N., Astrup, T.F., Tischberger-Aldrian, A., Pomberger, R., Vollprecht, D., Challenges and opportunities in managing biodegradable plastic waste: A review. *Waste Manage. Res.*, 43, 911–934, 2025.
30. Khatiwada, D., Golzar, F., Mainali, B., Devendran, A.A., Circularity in the Management of Municipal Solid Waste - A Systematic Review. *Environ. Clim. Technol.*, 25, 491–507, 2021.
31. Zeng, X. and Li, J., Emerging anthropogenic circularity science: Principles, practices, and challenges. *iScience*, 24, 102237, 2021.
32. Panel, I.R., Bringezu, S., Ramaswami, A., Schandl, H., O'Brien, M., Pelton, R., Nagpure, A., Assessing Global Resource Use: A systems approach to resource efficiency and pollution reduction, United Nations Environment Programme, 2017, Accessed: Jul. 31, 2025. [Online]. Available: https://experts.umn.edu/en/publications/assessing-global-resource-use-a-systems-approach-to-resource-effi.
33. Mathieux, F., Ardente, F., Bobba, S., Nuss, P., Blengini, G., Dias, P., Blagoeva, D., Torres De Matos, C., Wittmer, D., Pavel, C., Hamor, T., Saveyn, H., Gawlik, B., Orveillon, G., Huygens, D., Garbarino, E., Tzimas, E., Bouraoui, F., Solar, S., Critical Raw Materials and the Circular Economy – Background report. Report EUR 28832 EN, 2017, https://publications.jrc.ec.europa.eu/repository/handle/JRC108710.
34. Geissdoerfer, M., Savaget, P., Bocken, N.M.P., Hultink, E.J., The Circular Economy – A new sustainability paradigm? *J. Cleaner Prod.*, 143, 757–768, 2017.
35. Bertolazzi, S., Cuttitta, A., Pipitone, V., Addressing marine plastic pollution: a systematic literature review. *Curr. Opin. Environ. Sustainability*, 68, 101428, 2024.

36. Goel, M. and Tripathi, N.G., Plastic Pollution Challenges and Green Solutions, in: *Plastic Pollution Challenges and Green Solutions*, pp. 1–280, Springer, 2024.
37. Moshood, T.D., Nawanir, G., Mahmud, F., Mohamad, F., Ahmad, M.H., AbdulGhani, A., Sustainability of biodegradable plastics: New problem or solution to solve the global plastic pollution? *Curr. Res. Green Sustainable Chem.*, 5, 100273, 2022.
38. Shamsuddoha, M. and Kashem, M.A., Zero Plastic Drive: A Comprehensive Review on Unveiling Innovative Sustainable Solutions for a Circular Plastics Economy. *Sustainability*, 16, 10329, 2024.
39. Kurniawan, T.A., Mohyuddin, A., Othman, M., Goh, H., Zhang, D., Anouzla, A., Aziz, F., Casila, J., Ali, I., Pasaribu, B., Beyond surface: Unveiling ecological and economic ramifications of microplastic pollution in the oceans. *Water Environ. Res.*, 96, 2024.
40. Tumu, K., Vorst, K., Curtzwiler, G., Global plastic waste recycling and extended producer responsibility laws. *J. Environ. Manage.*, 348, 119242, 2023.
41. Adeyanju, G.C., Augustine, T., Volkmann, S., Oyebamiji S. Ran, U., Osobajo, O., Otitoju, A., Effectiveness of intervention on behaviour change against use of non-biodegradable plastic bags: a systematic review. *Discover Sustainability*, 2, 2021.
42. de A. Santos, R., Teles, E.O., Freires, F.G.M., Applying the Circular Economy Framework to Blockchain Agricultural Production. *Sustainability*, 16, 8004, 2024.
43. Castigliego, J.R., Pollack, A., Cleveland, C.J., Walsh, M.J., Evaluating emissions reductions from zero waste strategies under dynamic conditions: A case study from Boston. *Waste Manage. (Oxford)*, 126, 170–179, 2021.
44. Liu, M., Leping, C., Xuerou, S., Yue, X., Xueliang, Y., Qingsong, W., Qiao, M., Jian, Z., Zero-waste city pilot and urban green and low-carbon transformation: Quasi-experimental evidence from China. *Resour. Conserv. Recycl.*, 206, 107625, 2024.
45. Zaman, A.U. and Lehmann, S., Challenges and Opportunities in Transforming a City into a 'Zero Waste City'. *Challenges*, 2, 73–93, Nov. 2011.
46. Qian, Y., Jiang, J., Guo, B., Hu, F., The impact of 'zero-waste city' pilot policy on corporate green transformation: a causal inference based on double machine learning. *Front. Environ. Sci.*, 13, 1564418, 2025.
47. Swachh Bharat Abhiyan | Prime Minister of India. Accessed: Jul. 07, 2025. [Online]. Available: https://www.pmindia.gov.in/en/major_initiatives/swachh-bharat-abhiyan/.
48. Swachh Bharat Abhiyan (Clean India Mission) | Department of Economic and Social Affairs. Accessed: Jul. 13, 2025. [Online]. Available: https://sdgs.un.org/partnerships/swachh-bharat-abhiyan-clean-india-mission.

49. Financial and Economic Impacts of the Swachh Bharat Mission in India. https://www.unicef.org/india/media/1206/file/Financial-and-Economic-Impacts.pdf.
50. Hutton, G., Patil, S., Kumar, A., Osbert, N., Odhiambo, F., Comparison of the costs and benefits of the Clean India Mission. *World Dev.*, 134, 2020.
51. Varshney, K., Shet, P., Hoosenally, M., Cleaning Bharat (India): A Scoping Review of Perceptions Regarding How Swachh Bharat Can Continue to Improve Hygiene Across the Nation. *Hygiene*, 5, 17, 2025.
52. Siuta-Tokarska, B., Thier, A., Hornicki, K., The Concept of Extended Producer Responsibility in the Field of Packaging Industry and the Energy Sector in the Light of the Circular Economy-The Example of Poland. *Energies*, 15, 9060, 2022.
53. Tran, T.Y.A., Herat, S., Kaparaju, P., A Review on Extended Producer Responsibility Schemes for Packaging Waste Management and Research Gaps in the Field. *Nat. Environ. Pollut. Technol.*, 24, 2025.
54. Pani, S.K. and Pathak, A.A., Managing plastic packaging waste in emerging economies: The case of EPR in India. *J. Environ. Manage.*, 288, 112405, 2021.
55. Directive - 94/62 - EN - EUR-Lex. Accessed: Aug. 01, 2025. [Online]. Available: https://eur-lex.europa.eu/eli/dir/1994/62/oj.
56. Shennib, F., Eicker, U., Schmitt, K., OpenWasteAI - Open Data, IoT, and AI for Circular Economy and Waste Tracking in Resource-Constrained Communities. *IEEE Technol. Soc. Mag.*, 43, 39–53, 2024.
57. Kasulla, S., Zafar, S., Malik, S.J., Prakash Baxla, S., The Role of IoT in Waste Management and Sustainability. *Partn. Univers. Int. Res. J.*, 3, 76–88, 2024.
58. Lakhouit, A., Revolutionizing urban solid waste management with AI and IoT: A review of smart solutions for waste collection, sorting, and recycling. *Results Eng.*, 25, 104018, 2025.
59. Zhang, Q., Xujuan, Z., Xiaojun, M., Zhihe, W., Ran, T., Xiangwen, W., Xueyan, L., Recyclable waste image recognition based on deep learning. *Resour. Conserv. Recycl.*, 171, 105636, 2021.
60. Dong, Z., Chen, J., Lu, W., Computer vision to recognize construction waste compositions: A novel boundary-aware transformer (BAT) model. *J. Environ. Manage.*, 305, 114405, 2022.
61. Sirimewan, D., Bazli, M., Raman, S., Mohandes, S.R., Kineber, A.F., Arashpour, M., Deep learning-based models for environmental management: Recognizing construction, renovation, and demolition waste in-the-wild. *J. Environ. Manage.*, 351, 119908, 2024.
62. Carrillo-Hermosilla, J., Del Río, P., Könnölä, T., Diversity of eco-innovations: Reflections from selected case studies. *J. Cleaner Prod.*, 18, 1073–1083, 2010.
63. Wit, B., Dresler, P., Surma-Syta, A., Innovation in Start-Up Business Model in Energy-Saving Solutions for Sustainable Development. *Energies*, 14, 3583, 2021.

64. Bergmann, T. and Utikal, H., How to Support Start-Ups in Developing a Sustainable Business Model: The Case of an European Social Impact Accelerator. *Sustainability*, 13, 3337, 2021.
65. Yasmeen, R., Sarfraz, M., Shah, W.U.H., Ivascu, L., Cifuentes-Faura, J., The impact of public awareness, infrastructure, and technological development with economic growth on solid waste management of European countries: does governance quality matters. *Environ. Sci. Pollut. Res.*, 30, 53, 113442–113456, Nov. 2023, doi: 10.1007/S11356-023-30356-4/METRICS.
66. Farzadkia, M., Mahvi A, A., Norouzian Baghani, A., Sorooshian, A., Delikhoon, M., Sheikhi, R., Ashournejad, Q., Municipal solid waste recycling: Impacts on energy savings and air pollution. *J. Air Waste Manage. Assoc.*, 71, 737–753, 2021.
67. Abubakar, I.R., Maniruzzaman, K., Dano, U., AlShihri, F., AlShammari, M., Ahmed, S., Al-Gehlani, W., Alrawaf, T., Environmental Sustainability Impacts of Solid Waste Management Practices in the Global South. *Int. J. Environ. Res. Public Health*, 19, 12717, Oct. 2022.
68. Lee, E., Shurson, G., Oh, S.H., Jang, J.C., The Management of Food Waste Recycling for a Sustainable Future: A Case Study on South Korea. *Sustainability*, 16, 854, Jan. 2024.
69. Heng, K., Tan, K., Chan, A., Lee, C.C.C., Food Waste Valorization: Leveraging Singapore's Zero Waste Master Plan and 30-by-30 Goal. *Sustainability*, 16, 7321, 2024.
70. "indore report". https://uncrd.un.org/sites/uncrd.un.org//files/8th-3r_p7-2-background-paper.pdf.
71. Das, A.P. and John, A., Government Strategies for Achieving Circular Economy for Plastic Waste Management in Southeast Asia, in: *Sustainable Management of Urban Plastic Waste Through Circular Economic Approaches*, pp. 72–82, CRC Press, Jan. 2024.
72. Fatorachian, H., Kazemi, H., Pawar, K., Digital Technologies in Food Supply Chain Waste Management: A Case Study on Sustainable Practices in Smart Cities. *Sustainability*, 17, 1996, 20252025.

Index

Also of Interest

Check out these related titles from Scrivener Publishing

Solid Waste Management: Challenges, Sustainability and Advancements, Edited by Priyanka Singh, Pooja Agarwal, and V. Vivekanand, ISBN: 9781394267934. Tackle the global challenge of environmental sustainability with this essential book, which provides a critical review of innovative and sustainable solid waste management strategies by leveraging advanced technologies like AI and geographic information system-based predictive models.

Waste Management and Utilization for a Sustainable Environment, Edited by Amit Srivastava and Suman Lata Tripathi, ISBN: 9781394292974. *Waste Management and Utilization for a Sustainable Environment* is an essential guide to understanding how cutting-edge technologies and innovative engineering solutions can transform waste management into a cornerstone of a sustainable future.

Sustainable Production Innovations: Bioremediation and Other Biotechnologies, Edited by Alok Patel and Amit Kumar Sharma, ISBN: 9781119791904. Presenting the latest technologies and practices in this ever-changing field, this groundbreaking new volume covers the gambit for providing solutions and practical applications of smart and efficient energy systems.

Ecosystem Management: Climate Change and Sustainability, Edited by Arnab Banerjee, Manoj Kumar Jhariya, Abhishek Raj, and Taher Mechergui, ISBN: 9781394231218. This book is essential for anyone who wants to understand the challenges of environmental degradation and learn about the sustainable solutions needed to address these critical issues.

Sewage and Biomass from Wastewater to Energy: Possibilities and Technology, Edited by Inamuddin, Tariq Altalhi, Mohammad Luqman, and Joseph K. Bwapwa, ISBN: 9781394204311. Written and edited by a team of industry experts, this exciting new volume covers clean energy production from sewage and biomass while achieving a zero-waste strategy.

Biotechnological Processes for Industrial and Biomedical Waste Management, Edited by Naveen Dwivedi, Shubha Dwivedi, Ranjeet Kumar Mishra and Charles Oluwaseun Adetunji, ISBN: 9781394272006.

LAND AND ENVIRONMENTAL MANAGEMENT THROUGH FORESTRY, Edited by Abhishek Raj, Manoj Kumar Jhariya, Arnab Banerjee, Sharad Nema, and Kiran Bargali, ISBN: 9781119910404. Written and edited by a group of experts in the field, this groundbreaking reference work sets the standard for engineers, students, and professionals working in forestry, agriculture, ecology and environmental science, offering the scientific community a way towards combating climate change.

GREEN CHEMISTRY FOR ENVIRONMENTAL REMEDIATION, Edited by Rashmi Sanghi and Vandana Singh, ISBN: 9780470943083. Presents high quality research papers as well as in depth review articles on the new emerging green face of multidimensional environmental chemistry.

BIOREMEDIATION AND SUSTAINABILITY: Research and Applications, Edited by Romeela Mohee and Ackmez Mudhoo, ISBN: 9781118062845. Bioremediation and Sustainability is an up-to-date and comprehensive treatment of research and applications for some of the most important low-cost, "green," emerging technologies in chemical and environmental engineering.

EMERGENCY RESPONSE MANAGEMENT OF OFFSHORE OIL SPILLS: Guidelines for Emergency Responders, By Nicholas P. Cheremisinoff, PhD, and Anton Davletshin, ISBN: 9780470927120. The first book to examine the Deepwater Horizon disaster and offer processes for safety and environmental protection.

COMPLEX BIOLOGICAL SYSTEMS: Adaptation and Tolerance to Extreme Environments, By Irina R. Fomina, Karl Y. Biel, and Vladislav G. Soukhovolsky, ISBN: 9781119510406. Written and edited by some of the most well-respected authors in the area of the adaptation of plants and animals to climate change, this groundbreaking new work is an extremely important scientific contribution to the study of global warming.

GLOBAL CLIMATE CHANGE DEMYSTIFIED, By James G. Speight, ISBN: 9781119653851. Tackling one of the most controversial subjects of our time, one of the world's foremost environmental and petroleum engineers explores the potential causes and ramifications of global climate change.

DEATH OF THE CHESAPEAKE: A History of the Military Role in Polluting the Bay, By Richard D. Albright, ISBN: 9781118686270. This unique book focuses attention on the failure of current efforts to cleanup the Chesapeake Bay and suggests an approach often used in cleaning up environmentally damaged sites

Printed and bound by CPI Group (UK) Ltd, Croydon, CR0 4YY

24/08/2026

14947831-0006